Better Ceramics Through Chemistry

MATERIALS RESEARCH SOCIETY SYMPOSIA PROCEEDINGS   VOLUME 32

ISSN 0272 - 9172

MATERIALS RESEARCH SOCIETY SYMPOSIA PROCEEDINGS   VOLUME 32

Better Ceramics Through Chemistry

Symposium held February 1984 in Albuquerque, New Mexico, U.S.A.

EDITORS:

C. Jeffrey Brinker
Sandia National Laboratories, Albuquerque, New Mexico, U.S.A.

David E. Clark
University of Florida, Gainesville, Florida, U.S.A.

Donald R. Ulrich
Air Force Office of Scientific Research, Bolling Air Force Base, Washington, D.C., U.S.A.

NORTH-HOLLAND
NEW YORK · AMSTERDAM · OXFORD

This work was supported in part by the Air Force Office of Scientific Research, Air Force Systems Command, USAF, under Grant Number AFOSR-84-0060. The U.S. Government is authorized to reproduce and distribute reprints for Governmental purposes notwithstanding any copyright notation thereon.

Published by:

Elsevier Science Publishing Company, Inc.
52 Vanderbilt Avenue, New York, New York 10017

Sole distributors outside the United States and Canada:

Elsevier Science Publishers B.V.
P.O. Box 211, 1000 AE Amsterdam, The Netherlands

Library of Congress Cataloging in Publication Data

Main entry under title:

Better ceramics through chemistry.

   (Materials Research Society symposia proceedings, ISSN 0272-9172; v. 32)
   Sponsored by the Materials Research Society.

   Includes indexes.
   1. Ceramics—Congresses. I. Brinker, C. Jeffrey. II. Clark, David E.
   III. Ulrich, Donald R.   IV. Materials Research Society.
   V. Title. VI. Series.
TP786.B48 1984   666   84-18796
ISBN 0-444-00898-5

Manufactured in the United States of America

Contents

SCIENCE OF SOL-GEL PROCESSING

APPLICATIONS OF SOL-GEL PROCESSING

CHEMICAL SYNTHESIS OF CERAMIC POWDERS

CHARACTERIZATION OF GELS AND POWDERS

NOVEL MATERIALS THROUGH CHEMICAL SYNTHESIS

PREFACE

The papers contained in this volume were presented at the symposium entitled "Better Ceramics through Chemistry" at Albuquerque, New Mexico in February 1984. The symposium, sponsored by the Materials Research Society and funded by the Air Force Office of Scientific Research, was convened to explore chemical processing methods which are expected to have major impact on ceramics and glass fabrication.
Processing and its influence on structure and properties is one of the most rapidly evolving fields of study in materials science. It is increasingly apparent that processing must be better understood and controlled at the molecular level in order to optimize a material's performance. Sol-gel processing is one means of chemical synthesis which offers molecular level control of processing, and was the primary focus of the symposium.

For years it has been known that ceramics can be fabricated from solgels. In fact, several products based on this technology are already commercially available; these include ceramic fibers, abrasives and special coatings. However, it has only been during the last three to five years that interest in this field has become widespread.

Traditionally, the manufacture of glass and ceramics required high temperatures, but chemical processes such as sol-gel, pyrolysis, laser synthesis, and controlled precipitation involve lower processing temperatures and result in better material homogeneity. Perhaps of even greater significance, chemical processing may permit the synthesis of materials with unique properties not obtainable by other methods.

The objective of this symposium was to bring together the leading international researchers from government laboratories, universities, and industry who currently are investigating the chemical synthesis of materials. Fifty papers were presented on the various aspects of processing, characterization, and properties of chemically derived materials. The major emphasis was on understanding and controlling the chemistry of processing at the molecular level. A more comprehensive understanding of the chemistry of processing should enable us to develop new types of ceramics and improve existing ceramics as well.

The work reported herein will provide both an introduction and a reference for those who are studying chemical synthesis of ceramics. Additionally, the reader will find useful information on many new materials and processes currently under investigation.

SCIENCE OF SOL-GEL PROCESSING

STRUCTURE OF SOLUBLE SILICATES

DALE W. SCHAEFER and KEITH D. KEEFER,
Sandia National Laboratories, Albuquerque, NM   87185

ABSTRACT

Small angle x-ray scattering (SAXS) is the technique of
choice for the determination of structure on the 10–1000Å
scale. We have used this technique to study the growth and
topology of the macromolecules which precede gelation in
several chemical systems used in sol–gel glass technology.
The results show that branched polymers, as opposed to
colloids, are formed. The alcoholic silica system is akin
to organic systems where gelation occurs through growth and
crosslinking of chain molecules. Data are reported from
both the Porod and Guinier regions of the SAXS curve and
these data are interpreted in terms of geometrical structures
predicted by various disorderly growth processes. The results indicate that the degree of crosslinking can be controlled by catalytic conditions. The degree of crosslinking
may, in turn, control phase separation and processability
to a dense glass.

INTRODUCTION

Although there has been substantial research in the last decade on sol-
gel derived materials, little is known about the structure of solution
precursors. This paucity of knowledge extends to structures at all levels;
molecular, macromolecular, and microscopic. In this work we use small angle
x-ray scattering (SAXS) to probe the structure of the macromolecular precursors to the gel phase in systems of interest in glass technology.

The System

This work centers on the growth of macromolecules resulting from the
hydrolysis and condensation of silicon tetraethoxide ($Si(OC_2H_5)_4$, TEOS):

$$Si(OR)_4 + H_2O \rightarrow Si(OR)_3OH + ROH \tag{1}$$

$$2Si(OR)_3OH \rightarrow (RO)_3SiOSi(OR)_3 + H_2O \tag{2}$$

In these reactions R is a proton or an alkyl or silicate group. Reaction
(1) is either acid or base catalyzed. Details of the compositions are given
in Table 1. Table 2 gives some physical properties of the systems. The
solvent, n-propanol, is chosen to facilitate study of the reaction kinetics
by gas-liquid chromatography (GLC) [1]. Further details on the chemical and
physical conditions are published elsewhere [2]. The three compositions
studied differ in catalyst (acid vs. base) and in the H_2O/TEOS ratio used
for the hydrolysis. In order to eliminate the effects of phase separation
and precipitation, a two-step hydrolysis process was used to prepare all
the solutions. The first step consisted of mixing at 60°C tetraethylorthosilicate (TEOS), alcohol, water, and acid (HCl) in a molar ratio 1:3:1:
0.0007. This initial water addition equals one fourth the stoichiometric
amount required to fully hydrolyze the TEOS to monosilicic acid. After 1.5
h, additional water plus base or acid were added at 40°C for the second
hydrolysis step.

Mat. Res. Soc. Symp. Proc. Vol. 32 (1984) Published by Elsevier Science Publishing Co., Inc.

TABLE I. Compositions Investigated (mol%) [2].

Sample	n-prop	H_2O	TEOS	HCl	NH_4OH	H_2O/TEOS
A2	32.8	55.7	10.9	0.632	-----	5.1
A3	18.4	75.5	6.1	0.005	-----	12.4
B2	39.2	47.9	12.9	0.010	0.016	3.7

TABLE II. Summary of Physical Properties [2].

Samples	H_2O/Si	Apparent pH	Gel time (h)	Density of gel (g cm^{-3})	Surface Area (m^2 g^{-1})	Surface Area (m^2 cm^{-3})
A2	5.1	0.8	25	1.54	740	1139
A3	12.4	3.1	12	1.42	806	1144
B2	3.7	7.9	4	0.99	910	901

Structural Issues

The structure of sol-gel materials must be studied on several levels in order to develop a reasonable model relating chemistry to physical properties. At the molecular level, it is important to know the development of the local chemistry around the Si atom. Specifically, the kinetics of hydrolysis and condensation and the degree of branching are of central importance. R. Assink and B. Kay [3] use proton NMR to determine local structure on the systems studied here. Brinker, et. al. also addressed molecular issues using GLC [1].

SAXS is the technique of choice for structure determination at the macromolecular level. Because of the wide variety of possible structures, linear polymers to colloidal particles, it is important to determine the structure of silica condensation polymers as a function of chemical conditions. We are primarily concerned with the structure of a single macromolecule and we use the concept of fractal geometry [4] to describe the structure of silicates.

A further level of structure relates to collective phenomena and the morphology of the bulk gel and dried solid. We speculate on this problem by reviewing the factors which control phase separation in gels. As yet, we have performed only a limited number of scattering experiments on bulk materials, but C. J. Brinker [4] has studied the densification of the same systems studied here.

EXPERIMENTAL METHODS

Because SAXS from macromolecular solutions is relatively unfamiliar to ceramists, we briefly review this technique. First, we consider the familiar Bragg scattering law for crystals and develop scattering from amorphous materials by analogy.

Bragg's law relates the peaks in the angle dependence of the scattered x-ray intensity, I, to corresponding planes of spacing, d, in a crystal. For scattering angle θ

$$\frac{1}{d} = \frac{2}{\lambda} \sin(\theta/2) \tag{3}$$

where λ is the wavelength of the incident radiation. In amorphous systems distinct planes are not present. Scattering may occur, nevertheless, from correlated fluctuations of concentration or density. Scattering at an angle θ arises from fluctuations with the appropriate Fourier spatial frequency, K:

$$K = \frac{4\pi}{\lambda} \sin(\theta/2) \qquad (4)$$

Comparing (3) and (4), scattering at θ is due to fluctuations whose characteristic length is $2\pi/K$.

Structural Regimes

The form of the scattered intensity profile depends on the relationship between K^{-1} and other appropriate lengths in the sample. Figure 1 shows a schematic scattering curve for a typical dilute polymeric system.

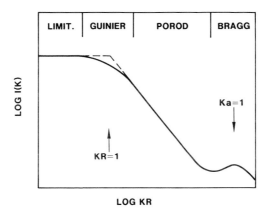

FIG. 1. Schematic small angle scattering curve from a dilute macromolecular solution.

Two lengths appear in this figure: R, the average radius of gyration of the macromolecule and a, the chemical length of the monomeric units. Typically $R \simeq 100A$ and $a \simeq 1A$.

Given the set of lengths (K^{-1}, R, a) it is convenient to divide the scattering curve in Fig. 1 into several regions. At large scattering angles ($Ka \simeq 1$) one probes distances comparable to chemical bonds. We call this regime Bragg scattering, because in crystalline systems we expect sharp diffraction lines when $Ka = 1$. In amorphous macromolecules, broad diffuse peaks may be observed and the shape of $I(K)$ in the Bragg regime can be used to determine the distribution of local conformational isomers [6]. Figure 2a shows a schematic picture of Bragg scattering for polymers, the lines represent the spatial Fourier component of spacing K^{-1} and the bond length is a.

4

 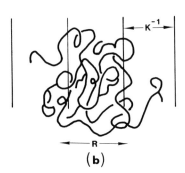

(a)   (b)

FIG. 2. Schematic representation of scattering in the Bragg (a) and Guinier (b) regimes. The bond length is a and R is the radius of gyration.

    We call the limit $KR \ll 1$ the limiting regime. In this regime the system behaves as a dilute gas of point scatterers. In this limit $I(K)$ is proportional to the osmotic compressibility which is proportional to cN, c being the concentration of monomers and N being the number-average degree of polymerization. For sol-gel problems the divergence of N near the gel point is of considerable interest, but we have not studied this problem yet.

    We analyzed our scattering curves in the Guinier and Porod regions. The Guinier regime occurs when $KR \leqslant 1$. From the initial decay of $I(K)$ one measures the Guinier radius, R_g,

$$I(K) \propto cN\left[1 - \frac{K^2 R_g^2}{3}\right] + \ldots \qquad (5)$$

The Guinier regime is shown schematically in Fig. 2b. In dilute solution R_g reduces to the Z-average radius of gyration of the scatterers. In interacting or overlapped systems R_g is a correlation length. In this paper we measure the concentration dependence of R_g during gelation. This information provides a qualitative measure of the degree of branching of the silica macromolecules.

    We call the regime $R \gg K^{-1} \gg a$ the Porod regime. Porod [7] showed that in this regime for systems with sharp boundaries (e.g., dense colloids) the scattered intensity decays as a power law:

$$I(K) \sim K^{-4} \; ; \; R \gg K^{-1} \gg a \qquad (6)$$

It turns out that power law decay is common for a wide variety of macromolecules and that the exponent is sensitive to the geometry of the scatterers [10,28].

    The Porod regime is shown schematically in Fig. 3. This figure illustrates K^{-1}, the wavelength of the appropriate Fourier component which is small compared to the size R of the macromolecule, but large compared to a, the chemical distance. In the Porod regime the scattering curve contains no information which depends on a, R_g, or N. Only geometric information remains so the power-law exponent must contain such information.

    If we assume power-law decay in the Porod regime, the exponent can be calculated by a simple scaling argument. The functional form

$$I(K) \propto cN(KR)^{-x} \qquad (7)$$

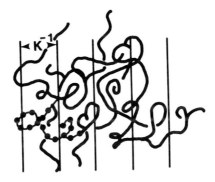

FIG. 3.   Porod regime scattering.   In this regime power-law decay is expected for I(K).

has all the properties required.   At KR = 1, I(K) = cN, the appropriate as-
ymptotic limit (K → 0).   In addition, eq. (7) has the assumed power law
form.   To calculate the exponent x, we make use of the observation of the
previous paragraph; that information about both N and R must drop out of
this equation.   To calculate x, however, we need a relation between R and
N.   It can be shown [8] quite generally that systems obeying (7) also obey

$$N \sim R^D \tag{8}$$

Note that for a uniform object D = 3 whereas for highly ramified objects
D < 3.   Random-walk linear polymers [10] give D = 2 whereas swollen linear
chains give D = 1.66.   Other values of D (usually near 2) are possible for
branched polymers depending on the growth laws.   D is called the fractal dimension [11].

Given D, the exponent x follows immediately from eq. (7) by requiring
that I(K) be independent of R and N.   Clearly, x = D.   The essential point
is that the power-law exponent in the Porod regime is the fractal dimension,
D.

Fractal Geometry

The concept of fractal geometry is a powerful new tool to describe random systems.   In the last few years there has been an explosion of literature on this subject so it is appropriate to review a few characteristics of
fractals.   The triumph of fractal theory is to quantify randomness and to
unify broad classes of problems.

Fractal objects are self-similar which means that the objects look the
same on all length scales.   If one imagines doubling the length K^{-1} in Fig.
3 and simultaneously clumping monomers together in pairs, then the picture
is essentially unchanged aside from the scale factor.   Self-similarity implies the power-law form assumed in eq. (7).   That is, I is homogeneous under a scale change λ,

$$I(\lambda K) = CONST.I(K) \tag{9}$$

Eq. (9) is consistent with and justifies the original power-law assumption,
eq. (7).

Dense objects, like colloidal particles, are not fractal because they have a distinct boundary. Consequently all scattering arises from the surface discontinuity. Thus, even though power-law decay (eq. (6)) is found for colloids, the exponent is not the fractal dimension.

Qualitatively, the fractal dimension is a measure of the compactness of an object. Objects with D substantially less than the dimension of space, are highly ramified, wispy structures whose density decreases with distance from the center of mass.

Development of Scattering Curves

Based on the above discussion of scattering curves, one of at least three patterns for development of the scattering curves during polymerization is expected. These patterns are shown schematically in Fig. 4. In all cases D is assumed to be unchanged during growth and that any background from solvent monomer is subtracted.

The increasing intercept in all three panels of Fig. 4 indicates growth. The intercept is proportional to cN (see Fig. 1), so panel (a) represents an increase in the number concentration of a given size whereas panel (b) represents growth in the size of a fixed number of scatterers. Panel (c) is characteristic of growth in both the number concentration and size of clusters.

The model curves in Fig. 4 should be compared with Fig. 5 which shows the development of the scattering curve for the acid catalyzed system, A2. In the later stages the curves develop as in Fig. 4b, indicating growth in the mean size at constant monomer concentration. The Porod slopes at large K are constant in the latter stages, showing that the local geometric struc- is unchanged as the clusters grow. Other systems studied showed similar behavior.

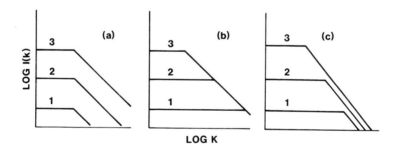

FIG. 4. Development of the scattering curves during polymerization: a) increasing number of scatterers without increase in size, b) increasing size of a constant number of scatterers; (big clusters growing at the expense of small; c) simultaneously increasing size and number of scatterers.

POROD REGIME SCATTERING

SAXS Results

The fractal dimension, D, of the macromolecules is obtained from the limiting slope of the curves in Fig. 6. Fig. 6 shows the scattering profiles [13] for the three systems studied as a function of normalized time increment Δ. These curves were recorded from samples diluted at least 10:1 and quenched to 7°C. A background of unpolymerized monomer solution was subtracted and the curves were corrected for detector sensitivity and linearity.

In all cases the fractal dimension (the Porod slope) is close to 2. This result contrasts with Fig. 7 which shows the scattering curves for colloidal silica and a colloidal ferrofluid. It is clear from the measured fractal dimensions that the clusters in the alcoholic silica systems are not colloidal particles but rather are polymeric. Under basic conditions (B2) and high water/TEOS ratio (A3), D is slightly greater than two whereas for the acid catalyzed system at low water content D is slightly less than 2. Although the deviations from 2.0 are small, we have observed systematic differences on several different samples studied over a one year period. We believe the differences show that A2 is a more open, linear polymer whereas A3 and B2 are more highly crosslinked. Evidence for this hypothesis occurs in the Guinier region as discussed below.

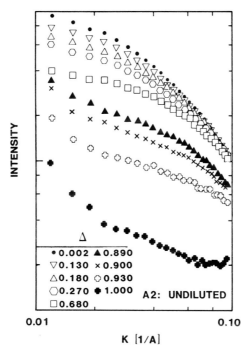

FIG. 5. Development of the scattering profile for the acid-catalyzed gel A2. Δ is the normalized time increment to gelation: $\Delta = t_{gel}-t/t_{gel}-t_0$ where t_{gel} is the gel time, t_0 is the initiation time and t is the observation time.

Growth Models

To interpret the observed D's, it is necessary to review the predictions of various growth models. Table 3 summarizes [12] the fractal dimension of numerous model macromolecules. Two types of linear polymers are listed in the table, ideal and swollen. The ideal chain results from a random walk process whereas the swollen chain results from a self-avoiding

walk [12]. Both extremes are observed in linear polymers [10]. The so-
called θ temperature [9] is the point at which ideal behavior is observed.
Most polymers swell as the temperature is raised above θ [10]. At sufficiently high temperature, the repulsive forces between monomers swell the
chain and self-avoiding statistics are realized.

   In analogy with linear chains, both swollen and ideal randomly branched
polymers are listed. The self-avoiding randomly branched structure is
called a lattice animal (LA) and its fractal dimension is the same as the
ideal linear polymer: D = 2. An LA at T = θ is somewhat collapsed with D =
2.17 according to the highly approximate calculation of Daoud and Joanny
[15].

   A third structure, which may in fact be Daoud and Joanny's θ-point
branched polymer, is a percolation cluster. Percolation clusters are gener-
rated by randomly adding sites to a lattice. deGennes suggests [17] that

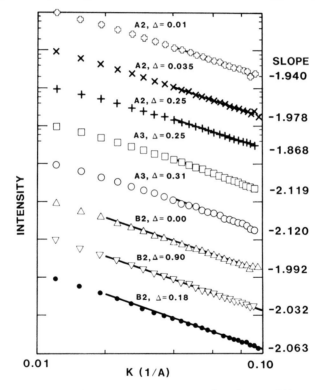

FIG. 6. Porod plots of the scattering curves for three gelling systems.
Δ is defined in Fig. 5.

percolation clusters are merely LA's in which the swelling forces are
screened by other clusters. If this idea is correct percolation clusters
are θ-point lattice animals.

   Cluster aggregates (CA) and diffusion limited aggregates (DLA) are
produced by kinetic growth processes. In DLA monomers approach a fixed
seed cluster by a random walk and bond on contact. Because of the random
walk of the approach, growth takes place almost exclusively on projections.

Once projections form, the interior of the cluster never fills in and the structure is highly dendritic. CA is basically the same as DLA except that the clusters themselves move and stick together on contact.

Since we know that branched polymers are produced in the silica system, the most reasonable interpretation of the results in Fig. 6 is that we have

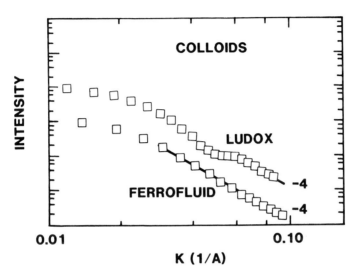

FIG. 7. Porod plots for colloidal particles.

TABLE III. Fractal dimension of various objects

Name	D
Swollen linear polymer [9,10]	1.66
Cluster aggregate [16]	1.75
Ideal linear polymer [9,10]	2.0
Swollen branched polymer [14]	2.0
Ideal branched polymer [15]	2.16
Diffusion limited aggregates [8]	2.5
Percolation cluster [12]	2.5
Dense particle [7,28]	3.0

generated lattice animals with D = 2. The slight deviations from D = 2 may be due to crossover to linear chains in the case of A2 and crossover to dense systems for A3 and B2 [10,19]. Family [19] shows that the limiting

$(R \to \infty)$ fractal dimension is independent of the degree of branching (excluding linear chains) so it is quite possible that the three systems studied are topologically quite different. At any rate, there is no question that we are dealing with branched polymers and not colloidal particles.

Keefer's [5] recent studies have identified oligomers (3-15 Si atoms, R_g = 5A) as important species in acid catalyzed systems at low water content. Since these conditions prevail during the first stage of the reactions studied here, the second stage of condensation actually takes place between oligomers of unknown and possibly cyclic species. In the present context, then, a "monomer" is ill-defined. In fact, systems which behave as linear polymers may even be a linear sequence of loops.

GUINIER REGIME SCATTERING

Although the Porod law analyses clearly demonstrate the existence of polymer molecules, the method is rather insensitive to the degree of branching. To gain further insight into branching, we performed a set of experiments to determine the Guinier radius before and after dilution. If the macromolecules are compact, highly branched clusters, then the Guinier radius is independent of the degree of dilution. If the chains are nearly linear, however, they should become strongly overlapped soon after condensation begins. In this case, the measured Guinier radius is substantially less than the radius of gyration of the chains, i.e., the polymer solution is semidilute [18,20]. Upon dilution the chains disentangle and the Guinier radius approaches the radius of gyration.

Figures 8, 9 and 10 show the effect of dilution on the Guinier radius. In most cases, the samples were diluted 10:1. Near the gel point dilutions

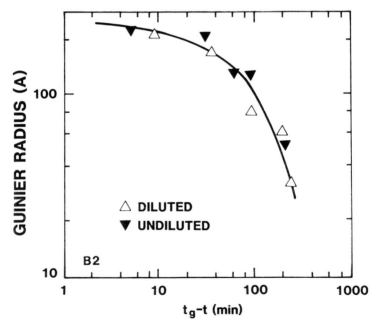

FIG. 8. Time and concentration dependence of the Guinier radius for B2. There is an error in the ordinate of this figure reported in a previous publication [28].

of 20:1 and 50:1 were also studied to assure that the samples were in the dilute limit. The data show that the A2 sample is strongly overlapped prior to dilution and that the B2 sample is essentially non-interacting up to the gel point. The A3 sample shows intermediate behavior.

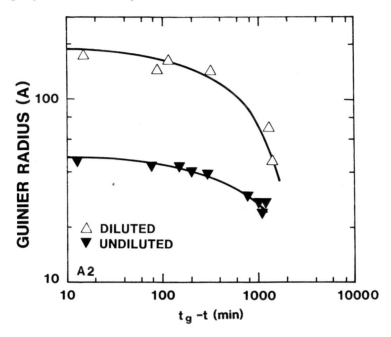

FIG. 9. Time and concentration dependence of Guinier radius for A2.

Consideration of both the Porod and Guinier regions leads to the conclusion that the degree of branching is controlled by the catalytic conditions and the H_2O/TEOS ratio. Linear chains are favored in acid catalyzed conditions and at low water content. Branching is favored in the opposite conditions. These conclusions are similar to those reached previously [2], but some of the chemical arguments presented in our earlier work [2] have not been substantiated [1,3,5].

The conclusions reached here are supported by other lines of evidence. It is well known [21], for example, that colloidal particles may be formed in aqueous solution under base catalyzed conditions. Colloidal particles represent the highly branched limit. By contrast, Sakka [22] finds that fibers can be spun from silicates formed in acidic-alcoholic solution. Spinability is a signature of linear chains. Sakka has showed that the molecular weight dependence of the viscosity, η, is consistent with linear chains. It should be noted, however, that $\eta \sim R^3/N \sim N^{(3D-1)}$. Since D is found to be quite insensitive to branching, viscosity is also only weakly dependent on branching.

The properties of the gels derived from the systems studied above are also consistent with the conclusions reached. The acid-catalyzed gels with low H_2O additions are clear and rubbery as expected for a weakly cross-linked system [23]. The base catalyzed gels, on the other hand, are often brittle and cloudy. In organic systems [24], brittle, cloudy gels are found in highly crosslinked systems.

PHASE SEPARATION

The possibility of controlling phase separation in the gel phase of sol-gel glass processing offers an interesting opportunity to tailor bulk material properties. Tanaka and coworkers [23-27] have studied collective phenomena in gels and have identified the parameters which control phase separation. It is worthwhile to review these ideas here and speculate on the consequences for sol-gel materials.

A highly schematic phase diagram for polymers is shown in Fig. 11 with temperature on the ordinate and mole-fraction monomer on the abscissa. Curve a represents the two phase region for monomer mixed with solvent. Compositions under the curve separate into two phases specified by a temperature

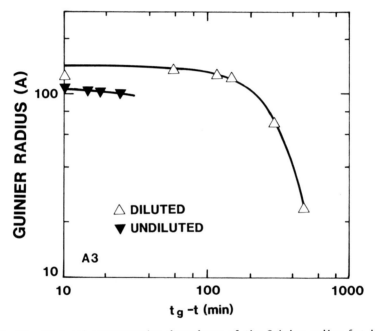

FIG. 10. Time and concentration dependence of the Guinier radius for A3.

tie line. Upon polymerization to linear chains, the two-phase region invades a large portion of the phase diagram as shown in curve (b). If the linear polymer is further crosslinked, curve (c) is expected. If a composition such as point P is chosen, a clear single phase product is expected if linear chains are formed, whereas the system should phase separate if the polymers are highly branched.

Hochberg, et al., have shown that chain stiffness and backbone charge also lead phase separation [27]. Since both the charge and stiffness of the silicate backbone can be controlled though ionization of silanol protons, there is promise for controlling phase separation with pH.

Once a gel has formed, phase separation occurs on a microscopic level. A sponge like structure is expected with randomly distributed patches of high and low concentration. Such a structure has enormous concentration fluctuations leading to the cloudiness associated with highly crosslinked gels.

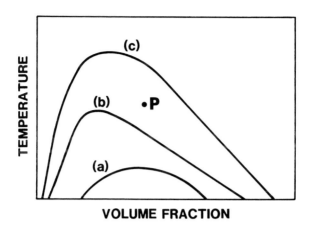

FIG. 11.   Schematic phase diagram of monomers, linear polymers and branched polymers.

     Although the above discussion is both qualitative and speculative, it is credible that phase separation can be exploited to control glass properties. For example, tough dense films are expected from weakly crosslinked gels or from dried linear chains. If porosity is desired, on the other hand, base catalyzed, highly crosslinked gels are indicated. To achieve very large pores, it may be possible to start with a weakly crosslinked system, to achieve high elasticity, and force the gel to phase separate by increasing backbone charge or changing temperature. Unfortunately, the silica system is very complex and it is impossible to independently control branching and the degree of ionization. In addition, no information exists on the statistics of silica polymers or on the thermodynamics of silica polymer solutions. In this situation it is impossible to make precise statements on the control of material properties. We hope, nevertheless, that the possibilities outlined here will encourage the required research.

ACKNOWLEDGEMENT

     We thank C. J. Brinker and C. S. Ashley for preparation of the solutions used in this work. We also thank J. S. Lin for supervision of the data acquisition. This work was performed at Sandia National Laboratories supported by the U. S. Department of Energy under Contract Number DE-AC04-76-DP00789. SAXS work was performed at the National Center for Small Angle Scattering at Oak Ridge National Laboratory.

REFERENCES

1.   C. J. Brinker, K. D. Keefer, D. W. Schaefer, R. A. Assink, B. D. Kay, and C. S. Ashley, J. Non. Cryst. Solids, 63, 45 (1984).
2.   C. J. Brinker, K. D. Keefer, D. W. Schaefer, and C. S. Ashley, J. Noncryst. Solids, 48, 47 (1982).
3.   R. A. Assink and B. D. Kay in Better Ceramics Through Chemistry, edited by C. J. Brinker, D. R. Ulrich, and D. E. Clark (Elsevier-North Holland, New York, 1984).

4. C. J. Brinker, Ibid.
5. K. D. Keefer, Ibid.
6. C. J. C. Edwards, R. W. Richards, R. F. T. Stepto, K. Dodgson, J. S. Higgins, and J. A. Semlyen, Polymer, 25, 365 (1984).
7. G. Porod, Kolbid Z., 124 (1951), 83.
8. T. Witten and L. M. Sander, Phys. Rev. Lett., 47, 1400 (1981).
9. P. J. Flory, Principles of Polymer Chemistry (Cornell University Press, Ithaca, NY, 1953).
10. D. W. Schaefer and J. G. Curro, Ferroelectrics, 30, 49 (1980).
11. B. B. Mandelbrot, Fractals, Form, Chance and Dimension (Freeman, San Francisco, 1977).
12. F. Family in Random Walks and their Applications in the Physical and Biological Sciences, Am. Inst. of Physics, Proceedings, 109, 33–72, (1984)
13. D. W. Schaefer and K. D. Keefer, submitted.
14. J. Isaacson and T. C. Lubensky, J. Phys. Lett. (Paris), 41, L–469, (1980).
15. M. Daoud and J. F. Joanny, J. Phys. (Paris), 42, 1359 (1981).
16. P. Meakin, Phys. Rev. Lett., 51, 1119 (1983); M. Kolb, R. Botet and J. Jullien, Phys. Rev. Lett., 51, 1123 (1983).
17. P-G. deGennes, C. R. Acad. Sci. Paris, 291, 17 (1980).
18. D. W. Schaefer, Polymer, 25, 387 (1984).
19. F. Family, J. Phys. A, 13, L325 (1980).
20. P. Wiltzius, I. Haller, D. Cannell, and D. W. Schaefer, Phys. Rev. Lett., 51, 1183 (1983).
21. R. K. Iler, The Chemistry of Silica, (John Wiley, New York, 1979).
22. S. Sakka in Better Ceramics through Chemistry, edited by C. J. Brinker, D. R. Ulrich, and D. E. Clark (Elsevier–North Holland, New York, 1984).
23. T. Tanaka, Phys. Rev. Lett., 40, 820 (1978).
24. R. Basil and M. K. Gupta, Ferroelectrics, 30, 63 (1980).
25. Y. Hirokawa, S. Katayama, and T. Tanaka, preprint, "Effects of Network Structure in the Phase Transition of Acrylamide–Sodium Acrylate Copolymer Gels."
26. T. Tanaka, S. Ishiwata, and C. Ishimoto, Phys. Rev. Lett., 38, 771 (1977).
27. A. Hochberg, T. Tanaka, and D. Nicoli, Phys. Rev. Lett., 43, 217 (1979).
28. D. W. Schaefer, K. D. Keefer, C. J. Brinker, Polym. Preprints, Am. Chem. Soc., Division of Polymer Chemistry, 24, 239 (1983).

THE EFFECT OF HYDROLYSIS CONDITIONS ON THE STRUCTURE AND GROWTH OF SILICATE POLYMERS

K. D. Keefer, Sandia National Laboratories, Albuquerque, New Mexico 87185

ABSTRACT

Small angle scattering experiments have demonstrated that the structure of the silicate species produced by the hydrolysis of silicon alkoxides in non-aqueous solvents ranges from extended, weakly cross-linked polymers to highly condensed, colloidal particles. In contrast, inorganic, aqueous silicate solutions yield primarily colloidal particles because the silicate species have a number of different silanol sites available and the preferred condensation reaction is that of weakly condensed species with highly cross-linked branch sites, such as those on an amorphous silica surface. It is proposed that in the alkoxide systems, however, the hydrolysis reaction may control the number and type of silanol sites available for condensation. In acid catalyzed reactions, the rate of hydrolysis of a silicate tetrahedron tends to decrease as alkoxide groups are removed. This favors the production of silanol sites on the end of chains, thus generating linear polymers. In base catalyzed reactions, it is argued that each subsequent hydrolysis of a tetrahedron should proceed more rapidly than the previous one, producing numerous branch points which are the preferred sites for condensation.

INTRODUCTION

The hydrolysis of silicon tetraethoxide $(Si(OC_2H_5)_4)$ and subsequent condensation of the resulting silanols ($\equiv SiOH$ groups) may be described by the following set of reactions:

$$\equiv SiOEt + H_2O \xrightarrow{\ OH^- \text{ or } H^+\ } \equiv SiOH + EtOH \qquad (1)$$

$$2\equiv SiOH \longrightarrow \equiv SiOSi\equiv + H_2O \qquad (2)$$

The gels which result from these reactions differ in several ways, depending on the water content of the system and whether acid or base was used to catalyze the hydrolysis. First, measurements of the small angle scattering[1] from these systems show that in acidic solutions or at low water concentration, the product of these reactions is weakly crosslinked and polymeric in nature. In alkaline solutions or at high water concentrations, silica forms more highly crosslinked polymers, even fully dense colloidal particles[2]. Second, analyses of the solutions show that in acidic solutions, hydrolysis tends to go to stochiometric completion, but in basic solutions it does not[9], the tetraethoxide monomer being particularly resistant[3]. Third, despite the degree of hydrolysis in solution, dried gels prepared from acidic solutions often have more alkoxy groups per unit surface area than gels prepared in basic solutions[4] indicating that reesterification (the reverse of hydrolysis) is more important in acidic systems. These differences can be accounted for by some simple chemical arguments steming from the mechanisms of the reactions. This reaction occurs via one of two different reaction mechanisms, depending on the type of catalyst and each mechanism favors a different degree of branching. In

Mat. Res. Soc. Symp. Proc. Vol. 32 (1984) © Elsevier Science Publishing Co.. Inc.

the initial stage of the reaction, even a small bias against full condensation of the silicate tetrahedra could radically alter the structure of the resulting polymers. The reaction mechanism also accounts for the extent of hydrolysis and the amount of organic residue in the dried gels.

SMALL ANGLE X-RAY SCATTERING EXPERIMENTS

The x-ray scattering measured in these experiments is Rayleigh scattering which arises from correlated, but not periodic, fluctuations in electron density. At very low angles, scattering from widely separated points interferes constructively giving information about the maximum dimensions of the particles. At higher angles scattering arises from fluctuations which are shorter range than the maximum dimension of the particle, but are longer than atomic dimensions. Small angle scattering measurements have several key virtues for studying silica polymers: the entities are studied in situ so no artifacts are introduced in sample preparation; the structure of solvated species may be studied; the technique provides definitive discrimination between single phase solutions and two phase sols.

Small angle scattering measurements were made with an Anton-Parr compact Kratky camera, which had been specially adapted to mount on a 12 kW Rigaku rotating anode x-ray generator. Graphite monochromatized CuK_α radiation was used with an evacuated beam path. Patterns were recorded with a TEC model 205 position sensitive proportional counter ($\sim$73 µm resolution) at sample to detector distances of 211 or 423 mm. The samples were contained in 1 mm diameter fused silica capillaries and measured at room temperature. Pure solvents were used as backgrounds. The combination of a very high intensity source and a position sensitive detector enabled scattering patterns to be recorded with 1% counting statistics in 100 to 1000 seconds. This permitted many solutions to be monitored at the extent of the reaction and also made dilution experiments feasible.

At low angles, the scattering was analyzed with Guinier's Law:

$$I(h) = I_e(h)N(\Delta\rho v)^2 \exp(-h^2 R_g^2/3)$$

where, $h = 4\pi (\sin \Theta/\lambda)$, 2Θ is the scattering angle, I_e the scattering from a single electron, N the number of particles, $\Delta\rho$ the electron density difference between particle and matrix, v the particle volume, and R_g is the electronic radius of gyration. No curvature terms or slit corrections were used and only the range $h < 1/R_g$ was fit. If interparticle interference effects were suspected (which cause an apparent reduction in R_g) the samples were diluted until the apparent R_g remained constant. It should be noted that the factor of $(\Delta\rho v)^2$ in the pre-exponential coefficient weights the result heavily towards the largest particles in the system.

At higher angles, in the range $1/R_g \ll h \ll 1/a$, where a is the molecular length scale of the system, the scattering curves were analyzed for power law behavior:

$$I(h) = Ah^{-D} \tag{4}$$

where A is an instrumental parameter and D is the power law exponent. If the curve obeys a power law, then a graph of $\log(I)$ vs. $\log(h)$ will be a straight line with slope $-D$. If, as in this case, infinitely long slits are used in the measurement, the value of the exponent will be exactly 1 less than would be observed with pinhole collimation, the latter being the case for which most theoretical work has been done.

A particular case of a power law is Porod's Limiting Law, in which D = 4 (3 if measured with slits) in the limit $h \to \infty$. Porod's Law is observed if in the range $1/R_g \ll h \ll 1/a$, the electron density fluctuation is effectively a step function, as is the case of colloidal particles, which have well defined surfaces.

The theory necessary to interpret quantitatively an arbitrary value of D does not yet exist, except in the case of self similar objects, in which case D is equal to the fractal dimension[1]. Qualitatively, however, D increases with increasing crosslinking and decreasing extension of the particle into its surroundings. For example, if long thin rods "crosslink" to form thin discs, D increases from 1 to 2. Similarly, an extended, self-avoiding random walk polymer has a D of 1.67 but a more compact random walk has D = 2. In this paper, therefore, increasing D is taken as indicating increased crosslinking, and decreasing extension in analogy with fractal dimension[1,5].

CONDENSATION OF SILICA

The tendency of silica to form dense, colloidal particles when polymerized in aqueous solutions above pH 2 may be explained by a condensation reaction mechanism proposed by Iler[6]. In this nucleophilic substitution mechanism, a deprotonated silanol reacts with a protonated silanol, displacing an OH⁻ (Fig. 1). The favored condensation, therefore,

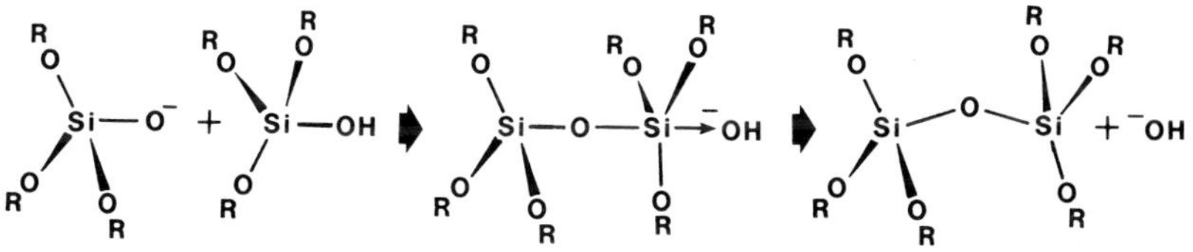

Fig. 1. Nucleophilic condensation reaction mechanism R = H, Et or $Si(OR)_3$

is between the most acidic and most basic silanols in the system. The acidity of a silanol proton depends on the other substituents on the silicon atom. When very basic substituents such as OH⁻ are replaced by less basic SiO⁻, the reduced electron density on Si increases the acidity of the protons on the remaining silanols. The most acidic silanols are, therefore, on the most highly condensed silicate units and the least on orthosilicic acid so the condensation reaction preferentially removes the least poly-merized species from solution and forms very highly condensed species[6].

Little work has been done in solutions with pH less than 2. The fact that the reaction rate increases as pH decreases indicates that protonated species are involved in this regime, although the reaction would be electrophilic rather than nucleophilic. The weakest acid, $Si(OH)_4$ would be the strongest base and most likely to be protonated to form an electrophilic species, and silanol on a triply condensed tetrahedron would be most subject to attack, again tending to form highly condensed species[6].

Power law analysis of small angle scattering from silicate polymers formed during the hydrolysis of silicon tetraethoxide in alcohol solution (Figs. 2 and 3) show that the resulting polymers can vary widely in form and are not fully dense. Since the hydrolysis step precedes the condensation reaction, under certain circumstances it controls the form of the polymers. In neutral solutions, the rate of hydrolysis of silicon tetraethoxide is extremely slow but is catalyzed by both acids and bases. Differences in the reaction mechanisms of the acid and base catalyzed reactions can account for the differences in the structure and chemistry of the resulting gels[7].

18

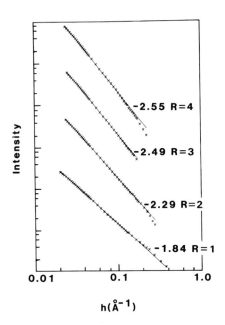

Fig. 2. Alkaline solution SAXS power law exponent change with molar ratio (R) of $H_2O:Si(OEt)_4$. 1 M $Si(OEt)_4$ 0.01 MNH_4OH in EtOH after 192 hours.

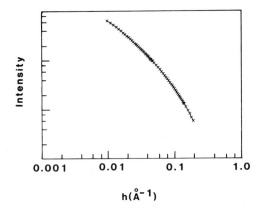

Fig. 3. Acidic solution power law analysis. 1 M $Si(OEt)_4$, 4 M H_2O, 0.0001 M HCl in ethanol.

HYDROLYSIS IN ALKALINE SOLUTION

Alkaline catalyzed hydrolysis of silicon tetraethoxide takes place by nucleophilic substitution[7]. A negatively charged hydroxide ion attacks the positively charged silicon forming a five coordinated intermediate (Fig. 4). In the model proposed here, the attacking and leaving groups are on opposite sides of the silicon, providing maximum separation of charge, analogous to similar reaction in organic systems. Back side attack and inversion of the molecule has important implications, as discussed below. The hydroxide ion would be regenerated by the reaction of the ethoxide ion with water in a subsequent fast step.

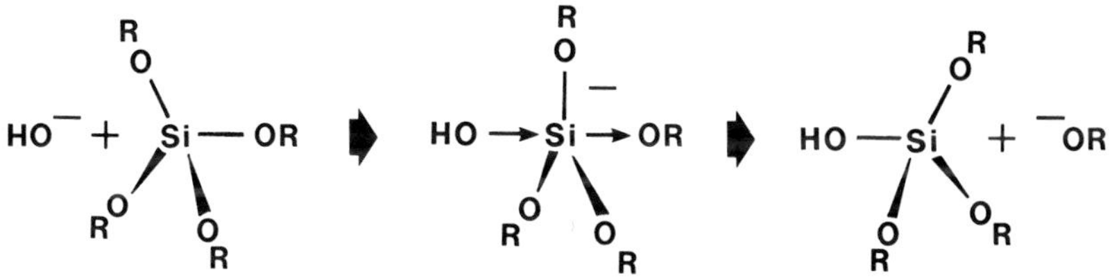

Fig. 4. Base catalyzed hydrolysis by nucleophilic reaction mechanism.
R = H, Et or Si(OR)$_3$

Nucleophilic attack will be sensitive to the electron density around silicon and to the steric effects of the substituents. The fewer bulky and highly basic alkoxy groups surround silicon, the more subject it is to attack by a hydroxide ion. Thus, silicon tetraethoxide should be hydrolyzed most slowly and monomeric hydrolysis products should tend to hydrolyze further at a faster rate, tending to produce orthosilicic acid. Such a system could then condense to form highly crosslinked, relatively compact polymers as described above.

The above factors indicate that alkaline catalysis only biases the system towards dense particles. Other factors are also important in determining how highly crosslinked the polymers are. For example, by LeChatelier's principle, water promotes hydrolysis and impedes condensation. As the water concentration is reduced, silicate monomers may start to condense before they are fully hydrolyzed forming less dense polymers. This effect can be seen from the scattering from base catalyzed solutions prepared at different water concentrations. A power law analysis (Fig. 2) indicates increasing crosslink density of the scattering species with increasing water concentration. Similarly, high catalyst concentrations promote hydrolysis (and can impede condensation) and bias the system again toward dense particles.

The production of highly condensed particles in alkaline solutions is often attributed to "ripening" by a dissolution/reprecipitation process. To check for this, the scattering from the base catalyzed solution with the highest water concentration was monitored as a function of time. A power law analysis of the results (Fig. 5) shows only a slow change of the exponent. This is probably due to the very low equilibrium solubility of silica in alcohol inhibiting dissolution/reprecipitation. This is not to suggest that ripening doesn't occur at higher water concentration or at longer times, but that under these conditions that it is not fast enough to account for the crosslink density of the growing polymers.

An often overlooked aspect of using base to catalyze the hydrolysis reaction is that acidic silanols tend to neutralize it, causing the reaction to slow with time if the concentration of silica is large with respect to the amount of base present. As a result, although hydroxide ion is a true catalyst for the hydrolysis reaction, it may not appear to be so because of its subsequent neutralization. This may also explain why observed rate laws for the hydrolysis reaction do not always appear to be first order in hydroxide[7], as this mechanism implies.

The inversion of the molecule postulated in the mechanism makes the

reaction sensitive to steric hindrance. If condensation occurs between
monomers which are incompletely hydrolyzed, the hydrolysis of the remaining
alkoxy groups would be retarded. As a result, polymeric species which were
not fully dense would likely be incompletely hydrolyzed.
     Both the inversion of the molecule and the deprotonation of silanols
inhibit the reverse reaction, i.e. reesterification. Since the species in
solution are highly crosslinked, the difficulty in forming the activated
complex inhibits reesterification as well. Further, the acidic silanols on
these highly crosslinked sites are likely to be deprotonated, and their
negative charge would prevent attack by a negatively charged attacking
species. The ramifications of the reversibility of the hydrolysis reaction
are discussed more fully in the next section.

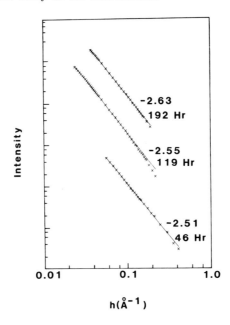

Fig. 5. SAXS power law exponent variation with time in alkaline
         solution. 1 M $Si(OEt)_4$, 4 M H_2O, 0.01 M NH_4OH in EtOH.

HYDROLYSIS IN ACIDIC SOLUTION

     The hydrolysis reaction in acidic solutions proceeds by an
electrophilic reaction mechanism[7]. In this mechanism, a protonated water
molecule is attracted to the oxygen atoms in the basic, alkoxide groups
(Fig. 6). In the proposed activated complex, partial bonds form between the
protons and the oxygen atoms in the alkoxides, and between the water oxygen
and the silicon while the original bonds weaken. One set of partial bonds
will strengthen fastest, forming the alcohol molecule and the silanol group.
The extra proton is given up to another water molecule to complete the
reaction in a fast step[7].
     It is generally accepted that if the charge on the attacking species is
positive, the reactivity of a tetrahedron should increase as the electron
density around the silicon increases. The more highly esterified silicate

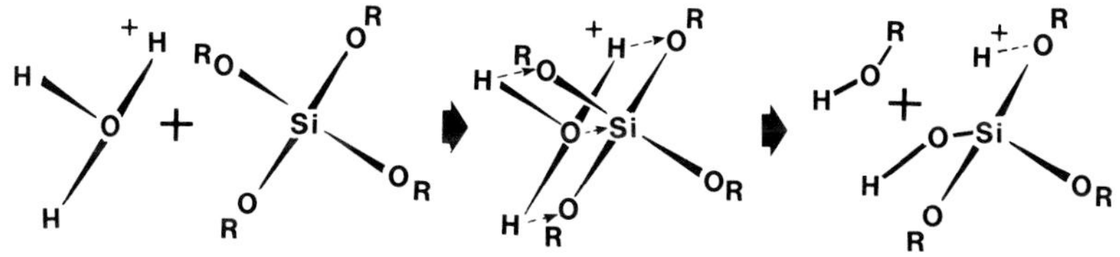

Fig. 6. Acid catalyzed hydrolysis by electrophilic reaction mechanism.

species should be the most prone to attack, rather than the least as in
alkaline solution. Also, if the attacking species tends not to be repelled
by the high electron density, highly esterified species are more subject to
hydrolysis simply because there are more alkoxy groups to attack. The
forgoing arguments imply that tetraalkoxide monomers should be more rapidly
hydrolyzed than end groups on chains, which in turn will be more rapidly
hydrolyzed than middle groups on chains, etc. As a result, acid catalyzed
hydrolysis tends not to produce orthosilicic acid monomer and condensation
could start between incompletely hydrolyzed species. The pattern of
hydrolysis rates would lead to less highly crosslinked polymers than result
in alkaline solution.

Again, the proposed reaction mechanism for acid solutions only biases
the system toward the production of weakly crosslinked polymers, a bias
which can be countered by other factors. For example, the rate of
hydrolysis increases monotonically as the pH decreases from 7, but the
condensation reaction rate has a local minimum at around pH 2[6]. In this
region, hydrolysis may well go to completion before any significant
condensation has occurred and polymerization would have the normal pattern.
As argued above, water concentration also promotes hydrolysis and tends to
inhibit condensation, so the resulting products would be more highly
crosslinked.

Acid catalysis contrasts with alkaline in several other ways. The
leaving group departs from the same side of the tetrahedron as the water
molecule attacks: no inversion of the molecule occurs. This front side
attack is much less sensitive to the degree of polymerization than is back
side attack, hence, even if condensation starts before all the alkoxy groups
are removed, the reaction can still go to completion. Also, if the pH of
the solution is above the isoelectric point, silanols will tend not to be
protonated and, therefore, not repel additional attack by hydronium ions.

The same factors which cause the reaction to go to completion in acidic
solutions also cause it to reverse readily. If the solution is not too
acidic, silanols will be neutral and not repel protonated alcohols and all
silanols will be susceptible to reesterification regardless of the degree of
condensation.

REESTERIFICATION

The importance of the reversibility of the hydrolysis reaction is that
many of the solvents used (ethanol, propanol, dioxane, but not methanol)
form azeotropes with water. When the gel is dried, the azeotrope is removed
first, since it has the highest vapor pressure. Depending on the exact
amounts of water and alcohol used in the reaction and the degree of
condensation, either all of the water or all of the alcohol may be removed
from the gel. If water remains, it will tend to complete the hydrolysis
reaction. However, if alcohol remains and the reaction is readily
reversible, the silanols will reesterify and the water produced will be

removed as the azeotrope and tend to drive the thermodynamically unfavorable reesterification reaction to completion. This phenomenon could account for the common and contradictory observation that although measurements in solution may show that the acid catalyzed hydrolysis goes to completion and base does not, the dried gels often have a higher organic content if acid catalyzed than if base catalyzed[8]. Because a small amount of water can cause the removal of a large amount of solvent, the gel would change from being completely hydrolyzed to completely esterified in a narrow range of water concentrations. For the common solvent ethanol, this effect is shown in Figure 7, in which the fraction of sites reesterified is calculated as a function of the molar ratio of water to silicon tetraethoxide at different molarities of the latter. This change occurs in a range of water contents

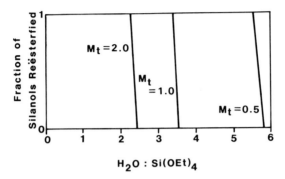

Fig. 7. Calculated fraction of silanols in gel reesterified upon drying ethanol solvent as a function of $H_2O:Si(OEt)_4$ at different molar concentrations of $Si(OEt)_4$, M_t. Calculated at $78^{\circ}C$, complete solution hydrolysis and a degree of polymerization of 3, neglecting ΔV of mixing.

often used in experiments and is very sensitive to concentration. Because water is released in the condensation reaction, the degree of polymerization is also important. In Figure 8, the water to silicon tetraethoxide ratio at which hydrolysis remains complete is calculated as a function of the degree of polymerization. For degrees of polymerization in the range expected for most gels (2.5 - 3.5), the amount of water required to prevent reesterification varies widely and covers the range often used in experiments.

Although reesterification is not likely to occur in alkaline solution, the extent of hydrolysis measured in solution may still be a poor indicator of the organic content of a dried gel. Since the species most likely to account for residual alkoxy groups in basic solution is the volatile tetraethoxide itself, the organic content of a dried base catalyzed gel may well be lower than predicted from solution analysis.

Polymer Structure and Aggregation

To assess the effect of pH and water concentrations on polymer growth rate and gel formation, small angle scattering measurements were made on a variety of compositions after 48 hours of reaction at $25^{\circ}C$ (Table 1). Two interesting observations can be made. First, in alkaline solution, polymers got large regardless of the water concentration, whereas in acidic solutions, a large concentration (more than enough to assure complete hydrolysis) was required for growth, and even then only at a nominal pH of 4. Second, although the scattering polymers are small, extensive reaction

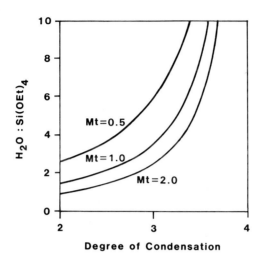

Fig. 8. Calculated $H_2O:Si(OEt)_4$ required to prevent reesterification
of dried gel is a function of degree of polymierization at
different molar concentrations of $Si(OEt)_4$, M_t. As the range
over which the gel would be partially esterified as narrower
than the line, gels prepared above each line would remain completely hydrolyzed, those below completely reesterified.

has occurred in many of the acidic solutions, as an R_g of 5A would
correspond to 5-15 tetrahedra. These solutions were monitored for 20 days
and although the polymers in one solution continued to grow, resulting in a
gel after 100 hours, little or no growth was seen in any of the other acidic
solutions and they did not gel. The polymers in the basic solution

TABLE 1

$R_g(A)$ after 48 hr. at $25^{\circ}C$ at different water and catalyst concentration

$H_2O:Si(OEt)_4$	HCl (M/ℓ) (2M $Si(OEt)_4$)				NH$_4$OH (M/ℓ) (1M $Si(OEt)_4$)
	10^{-2}	10^{-3}	10^{-4}	10^{-5}	10^{-2}
4	5.9	6.7	41.0	0	61.0
3	4.8	4.2	0	-	61.0
2	3.4	2.3	0	-	60.0
1	-	-	-	-	58.0

continued to grow slowly, but gelation did not occur. These phenomena can
be accounted for by the reaction mechanism proposed here.
      In the acidic solutions it appears that the oligomers of R_g ∿5A are
forming relatively quickly at all pHs and water concentrations, but further
particle growth occurs by agglomeration of these oligomers. Growth is slow
and is sensitive to pH and water concentration. The power law analysis of
the x-ray scattering of this system (Fig. 3) might then be interpreted as a
crossover from a low D in the range h = 0.02 to 0.04 arising from a random
growth process [1] to a larger D arising from the structure of the oligomer.
The very broad transition range might be due to polydispersity.

Condensation occurs rapidly between species which differ in degree of polymerization but is slow between species which are similar[6]. This implies the initial growth of these oligomers could be due to monomers adding to ends of chains, but if the ends of the chains reacted to form a ring, all of the species would have the same degree of polymerization and the reaction rate between fully hydrolyzed rings would be slow. Once formed separately it would be difficult for two rings to form a highly condensed entity. This is not to say that denser oligomers couldn't form but that they would require branching to occur very early in growth. In addition, rings and similar structures require a greater degree of hydrolysis and condensation to form a gel than would linear chains. For example, at least 2 more condensations per 6 member ring are required to form a "chain" of rings, a system which is still a liquid. However, at this degree of condensation, a system of long linear polymers could be crosslinked every 3 monomers and form a very stiff gel.

In alkaline solutions, the polymers continue to grow to a fairly large size even at very low water concentration (Table 1). This is consistent with the hypothesis that the hydrolysis of a single monomer tends to go to completion, allowing weakly crosslinked species to react with highly condensed species. Hence, at low water concentration, relatively few, but large, highly condensed species form, whereas in acid, under similar conditions, many more small, weakly condensed species are formed. This implies also that in alkaline solutions, gelation may occur at relatively low overall extents of hydrolysis as the large, hydrolyzed species could form a gel with unhydrolyzed monomer remaining in the solvent. Thus, the average chemical state of these systems is a poor predictor of polymer structure and gel formation.

CONCLUSIONS

The difference in the structure and chemistry of silica gels resulting from acid versus base catalyzed hydrolysis of silicon tetraethoxide can be accounted for by a series of reaction mechanisms discussed here. Nucleophilic hydrolysis and condensation in alkaline solutions tends to produce highly crosslinked species. The gels which form may not be completely hydrolyzed, but may not to reesterify during drying. In contrast, in acidic solutions, the electrophilic reaction mechanism tends to produce weakly crosslinked species which tend to be completely hydrolyzed in solution, but which under appropriate conditions are predicted to reesterify during drying. Both of these mechanisms only bias the system; other factors such as water concentration also affect the result.

REFERENCES

1. D. W. Schaefer, K. D. Keefer, and C. J. Brinker, this volume.
2. W. Stober, A Fink and E. Bohn, J. Coll. Interface Sci. 26, 62 (1968).
3. R. A. Assink and B. D. Kay, this volume.
4. C. J. Brinker, W. D. Drotning and G. W. Scherer, this volume.
5. J. E. Martin, to be published, Macromolecules.
6. R. K. Iler, The Chemistry of Silica (John Wiley, New York, 1979).
7. R. Aelion, A. Loebel and F. Erich, J. Am. Chem. Soc. 72, 5705 (1950).
8. C. J. Brinker, K. D. Keefer, D. W. Schaefer and C. S. Ashley, J. Non-Cryst. Solids, 48, 47 (1982).
9. C. J. Brinker, K. D. Keefer, D. W. Schaefer, R. A. Assink, B. D. Kay, C. S. Ashley, J. Non-Cryst. Solids 63, 45 (1984).

A COMPARISON BETWEEN THE DENSIFICATION KINETICS OF COLLOIDAL AND POLYMERIC
SILICA GELS

C. J. BRINKER,* W. D. DROTNING,* AND G. W. SCHERER,**
*Sandia National Laboratories, P. O. Box 5800, Albuquerque,
New Mexico 87185; **Corning Glass Works, Corning, New York

ABSTRACT

    Silica gels were prepared by three methods in which the
original silicate species varied from extended linear or
randomly branched polymers to more highly crosslinked clusters to colloidal particles of anhydrous silica. During isothermal sintering experiments, the viscosities of the two
polymer gels increased significantly (up to 3 orders of
magnitude) while the isothermal viscosity of the colloidal
gel was constant. Viscosity increases were explained by
crosslinking and structural relaxation of the polymeric gels.

INTRODUCTION

    Most researchers (e.g. [1]) who have attempted to describe the
gel → glass conversion for alkoxide-derived gels have utilized structural
models based on Iler's representations of colloidal silica gels formed in
aqueous solutions [2], i.e. the gels are considered to be composed of fully
polymerized (anhydrous) particles (Fig. 1a and b). Gel densification has
been considered to be essentially a sintering process [1], and differences
in densification behavior have been attributed to differences in texture,
e.g. the characteristic particle size, and/or differences in oxide composition, e.g. the densification temperature generally increases with T_g of
the corresponding melted composition.
    If these views are correct, it should be possible to quantitatively
describe gel densification by application of a viscous sintering model. The
purpose of the present investigation, therefore, is to compare the densification behavior of "polymer" gels (derived from metal alkoxides) to colloidal gels (composed of anhydrous, oxide particles) whose densification has
previously been described by a viscous sintering model [3]. Differences in
densification behavior between polymer and colloidal gels will be explained
in terms of gel structure as determined by this and previous investigations
of the sol → gel → glass conversion [4,5,6]. These differences will show
that it is seldom appropriate to strictly employ Iler's models of aqueous
silicate gels to describe the structures of metal-alkoxide-derived gels
prepared from alcoholic solutions.

GEL SYNTHESIS AND STRUCTURE

    Colloidal silica was prepared by flame oxidation of $SiCl_4$ as described
in reference 7. This technique results in fully dense, spherical SiO_2
particles 10-100 nm in diameter which may be single or clustered. Gels were
prepared by dispersing the particles in a non-polar solvent, e.g. chloroform, using long chain alcohols as a steric barrier to gelation. The
addition of a strong base such as an amine deprotonates the surface silanols
generating charges that cause gelation. The gel microstructure obtained,
after drying for 72 h at room temperature, is shown in FIG. 1c. This
xerogel has a density, ρ_0, of 0.54 g cm^{-3} (relative density, $\rho_0/\rho_s = 0.25$
where ρ_s = the density of the melted glass), a surface area of 80 m^2/g, and
the average pore diameter ∿60 nm. From FIG. 1c, it is difficult to resolve

26

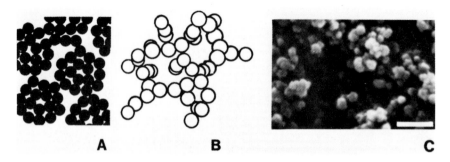

FIG. 1. a) hierarchical packing, b) ramdom packing coordination number =
3, c) SEM micrograph of colloidal gel, bar = 500 nm

whether this low density results from more open packing (coordination of
3-4) or from a hierarchical packing, e.g. in which primary particles are
loosely randomly packed (coordination ∿6) into agglomerates which are in
turn loosely randomly packed. However, it is likely that because of the low
surface tension of the solvent and the large pore size of the gel, the
capillary pressures during drying are minimized resulting in a weakly
compacted xerogel of low coordination number.

Acid (A2) and (B2) base catalyzed polymer gels were prepared using a
2-step hydrolysis process as described in detail in references 4 and 5
(samples A2 and B2 respectively). The first step consisted of hydrolyzing
TEOS with 1 mol H_2O/mol tetraethylorthosilicate (TEOS) under acidic
conditions. The second step, performed after 90 min. of reaction time,
consisted of the addition of water plus acid, A2, or base, B2 (H_2O:Si ∿4-5).

Small and intermediate angle x-ray scattering (SAXS) performed on these
solutions during the second hydrolysis step up to the gel point proved that
colloidal silicates were not formed in either the acid or base catalyzed
systems [5,6]. Instead (under acidic conditions), extended linear or
randomly branched polymers were formed which were highly overlapped prior to
gelation. ^{1}H NMR and gas chromatographic (GC) solution analyses [5] showed
that these polymers were completely hydrolyzed long before the gel point.
The base-catalyzed system resulted in discrete polymeric clusters which were
more highly condensed or collapsed compared to the acid-catalyzed polymers
(Figs. 2a and 3a). NMR and GC analyses indicated that gelation in B2
occurred before hydrolysis was complete. Incomplete hydrolysis was
attributed to unhydrolyzed monomer and perhaps only partial hydrolysis of
the polymers themselves.

Whereas desiccation does not change the skeletal structure of colloidal
gels, polymeric gel structures can change considerably during solvent
removal due to continued crosslinking, reesterification, dissolution and
repolymerization, and perhaps phase separation [8]. Additional crosslinking
occurs during desiccation of polymeric gels, because the increasing polymer
concentration forces reactive terminal OH groups to come within close
proximity. When the gel network becomes sufficiently crosslinked that it
can resist the compressive force of surface tension, additional solvent
removal results in the formation of internal porosity.

Under acidic conditions, silanols tend to be protonated, and the rate
of condensation according to: ≡SiO⁻ + HOSi≡ → ≡Si-O-Si≡ + OH⁻ is low. In
addition, reesterification is predicted to be enhanced under acidic
conditions [9] further reducing the condensation rate (in fact, as water
evaporates with alcohol as an azeotrope, reesterification may be complete).
When the extent of crosslinking is low, the tendency toward phase separation
is minimized [8]. Thus, as solvent is removed, the weakly crosslinked

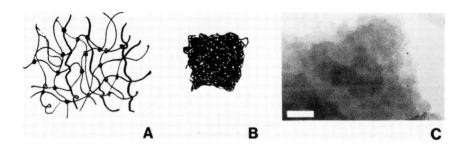

FIG. 2. a) schematic representation of acid catalyzed gel, b)
         desiccated xerogel and c) TEM micrograph of A2, bar = 25nm

network can uniformly compact to a relatively high density prior to pore
formation (FIG. 2).

    Conversely, lower density xerogels result from more basic systems in
which there exists greater concentrations of deprotonated silanols and in
which the rate of reesterification is suppressed [9]. Under these
conditions, the rate of condensation is increased causing an increased
tendency toward phase separation [8]. In addition, because of the greater
rate of depolymerization of the network at higher pH, ripening and neck
growth occur causing further fluctuations in polymer concentration and
reinforcing the gel network against compaction. Thus, porosity develops at
an early stage of desiccation and the resulting microstructure exhibits both
lower density and greater fluctuations in density than the acid-catalyzed
xerogel (FIG. 3).

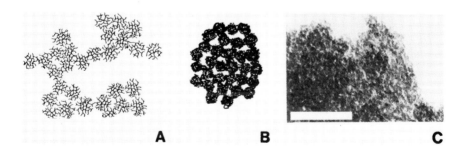

FIG. 3. a) schematic representation of base catalyzed gel, b) desiccated
         xerogel and c) TEM micrograph of B2, bar = 100 nm

    As observed by TEM, A2 and B2 xerogels exhibit textural differences
consistent with the above descriptions (FIGS. 2c and 3c). After drying at
$50^{\circ}C$ for 2 weeks, the initial relative densities were 0.50 and 0.29, and
from analyses of N_2 adsorption-desorption isotherms after degassing at
$150^{\circ}C$, the surface areas and average pore diameters were determined to be
929 and 942 m^2/g and 2.3 and 6.7 nm for A2 and B2, respectively.

WEIGHT LOSS AND SHRINKAGE

Weight loss and shrinkage were measured in desiccated air during constant heating rate (2-23°C/min) and isothermal treatments using a DuPont 1090 Thermal Analyzer and Theta dilatometer, respectively. Weight loss and shrinkage results were combined with the original xerogel density to calculate density as a function of temperature or time. For the colloidal gels, density was also determined by mercury porosimetry.

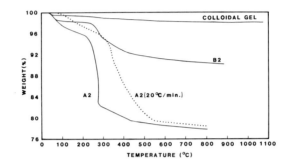

FIG. 4. Percent of original weight for colloidal gel, B2, and A2 heated at 2°C/min. Dotted line equals A2 heated at 20°C/min.

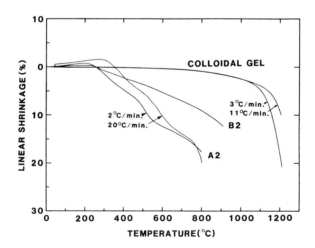

FIG. 5. Linear thermal shrinkage at constant heating rates for A2, B2, and colloidal gel.

Figures 4 and 5 show weight loss and shrinkage during constant heating rate experiments. Below 150°C, the weight loss is due primarily to desorption of physisorbed water and solvents. Above 150°C, weight loss is attributed to the removal of chemically bound water and (for A2 and B2) the pyrolysis of unhydrolyzed alkoxy radicals. The colloidal gel loses less than 1.5 wt% above 150°C substantiating the proposed model that this gel is composed of essentially fully polymerized (anhydrous) silica. By comparison B2 and A2 lose an additional 8 and 18 wt%, respectively. Because all of this weight loss is attributable either to Si-OR or Si-OH originally present in the desiccated gels, this proves that the polymer gels are initially less highly crosslinked (on average, fewer bridging oxygen atoms/silicon) than colloidal silica. Further evidence supporting this hypothesis was obtained from Raman spectroscopy: spectra of desiccated A2 gels showed only weak absorptions due to silica in four-fold coordination with bridging oxygens. However, there was strong evidence for chain-like structures [10].

The weight loss between 200 and 300°C for gels A2 and B2 heated at 2°C/min is accompanied by an exotherm and is attributed primarily to combustion of alkoxy radicals [4]. Weight loss above 300°C is attributed to dehydration [11]. A2 and B2 lose equal amounts of weight above 300°C, however, they differ significantly between 200 and 300°C. This difference indicates that B2 gels are initially more highly crosslinked and/or less highly esterified (terminal OR groups) compared to A2 gels, in keeping with the SAXS results and the predicted rates of reesterification. Further evidence in support of this idea is obtained from the shrinkage curves (FIG. 5). Below ∿150°C these curves represent the linear thermal expansion of the skeleton. A2 exhibits increased expansion compared to B2 indicative of a less highly crosslinked structure. Both A2 and B2 exhibit large expansion compared to fully crosslinked, colloidal silica.

Gel shrinkage has generally been attributed to viscous sintering. If this is correct, then the shrinkage rate is proportional to $\gamma/\eta d$ (where γ= surface energy, η = viscosity and d = pore diameter) and there is no associated weight loss. The colloidal silica gel used in the present investigation shrinks abruptly at $T > 1100°C$ where $\eta < 10^{12}$ poises [7]. Because of the much smaller pore sizes of A2 and B2, comparable shrinkage rates due to viscous sintering are expected at $\eta < 3 \times 10^{13}$ poises. As will be shown in the later discussion, this corresponds to $T > 700°C$ for A2 and $T > 900°C$ for B2. It is unlikely, therefore, that much of the shrinkage observed below ∿700°C for A2 and ∿900°C for B2 can be attributed to viscous sintering.

For temperatures less than 700°C, FIG. 4 and 5 suggest a relationship between shrinkage and weight loss, i.e. greater weight loss is accompanied by greater shrinkage. This implies that some of the shrinkage in A2 and B2 results from condensation reactions: $\equiv$Si-OH + HO-Si$\equiv$ → $\equiv$Si-O-Si$\equiv$ + H_2O. Compared to B2, A2 exhibits increased shrinkage between 400 and 550°C which is not accompanied by increased weight loss. We attribute this shrinkage to structural rearrangements. From the absence of any microstructural features which could be interpreted as primary particles (FIG. 2c), we feel it is unlikely that these rearrangements involve macroscopic movements of particles to higher coordination sites. Instead, we propose that rearrangement occurs by structural relaxation, i.e. the microscopic motions of atoms or polymer fragments which result in reduced excess free volume and increased skeletal density. Skeletal density measurements confirm this hypothesis: the skeletal density increased by 25% over this temperature interval. Preliminary DSC analyses showed an exotherm associated with this skeletal densification which is consistent with a structural relaxation mechanism [12].

To determine how the polymeric nature of gels A2 and B2 affect the kinetics of viscous sintering, isothermal shrinkage experiments were performed at temperatures where viscous sintering was expected to be the predominant shrinkage mechanism. Isothermal data were analyzed using a viscous sintering model based on a cylindrical cubic array geometry [3]. This is the only available model which describes initial and intermediate stage sintering of materials with continuous porosity. According to this model, the quantity K is calculated:

$$K \equiv \frac{\gamma}{\eta \ell_o} \ (\rho_o/\rho_s)^{1/3}$$

where ℓ_o is the original length of the cube edge and ρ_o/ρ_s is the original relative density. K is inversely proportional to viscosity and is normally a constant at constant temperature. This is demonstrated in FIG. 6 where K^{-1} is plotted versus density for the colloidal gel. By comparison the value of K^{-1} calculated for A2 is observed to increase by over 3 orders of magnitude at $708^\circ C$ (FIG. 7). Because ℓ_o does not increase, and γ does not

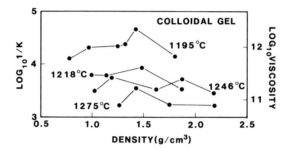

FIG. 6. Isothermal plots of K^{-1} (and viscosity) for the colloidal gel.

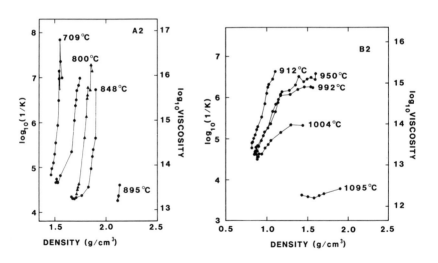

FIG. 7. Isothermal plots of K^{-1} (and viscosity) for A2 after heating at $2^\circ C/min$. ▲'s denote an initial heating rate of $20^\circ C/min$.

Fig. 8. Isothermal plots of K^{-1} (and viscosity) for B2 after heating at $2^\circ C/min$.

decrease, the increase in K^{-1} is a result of increasing viscosity. Both crosslinking [13] and structural relaxation [14] are known to increase the viscosity. Therefore, the isothermal sintering results (FIGS. 7 and 8) show that for every temperature investigated the viscosities of A2 and B2 increase isothermally as the gel structures progressively change (by crosslinking and structural relaxation) toward the equilibrium structures characteristic of the given temperatures. At temperatures $\geq 950°C$, the viscosities of B2 gels reached "plateau" values. Plotting these plateau viscosities according to the Arrhenius equation resulted in an activation energy of 155 Kcal/mole which is in the range normally reported for fused silica. This suggests that after sufficient heating, the skeletal structure of B2 becomes similar to the structure of fused silica.

The differences in structure between A2 and B2 are evident from the comparison of the temperature dependence of viscosity. Whereas the initial viscosity of B2 was 2×10^{13} poises at $912°C$, this same initial viscosity was obtained at $800°C$ for A2. This implies that compared to B2, A2 is less highly crosslinked and/or contains additional excess free volume so that its initial viscosity is reduced and correspondingly larger increases in η are observed during the isotherms.

The effects of reduced crosslinking and/or additional free volume are further illustrated by comparisons of the heating rate dependence of η. Crosslinking and structural relaxation, both of which are kinetically limited processes, are postulated to be the predominant shrinkage mechanisms below $700°C$ for A2. Therefore, as the heating rate is increased, the amounts of crosslinking and structural relaxation which occur while heating to a particular isothermal temperature are reduced (as evidenced by the initial reduction in shrinkage at $T < 700°C$ (FIG. 5) causing a corresponding reduction in η. As shown in FIGS. 5 and 7, where the values of shrinkage and η for A2 heated to $800°C$ at $20°/min.$, are compared to those values obtained at $2°C/min.$, the increased heating rate causes viscous sintering to commence at a lower temperature (resulting in increased density by $800°C$) and reduces the initial value of η at $800°C$. In contrast to these results, no heating rate dependence of shrinkage is observed below $1000°C$ for the colloidal gel (FIG. 5) and, because viscous sintering also is a kinetically limited process, the increased heating rate reduces the amount of shrinkage occurring between 1000 and $1225°C$. This result implies that, for this colloidal gel shrinkage occurs only by viscous sintering and indicates that the reduced viscosity resulting from the increased heating rate for A2 more than compensates for the expected reduction in shrinkage (when viscous sintering is the only shrinkage mechanism.)

CONCLUSIONS

Silica gels were prepared by 3 methods in which the original silicate species varied from extended linear or randomly branched chains (A2) to more highly crosslinked clusters (B2) to dense particles of colloidal silica. Weight loss measurements showed that the polymeric gels A2 and B2 were less highly crosslinked than the colloidal gel, and viscosity determinations showed that A2 was less highly crosslinked and/or contained additional excess free volume compared to B2. It could be argued that the differences between the polymer and colloidal gels are due only to the difference in size of the units which comprise the gels, i.e. compared to colloidal gels, polymer gels are composed of much smaller, anhydrous particles which have much higher surface areas and thus correspondingly higher surface hydroxyl contents. However, this argument cannot explain the substantial differences in densification behavior between the acid and base catalyzed polymer gels which have nearly identical surface areas.

Our results suggest that, because they are derived from weakly crosslinked polymeric silicates, polymer gels contain a very uniform distribution of non-bridging hydroxyl or alkoxy radicals which are thought to exist both within and on the surfaces of the skeletal phase. Compared to the colloidal gel in which OH exists only on the surfaces of larger, anhydrous particles, polymer gels contain no large-scale regions of fully polymerized silica. Therefore, on heating the skeletal phase continually evolves as polymerization and structural relaxation occur causing the skeletal structure to approach the structure of anhydrous silica. This skeletal evolution is manifested as increased viscosity during isothermal experiments.

ACKNOWLEDGEMENTS

The technical assistance of C. S. Ashley and D. L. Kirby is greatly appreciated.

REFERENCES

1. Z. Zarzycki, M. Prassas, J. Phalippou, J. of Mat. Sci. 17 3371-3379 (1982)
2. R. K. Iler, The Chemistry of Silica, (John Wiley & Sons, New York, 1979).
3. G. W. Scherer, J. Am. Ceram. Soc. 60 236-239 (1977).
4. C. J. Brinker, K. D. Keefer, D. W. Schaefer, and C. S. Ashley, J. Non-Crystl. Solids 48 47-64 (1982).
5. C. J. Brinker, K. D. Keefer, D. W. Schaefer, R. A. Assink, B. D. Kay and C. S. Ashley, J. Non-Cryst. Solids 63 45-59 (1984).
6. D. W. Schaefer, K. D. Keefer, C. J. Brinker, Polymer Preprints Am. Chem. Soc., Div. of Polym. Chem. 24 239 (1983).
7. G. W. Scherer and J. C. Luong, J. Non-Cryst. Solids 63 163 (1984).
8. D. W. Schaefer, K. D. Keefer, and C. J. Brinker, these Proceedings, p. 1.
9. K. D. Keefer, these Proceedings, p. 15.
10. C. J. Brinker and L. C. Klein, unpublished results.
11. T. A. Gallo, C. J. Brinker, L. C. Klein, G. W. Scherer, these Proceedings, p. 85.
12. C. J. Brinker, E. P. Roth, G. W. Scherer, to be presented at Conference on Effects of Modes of Formation on the Structure of Glass, Nashville, TN, July 9-12, 1984.
13. G. Hetherington, K. H. Jack, and J. C. Kennedy, Phys. Chem. Glasses 5 130 (1964).
14. S. M. Rekhson et al., Sov. J. Inorg. Mat. 7 622-623 (1971).

EFFECT OF WATER ON ACID- AND BASE-CATALYZED HYDROLYSIS OF TETRAETHYLORTHOSILICATE (TEOS)

L. C. KLEIN* AND G. J. GARVEY**
*Rutgers-The State University of New Jersey, Ceramics Department, P.O. Box 909, Piscataway, NJ 08854; **Room 12-007, MIT, Cambridge, MA 02139

ABSTRACT

Two series of gels were prepared by hydrolyzing TEOS with HCl or NH_4OH catalyst. In the first series, the molar ratio of water to TEOS was 4:1 and the catalyst addition was (1×10^{-3}), (4×10^{-3}), (1×10^{-2}), (4×10^{-2}), and (1×10^{-1}). In the second series, the catalyst addition was (1×10^{-3}) moles and the molar ratio of water to TEOS was 2:1, 4:1, 8:1, 16:1, and 32:1. Dried samples were characterized for surface area, porosity and C and H content. Samples were heated to $800^\circ C$ and thermogravimetric weight loss was recorded. Both HCl and NH_4OH are catalysts for hydrolysis, but during polymerization, the relative rates of hydrolysis and condensation determine resulting dried microstructures.

INTRODUCTION

The mechanisms for hydrolyzation in acid- and base-catalyzed solutions of ethanol-water-tetraethylorthosilicate (TEOS) have been studied directly with titration and gas chromatography [1,2,3,4] and indirectly by monitoring changes in physical properties, most often viscosity, [5,6,7,8]. The mechanism for acid-catalyzed hydrolysis is electrophilic attack and the mechanism for base-catalyzed hydrolysis is nucleophilic attack [9]. Not only are the mechanisms different, the kinetics are different. For acid-catalysis, hydrolysis is complete and the number of unreacted hydroxyls per silicon decreases with decreasing acid concentration [1]. For base-catalysis, hydrolysis is incomplete and the effect of polymerization is large. At the same time, the solubility of silicic acid in basic solutions makes dissolution and rearrangement of the growing polymer possible [10].

If acid and base had equivalent catalytic effect, it would be easier to compare the results using a parameter such as pH. However, the catalysts commonly used, HCl and NH_4OH, do not have equal strength and an alcohol-water mixture is a further complication. Several attempts have been made to compare the effects of acid and base. One scheme that qualitatively predicts physical properties of dried gels is that acid-catalyzed solutions yield linear polymers which become tangled at the sol-gel transition, while base-catalyzed solutions yield branched clusters which coalesce [11]. Accordingly, the dried gels can be classified by texture, either fine or coarse. Likewise, acid-catalyzed gels have higher bulk densities [12], and base-catalyzed gels are very friable. Acid-catalyzed gels should have hydroxyl contents that scale with surface area, while base-catalyzed gels do not, since the solubility of silicic acid leads to internal condensation [10]. In the course of rearrangement in base-catalyzed gels, the NH_4OH content is reduced and the polymer does not gel as a unit. Rather, what appears to be particles sediment from the solution, though in detail the particles are not colloids [9]. Of course, the level of water in the initial solution makes distinctions between acid vs base and high vs low level of catalyst even more uncertain.

Mat. Res. Soc. Symp. Proc. Vol. 32 (1984) Published by Elsevier Science Publishing Co., Inc.

In an effort to study systematically the role of catalyst in the hydrolysis of TEOS, this study was undertaken. A series of solutions with constant molar ratio water to TEOS and varying catalyst addition was prepared, along with a series of solutions with the same catalyst addition and varying molar ratio water to TEOS. By characterizing surface area and porosity of dried gels, C and H content of dried gels and thermogravimetric weight loss for fired gels, it was hoped that a better understanding of the role of the catalyst would emerge in the processing of monolithic silica shapes [12].

Experimental Techniques

Two series of acid and base-catalyzed solutions were prepared. In the first series, the molar concentration of catalyst was (1×10^{-3}), (4×10^{-3}), (1×10^{-2}), (4×10^{-2}), and (1×10^{-1}) for both HCl and NH_4OH, with a water to TEOS molar ratio of 4:1. TEOS and ethanol were added in equal volumes for all samples. In the second series, the water to TEOS ratio was varied. The ratio was 2:1, 4:1, 8:1, 16:1 and 32:1 for both HCl and NH_4OH with molar concentrations of 1×10^{-3}. The volume ratio of ethanol to TEOS for this series was 4:1 to permit solubility of more water.

Twenty ml Pyrex^{TM} scintillation vials with airtight polyethylene lids were used to mix, gel and dry the samples. Each vial was filled with 8 ml of TEOS and the appropriate volume of ethanol. Then the aqueous electrolyte was added, and finally the water was added. The vials were shaken and placed in a drier at $80^{\circ}C$ to react for two days. In this time all acid-catalyzed solutions gelled. No base-catalyzed solutions had gelled but most had sediment. After two days of reacting at $80^{\circ}C$, samples were uncapped and placed in a bell jar into which dry nitrogen was flowed. The bell jar was maintained at $80^{\circ}C$. After 24 hours the bell jar was heated to $200^{\circ}C$. After another 24 hours, samples were withdrawn, capped and allowed to cool. The high drying temperature eliminated most physically trapped water and ethanol.

Samples for nitrogen adsorption treatments were ground in an alumina mortar and pestle. For samples, with water to TEOS ratios equal or greater than 4:1 a sample size of 20 mg was used. For a water to TEOS ratio of 2:1 a 500 gm sample was required because of its low surface area. Samples (-200 to +325 mesh) were weighed into the sample cell. The cell was then attached to the outgassing station of the Quantasorb^{TM} Surface Area Analyser (Quantachrome Corp., Greenvale, NY). High purity dry nitrogen was flowed over the sample at a rate of 5 ml/min. Samples were outgassed for 12 hours at $200^{\circ}C$. Standard procedures recommended by the manufacturer for obtaining nitrogen adsorption-desorption isotherms were used. The volume of pores with radius less than 50 nm was determined by first adsorbing pure nitrogen then switching to a 98% N_2-2% He mixture at 77K. As the sample was heated to room temperature, the desorption signal was integrated and calibrated. Isotherms were obtained with ten prepared mixtures of helium and nitrogen. Surface area was estimated from the isotherms.

Quantitative chemical analyses for C and H in dried gels were carried out using combustion analysis (Robertson Laboratory, Florham Park, NJ). Also, samples were heated in a Mettler thermal analyzer to $800^{\circ}C$. Total weight loss between 120 and $800^{\circ}C$ was recorded during thermogravimetric analysis.

Results

All samples had reacted by the third day. At that time, there was a sharp contrast in the appearance of acid-catalyzed and base-catalyzed gels. All acid-catalyzed gels were transparent. Cracks developed when they were gelled in the capped vials and conchoidal patterns could be seen. All

base-catalyzed gels were cloudy. The samples prepared with 4 moles water
showed a tendency to sediment. The thickness of the sediment layer increased with increasing base addition. For the same molar addition of
base, the cloudiness decreased as the water level increased. When the
acid-catalyzed gels were dried, the result was large, transparent fragments. When the base-catalyzed gels were dried, the result was weakley
coalesced powder.

The adsorption-desorption isotherms for acid-catalyzed gels at all
water levels showed little or no hysteresis. This implies cylindrical
pores. According to the desorption isotherms for the five water levels
with constant acid addition, the volume of liquid nitrogen per gram gel
(V_s) increased for all nitrogen partial pressures as the water level increased. The pore volume was read directly from the isotherm at the point
corresponding to 98% N_2. The single point BET value was obtained using the
V_s value corresponding to 10% N_2. Each isotherm had an inflection point at
intermediate values of nitrogen partial pressure which became more pronounced with increased water. The plot of the isotherm approached the y-
axis at right angles at high nitrogen partial pressures. The isotherms
for base-catalyzed gels showed increasing hysteresis at high nitrogen
partial pressures with increasing water level. This implies non-uniform
cross-section pores. For the five water levels with constant base
addition, lower V_s values were measured for all nitrogen partial
pressures as the water level increased. This is the opposite trend from
that seen in acid-catalyzed gels. For a constant water level, the V_s value
decreases with increased base addition, as well.

The measured surface area for the constant water level series, with
the corresponding porosity, is plotted in Figure 1a. The surface area and
porosity for the constant catalyst addition is plotted in Figure 1b. For
the gels prepared with 4 moles water per mole TEOS, the acid level has
little effect on surface area or porosity, while an increase in the base
level reduces surface area and porosity. For the gels prepared with .001
moles catalyst, the acid-catalyzed gels show an increase in surface area
and porosity with an increase in water, while base-catalyzed gels show a
decrease, except that the base-catalyzed gel with 4 moles water per mole
TEOS has a higher porosity than ratios 2:1 or 8:1.

The analyzed C and H contents of dried gels (all in weight %) are
given in the Table. For the gels prepared with 4 moles water per mole
TEOS, there is little variation in C or H with increasing acid addition,
while the hundred-fold increase in base reduces the C and H by half. For
the gels prepared with .001 moles catalyst, both acid- and base-catalyzed
gels show a reduction in C and H for increased water levels. At the
highest water level, the analyses are similar, though the low water acid-
catalyzed gels have higher C and H contents than low water base-catalyzed
gels.

The total weight loss between 120 and 800°C was recorded. Thermogravimetric analysis was accompanied by thermal analysis. The only event
recorded with thermal analysis was oxidation of residual organics at 410°C.
The height of the exothermic peak scaled with the C and H analysis, as
expected. The fraction gravimetric weight loss is plotted in Figure 2a for
the constant water level series and in Figure 2b for the constant catalyst
addition series. For the gels prepared with 4 moles water per mole TEOS,
there is little change in fraction weight loss with increasing acid addition, but a noticeable decline in fraction with increasing base. For the
gels prepared with .001 moles catalyst, both acid- and base-catalyzed gels
show a decline in fraction weight loss for increased water levels, except
that the base-catalyzed gel with 4 moles water per mole TEOS has a higher
weight loss.

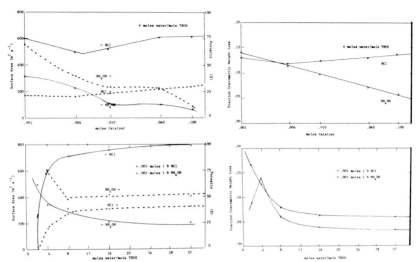

Fig 1a. Surface area (right) and % porosity (left) vs moles catalyst for 4 moles water/mole TEOS.
Fig 1b. Vs moles water for (1×10^{-3}) moles catalyst.

Fig 2a. Fraction gravimetric weight loss vs moles catalyst for 4 moles water/mole TEOS.
Fig 2b. Vs moles water for (1×10^{-3}) moles catalyst.

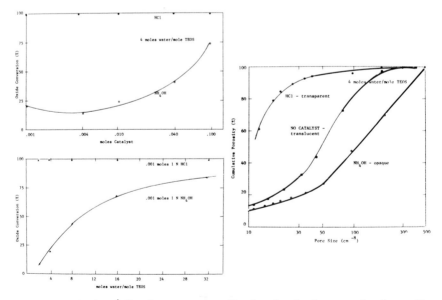

Fig 3a. Calculated %oxide conversion vs moles catalyst for 4 moles water/mole TEOS.
Fig 3B. Vs moles water for (1×10^{-3}) moles catalyst.

Fig 4. Cumulative porosity for acid- (filled circles), base- (open circles) and un-catalyzed (boxes) solutions normalized to 50 nm.

TABLE I. Effluent Analysis for C and H (all in weight %).

Moles Catalyst Addition	4 moles water/mole TEOS (HCl)		(NH$_4$OH)	
	C	H	C	H
.001	6.11	1.68	8.52	1.92
.004	5.97	1.68	7.10	1.70
.010	6.29	1.70	4.01	1.01
.040	5.60	1.49	3.24	0.96
.100	5.92	1.62	3.04	0.93

$\dfrac{\text{moles H}_2\text{O}}{\text{moles TEOS}}$	.001 mole 1 N Catalyst (HCl)		(NH$_4$OH)	
	C	H	C	H
2	13.15	2.83	4.25	1.46
4	6.11	1.68	8.52	1.95
8	3.48	1.14	2.36	0.88
16	2.14	0.95	2.02	0.69
32	1.79	0.95	1.80	0.63

Discussion

Using the fraction weight loss from gravimetry, it is possible to estimate the % oxide conversion. The oxide conversion is the net dry weight of the sample times one minus the fraction weight loss divided by the calculated oxide content of TEOS used. TEOS is 28.85% by weight silica. The estimated oxide conversion for all acid-catalyzed gels is 100%. The oxide conversion for base-catalyzed gels increases with increasing catalyst addition (Figure 3a) or increasing water (Figure 3b) but never reaches 100%. The chemistry of dried and fired gels is a result of both reaction kinetics and microstructure. From the start, the acid-catalyzed solutions undergo more complete hydrolysis than base-catalyzed solutions, even though the rate of hydrolysis with respect to the rate of condensation is slower. With acid-catalyst, hydrolysis is reversible, so some reesterification occurs. This explains the C and H content of dried acid-catalyzed gels.

The base-catalyzed solutions do not undergo complete hydrolysis. The rate of hydrolysis may be faster than the rate of condensation, but the strong crosslinking during polymerization makes elimination of alkoxy and hydroxy groups within clusters difficult. The conversion efficiency in low water base-catalyzed is only 25%. Since the entire container of solution does not gel, it is likely that unreacted TEOS remains in the liquid above the sediment layer. As the sediment is heated to convert the gel to oxide, much of the TEOS has been discarded with the solvent.

The microstructure of dried and fired gels has an effect on the chemistry of gels, especially when bulk samples are considered. Continuous open porosity allows oxidation of residual organics, where closed porosity usually gives bloating. The adsorption-desorption isotherms can be used to plot a pore size distribution. The result for acid-catalyzed gels is a narrow distribution, with the width and mode of the peak decreasing in radius and the height increasing as the water level is increased. The absence of hysteresis in the isotherm indicates cylindrical pores. The result for base-catalyzed gels is a broad distribution at low water levels which becomes bimodal at intermediate water levels. The distribution peaks are located at 1.5 and 5 nm. The 5nm peak increases and the 1.5 nm peak decreases as the water level is increased further.

Using the desorption portion of the isotherm, the cumulative pore volume was calculated. The cumulative porosity is plotted vs pore size in Figure 4. The calculation was normalized to 50 nm. The three curves in Figure 4 are a comparison of an acid-catalyzed gel with .001 moles HCl, a gel with no catalyst addition and a base-catalyzed gel with .001 moles NH_4OH. Normalizing the cumulative porosity to 50 nm is a fair assumption for the acid-catalyzed gel because the curve approaches 100% asymptotically. For the base-catalyzed gel, the curve is inclined at 100% which means the representation of cumulative porosity between 1 and 50 nm is not accurate. However, the base-catalyzed gels with water levels 16 and 32 moles give cumulative porosity curves with an inflection lower than 50 nm so that the curve approaches 100% more nearly horizontal. The shape of these curves predicts the optical appearance of samples. Acid-catalyzed gels are transparent, uncatalyzed gels are translucent and base-catalyzed gels scatter light to produce opacity.

There is a noticeable difference in appearance in the newly gelled solutions which can be traced to the hydrolysis mechanisms. In the acid-catalyzed solution, the first hydrolysis of the TEOS monomer is easier than the second, so a growing polymer will have an even distribution of hydroxyls. The polymer can form an occasional crosslink by a condensation reaction, and any structural features which might scatter light remain too small during gelling or drying. In the base-catalyzed solution, the successive hydrolysis of the TEOS monomer becomes faster, so that condensed polymers co-exist with unreacted monomer. Structural features develop in the overall polymer network that scatter light, both in solution while gelling and in the drying sediment.

Lacking, for the moment, direct evidence of the size of features from techniques such as electron microscopy, the molecular structure upon gelling has to be inferred from the microstructure of dried gels. Using surface area and pore volume measurements, it is possible to suggest that the uniform interconnected porosity of acid-catalyzed gels derives from a molecular structure largely composed of linear chains. In the base-catalyzed gels, it would seem that the polymers are branched, except that caution must be exercised when inferring the structure of base-catalyzed solutions from their dried gels. Silicic acid is more soluble in basic solutions. This is shown by the trend in surface area decreasing with base (Figure 1a) and increasing water (Figure 1b). With this increased solubility, silica dissolves and reattaches far more easily in basic than acidic aqueous medium.

At the same time, the base catalyst is being consumed in the nucleophilic attack. Both HCl and NH_4OH serve as catalysts to promote hydrolysis, but HCl retains its catalytic role through polymerization as NH_4OH becomes a reactant. This difference in catalytic role is exaggerated when considering effect of water. High water base-catalyzed solutions appear more like acid-catalyzed solutions than low water base-catalyzed solution, probably because an abundance of water, as a source for protons and OH, makes the consumption of base-catalyst slower. As a result, the base-catalyzed high water solution is largely hydrolyzed, with condensation being mildly effected. In comparing the effect of catalyst addition vs water level, first in acid-catalyzed solution, the acid addition has little effect while the water level has a strong effect, on surface area, porosity and oxide conversion. The low water solution gels to a polymer which is so weakly reacted that it continues to shrink after 12 hours at $200°C$. The high water solution gels to a polymer with a uniform skeleton and interconnected cylindrical pores, as measured by BET. In base-catalyzed solutions, both the base and the water level have strong effects. For example, the high water solutions are more completely hydrolyzed and condensed, as are the high base solutions, than low water or low base. In fact, dilution with ethanol allows for even more complete hydrolysis because reactants remain in solution longer. In contrast to the effect of water in acid-catalyzed solutions, which give some of the highest surface areas measured in dried gels, an increase in water level leads to lower surface area dried gels from base-catalyzed solutions. The lower surface area can be explained by the condensation mechanisms [2] or the increased solubility of silicic acid [10].

Reviewing the trends observed in acid and base-catalyzed gels, a high water acid-catalyzed solution should produce a gel which has high surface area, interconnected pores and low residual C and H which will oxidize during firing. These features are important when considering formation of silica monoliths and converting these to dense shapes. There are drying effects and temperature effects, as well, but the effect of catalyst is some what clearer in this system as a result of this systematic study.

Conclusions

Catalyst concentration and water level have been varied in the TEOS-water-ethanol system. HCl and NH_4OH serve to catalyze hydrolysis, but NH_4OH is consumed during condensation. Increased water levels increase the rate of hydrolysis in both cases. The acid addition has little effect on surface area, porosity or oxide conversion, while the base addition has a large effect. The effect of base is due to the condensation scheme or the solubility of silicic acid, which in either case gives a more condensed polymer at the time of gelling. Increasing base addition decreases surface area and porosity, and the oxide conversion is never complete. In appearance, acid-catalyzed solutions give transparent gels, while base-catalyzed solutions are translucent to opaque.

Acknowledgement--The financial support of NSF Division of Materials Research (DMR80-12902) is greatly appreciated.

REFERENCES

1. R. Aelion, A. Loebel and F. Eirich, Am. Chem. Soc. 72 (1950) 5705-5712.
2. C. J. Brinker, et. al., J. Non-Crystal. Solids 48 (1982) 47-64.
3. C. J. Brinker, et. al., J. Non-Crystal. Solids (1983) to be published.
4. L. C. Klein and G. J. Garvey, Am. Chem. Soc. Ind. Eng. Chem. No. 194 Symposium (1982) Chapter 18.
5. K. Kamiya, S. Sakka and M. Mizutani, Yogyo Kyokai Shi 86 (1978) 553-559.
6. M. Nogami and Y. Moriya, J. Non-Crystal. Solids 37 (1980) 191-201.
7. P. Yu, H. Liu and Y. Wang, J. Non-Crystal. Solids 52 (1982) 511-520.
8. W. C. LaCourse, et. al., J. Canadian Ceramic Soc. (1984) to be published.
9. K. D. Keefer. This volume.
10. R. Iler, The Chemistry of Silica, John Wiley and Sons, Inc. 1979.
11. C. J. Brinker and G. W. Scherer, Submitted to J. Non-Crystal. Solids, 1983.
12. L. C. Klein and G. J. Garvey, Ultrastructure Processing, John Wiley and Sons (1984).

GEL STRUCTURES IN LEACHED ALKALI SILICATE GLASS

Bruce C. Bunker, Thomas J. Headley, and Sally C. Douglas
Sandia National National Laboratories, P. O. Box 5800,
Albuquerque, New Mexico 87185

ABSTRACT

Transmission electron microscopy of leached $Na_2O \cdot 3SiO_2$ and $K_2O \cdot 3SiO_2$ glasses showed that the hydrated glass surface is a silica rich gel which phase separates during leaching. Phase separation is initiated when pockets of an aqueous phase nucleate and grow in the hydrated silica matrix. The aqueous pockets eventually become a network of interconnected pores with a pore diameter of ~ 30 nm. In advanced stages of leaching, the remaining interconnected silica rich phase resembles an aggregation of colloidal silica particles. The observed phase separation is consistent with literature models concerning the chemistry of silica in aqueous solutions and in leached glass.

INTRODUCTION

When alkali silicate glasses react with water, an alteration layer which is depleted in alkali cations and enriched in water usually forms on the glass surface. The layer is referred to as the gel layer, since its composition and physical properties resemble those of silica gel. Recently, small angle x-ray scattering results[1] reported for leached $Na_2O \cdot 3SiO_2$ (mole ratio) and $4BaO \cdot 28Na_2O \cdot 68SiO_2$ (mole ratio) glass suggested that the gel layer is not homogeneous, but contains either voids or particles as large as 15 nm in diameter, indicative of phase separation. Raman spectroscopy analyses[2] on leached $Na_2O \cdot 3SiO_2$ glass suggest that the phases which form in the surface gel consist of a phase resembling fused silica and an aqueous phase containing silicic acid. Glass dissolution studies indicate that although the surface gel functions as a diffusion barrier in the early stages of leaching, the barrier is ineffective at longer times, leading to linear leaching kinetics[3]. A change in gel structure, such as phase separation, might account for the observed changes in leaching kinetics.

In order to elucidate the structures of surface gels and to explore possible relationships between gel structures and alkali leaching kinetics, transmission electron microscopy (TEM) was performed on leached alkali-silicate glass as a function of solution pH, temperature, and leaching time. TEM confirmed that a silica-water phase separation occurs in leached surface gels. Observed structures are shown to be consistent with the aqueous chemistry of silica and silica gels.

EXPERIMENTAL

Two glasses ($Na_2O \cdot 3SiO_2$ and $K_2O \cdot 3SiO_2$, mole ratio) were prepared by mixing desired amounts of reagent grade alkali carbonates and silica. These were melted at $1500^{\circ}C$, stirred for 18 hours, cast into slabs, and annealed at appropriate temperatures, depending on the glass composition. The slabs were cut into 1.0 cm x 0.5 cm x 0.15 cm samples, polished to a 1 μm finish with diamond paste, and ultrasonically cleaned in methanol.

Glass samples were leached for various times in 100 ml of deionized water, HCl, or NaOH, depending on the desired solution pH. For intermediate solution pH values, the pH of the deionized water was held constant using pH stat titration techniques. A constant temperature water bath was used to control the temperature of the leachate. After leaching, samples were rinsed in methanol and air dried. For TEM examination, leached surface layers were scraped off, ground into thin flakes, and dispersed onto a carbon substrate using a butanol slurry. Transmission electron micrographs were obtained at 200kV in a JEM 200CX using low beam intensity to avoid beam alteration of the sample.

RESULTS

TEM revealed that phase separation occurs in surface gels for both $Na_2O \cdot 3SiO_2$ and $K_2O \cdot 3SiO_2$ under all leaching conditions investigated ($T = 20^\circ C - 80^\circ C$, $pH = 1 - 11$). The kinetics of phase separation depends on glass composition, temperature, and pH, and in general follows alkali leaching kinetics.

The first stage of gel formation involves alkali leaching without any apparent change in the amorphous structure of the glass. Next, small (5 nm) droplets of an aqueous phase begin to nucleate within the gel (Fig. 1). The nucleated droplets grow in size and decrease in number and can attain an average droplet size as large as 30 nm (Fig. 2). During coarsening, the pores (or droplets) begin to interconnect (Fig. 3), eventually resulting in a totally interconnected aqueous phase (Fig. 4). Finally, the silica rich phase collapses into a morphology which resembles an aggregation of colloidal silica particles (Fig. 5). In fact, the only discernible difference between leached surface gels in the final stage of phase separation and gels made at similar pH values from colloidal silica particles (Ludox HS-40[a])(Fig. 6) is the greater degree of interparticle necking observed in the leached alkali silicate gels.

The above progression as a function of time was deduced from an examination of many thin flakes from the gel layer, since each flake generally revealed only one stage in the morphological development and its relative position within the gel layer was unknown. However, in a few instances, flakes were observed which apparently had been located near the interface between the surface gel and unleached glass. For $K_2O \cdot 3SiO_2$, the entire phase separation sequence has been observed in a single flake (Fig. 7), going from isolated water droplets in a silica rich matrix (on the left) to a colloidal silica aggregate in water (on the right). For example, the interface between leached and unleached glass has apparently moved to the left faster than the phase separated morphology was able to develop in the freshly leached material. For $Na_2O \cdot 3SiO_2$, a flake thought to represent the glass:gel interface (Fig. 8) exhibits a much sharper phase transition zone. Here, the phase separation process has progressed all the way to the colloidal particle stage before the glass:gel interface has moved an appreciable distance. The sharpness of the transition zone varies with glass composition and pH, and appears to be controlled by the relative rates of alkali leaching and phase separation. While alkali leaching occurs at a much slower rate for $Na_2O \cdot 3SiO_2$ than for $K_2O \cdot 3SiO_2$[3], phase separation rates may be comparable for the two glasses.

More complete mechanistic details of phase separation in leached surface gels could be obtained if fully intact leached layers could be examined in the TEM. To our knowledge, ion beam milling is the only way to produce suitable thinned specimens. Unfortunately, it appears that the ion milling step destroys the phase separated gel morphology. All ion milled layers examined to date appear to be TEM amorphous. Figure 9 illustrates that even the TEM electron beam is capable of amorphotizing phase separated gels and that care must be taken in examining gels by TEM techniques.

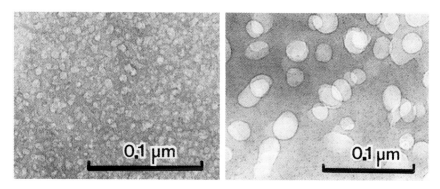

FIG. 1. K$_2$O·3SiO$_2$ leached 1/2 hr.,
T = 80°C, pH 1.5

FIG. 2. Na$_2$O·3SiO$_2$ leached 4 hr.
T = 80°C, pH 10.7

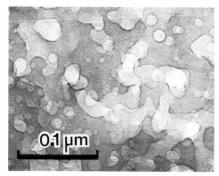

FIG. 3. K$_2$O·3SiO$_2$ leached 1/2 hr.,
T - 80°C, pH 10.5

FIG. 4. K$_2$O·3SiO$_2$ leached 8 hr.,
T - 22°C, pH 1.4

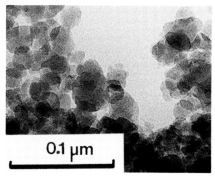

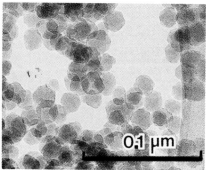

FIG. 5. Na$_2$O·3SiO$_2$ leached 2 days,
T = 80°C, pH 10.5

FIG. 6. Ludox HS-40 gel, gelation
pH = 5.4

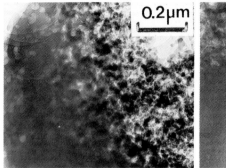

FIG. 7. $K_2O \cdot 3SiO_2$ leached 1/2 hr.,
         T = 80°C, pH 10.5

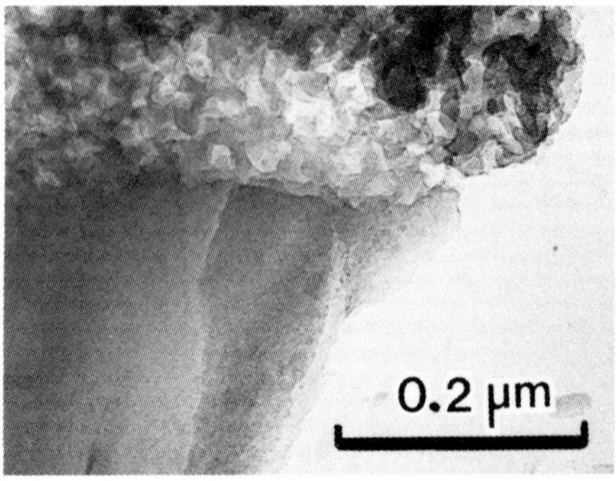

FIG. 8. $Na_2O \cdot 3SiO_2$ leached 4 hr.,
         T = 80°C, pH 1.5

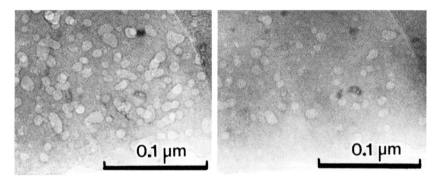

FIG. 9. TEM electron beam damage in leached $K_2O \cdot 3SiO_2$ after
         (a) 10 sec. and (b) 45 sec. of beam time.

DISCUSSION

We present here a plausible mechanism which can explain the formation
of phase-separated leached layers on alkali silicate glasses. The chemical
reaction sequence is given below:

1) $\equiv Si-O^- Na^+ + H_3O^+ \longrightarrow \equiv Si-OH \cdot H_2O + Na^+$

2) $\equiv Si-O-Si(OH)_3 \cdot H_2O \longrightarrow \equiv Si-OH + Si(OH)_4$

3) $\equiv Si-OH \cdot H_2O \longrightarrow \equiv Si-O-Si \equiv + 3 H_2O$

Bulk alkali silicate glasses contain a random distribution of anionic
non-bridging oxygens which are charge compensated by alkali cations. During
leaching, an ion exchange reaction occurs (reaction 1) in which alkali
cations are replaced by protons to make silanol groups. The silanol groups
are hydrophilic, and readily hydrogen bond to water molecules. The water
molecules can readily hydrolyze adjacent Si-O bonds to depolymerize the
silicate network and ultimately produce dissolved silicates such as silicic
acid (reaction 2) some of which is leached from the glass. Simultaneously,

repolymerization reactions can also occur (reaction 3), both between adjacent silanol groups and between network silanols and free silicate species dispersed in the gel.

The key to phase separation is that the depolymerization and repolymerization reactions do not occur in a random fashion. Silanol groups are hydrophilic and attract water, while siloxane bridging bonds (Si-O-Si) are hydrophobic and repel water. Therefore, as repolymerization of the silicate network proceeds, silicate clusters are produced in which all silanol groups and their associated water molecules occupy the cluster surface, while the hydrophobic siloxane bridges reside in the cluster interior, creating a highly crosslinked network containing few silanol groups or entrapped water molecules[4]. As the clusters grow with time, water which is expelled to cluster surfaces eventually nucleates into the aqueous droplets observed in the TEM. After the aqueous phase interconnects, the silicate clusters continue to grow until they resemble interconnected colloidal silica particles.

The ultimate size of colloidal silica particles grown in aqueous solutions via polymerization reactions is reported[4] to vary from 2-4 nm, for colloids grown in acidic solutions, to over 200 nm, for colloids grown in basic media. The microstructures observed in the TEM are consistent with colloid growth in a medium having a pH of 9-11 regardless of the solution pH. The TEM results are consistent with pH measurements[3] which show that the pH within the gel varies only between pH9 and pH11 as the solution pH is varied from pH1 to pH12. The relatively constant gel pH is due to the high concentration of silanol groups in the leached gel which buffers the gel pH. Changes in colloid morphology with pH might occur in gels having a lower silanol concentration, or at the immediate solution/gel interface, where the solution pH can dominate the capacity of the silanol buffer.

The molecular structure of leached gel layers is in sharp contrast to the structures reported for gels formed in ethanol by the hydrolysis and subsequent polymerization of tetraethylorthosilicate $[Si(OC_2H_5)_4][5]$. In ethanol, the tendency for phase separation is less pronounced than it is in water, since differences in solvation of silanol groups and siloxane bridges are less pronounced. In basic ethanol solutions, where rapid hydrolysis occurs, colloidal silicate particles are observed, but small angle x-ray scattering results show that the solid/liquid phase boundary of the particle surfaces are much more diffuse than for the alkali silicate gels, suggesting that much of the particle interior is not fully polymerized. In acidic ethanol solution, the gel polymerizes as an amorphous network of linear polymeric silicate chains. For gels formed in water, silicate chains are inherently unstable relative to highly crosslinked colloidal silica particles regardless of the solution pH.

The phase separation sequence depicted in Figure 1-6 should influence the ability of the leached surface gel to function as a diffusion barrier[3]. If one assumes that diffusion through the aqueous phase in the phase separated material is rapid relative to diffusion through the various silicate phases, it becomes clear that diffusion through the gel should become more rapid as a function of time. Leaching kinetics suggest that repolymerization reactions can accelerate diffusion prior to the observation of phase separation by TEM. Leaching data indicate that the surface gel ceases to function as a diffusion barrier even before it reaches the colloidal particle stage of phase separation (Fig. 6). The direct correlation between phase separation and alkali leaching kinetics will be detailed in a subsequent publication.

CONCLUSIONS

TEM has shown that phase separation occurs in gel layers formed on leached alkali silicate glasses, indicating that the gel structure breaks down and repolymerizes after the removal of alkali cations. It is proposed that during repolymerization, the silicate network is converted from a

nearly homogeneous structure containing a random distribution of
non-bridging oxygens and water molecules into a phase separated structure
consisting of a network of fused silica particles (containing few internal
silanol groups or water) surrounded by an aqueous phase containing dissolved
silicate species. Such a process can lead to the breakdown of the gel as an
effective diffusion barrier.

REFERENCES

1. M. Tomozawa and S. Capella, J. Am. Ceram. Soc., 66, C-24 (1983).
2. G. J. Exarhos and W. E. Conaway, J. Non-Cryst. Solids, 55, 445
   (1983).
3. B. C. Bunker, G. W. Arnold, E. K. Beauchamp, and D. E. Day, J.
   Non. Cryst. Solids, 58, 295 (1983).
4. R. K. Iler, The Chemistry of Silica, John Wiley & Sons, New
   York, 1979.
5. C. J. Brinker, K. D. Keefer, D. W. Schaefer, and C. S. Ashley,
   J. Non. Cryst. Solids, 48, 47 (1982).

THE PROCESSING AND CHARACTERIZATION OF DCCA MODIFIED GEL-DERIVED SILICA

S. WALLACE AND L. L. HENCH
Department of Materials Science and Engineering, University of Florida,
Gainesville, FL 32611

INTRODUCTION

The production of small silica gel monoliths by the hydrolysis and polycondensation of tetramethyl orthosilicate (TMS) in a methanol solution is now a common procedure. Drying is generally done slowly in a methanolic atmosphere [1] or under hypercritical conditions [2]. The potential application of sol-gel technology in the production of materials for large structures requires a rapid processing time, which causes a problem due to cracking. For large scale space structures the materials produced also require a low molecular weight, a low densification temperature, control of devitrification and a wide range of physical properties and gel densities. These requirements are potentially met by multicomponent silica based gels [3]. To decrease the drying time, which is the longest part of gel processing, the gel strength needs to be increased to resist cracking and the drying stress reduced. The gel strength can be improved by optimizing the variables in the gel manufacturing process.

During the initial stages of drying, the capillary force caused by the evaporation of solvents from the micropores in the gel creates an overall drying stress and local differential stresses due to non-uniform pore size distributions [2]. The capillary force depends on the rate of liquor evaporation which is a function of solvent vapor pressure, and is inversely proportional to the pore size. During the final stages of drying, cracking is the result of non-uniform shrinkage of the drying body [2]. This can be due to temperature gradients, compositional inhomogeneities and different local rates of reaction. It has been observed that for a given volume of TMS, there is a critical ratio S* (S = Volume of Solvents/Volume of TMS) above which cracking will not occur during the final stages of drying. The drying stress is therefore a function of the pore size and the rate of evaporation of the liquor, which depends on the liquor vapor pressure. The pore size can be controlled by the processing variables and the vapor pressure can be controlled by adding an organic solvent, called a drying control chemical additive (DCCA), to the sol.

In this study we examine the role of formamide (NH_2CHO) as a DCCA in TMS gel-derived silica, as a basis for the further investigation of multicomponent silica gels. The vapor pressure of NH_2CHO is 0.1 Torr at 40°C, compared to 100 Torr at 21°C for CH_3OH. The combination of the two liquids reduces the rate of evaporation during the initial stages of drying, thus reducing the drying stress.

GEL PROCESSING OPTIMIZATION

Base catalyzed gels gelate quickly due to OH^- catalysis of the polycondensation reaction leading to a particulate structure with many unreacted $SiOCH_3$ radicals. The addition of an acid causes rapid hydrolysis of the TMS, catalyzed by H^+ ions [4], as shown by the instantaneous appearance of the SiOH absorbance band at 948 cm^{-1} in an FTIR spectra. This causes longer gelation times (see Fig. 1) but allows for the formation of a larger percentage of Si-O-Si bridging bonds, a high connectivity polymeric structure, fewer unreacted $SiOCH_3$ radicals and subsequently more

Mat. Res. Soc. Symp. Proc. Vol. 32 (1984) Published by Elsevier Science Publishing Co., Inc.

48

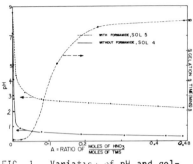

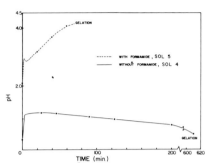

FIG. 1. Variation of pH and gelation time with A, the molar
ratio of HNO_3 to TMS.

FIG. 2. Variation of pH with time
of gelation (R=10).

Si-OH bonds. This means that for a given aging time, a gel will be stronger under acid rather than base catalysis.

Consequently the use of NH_2CHO in a concentrated HNO_3 catalyzed TMS-CH_3OH-H_2O was investigated. Table I shows the influence of progressive additions of NH_2CHO and HNO_3 on gelation times. NH_2CHO accelerates the gelation of gels both with (see Fig. 2) and without HNO_3, in each case raising the pH and presumably increasing the rate of polycondensation. This is confirmed by the observation that the rate of shrinkage during aging increases with NH_2CHO addition, and is fastest when HNO_3 is also present, due to the resultant rapid Si-OH formation leading to the strongest gel in the shortest time.

TABLE I. Gelation time for different sols.

No	Sol(R=10, A=0.475)	pH	Gelation Time (60°C)
1	TMS + H_2O + HNO_3	~ 1	50 mins
2	TMS + H_2O + CH_3OH	~ 7	6 hrs
3	TMS + H_2O + CH_3OH + NH_2CHO	~ 9	1 min
4	TMS + H_2O + CH_3OH + HNO_3	~ 1	12 hrs
5	TMS + H_2O + CH_3OH + HNO_3 + NH_2CHO	~ 4	1 hr

Decreasing the pH leads to a higher surface area (Fig. 3) and a corresponding decrease in pore size, which increses the drying stress. Consequently to prevent cracking in the final stages of drying this pore size decrease must be reversed by increasing S (Fig. 4), which also ensures complete miscibility during aging and drying.

Figure 5 shows the optimized flow diagram for the production of silica monoliths for 15 cm^3 TMS, with R = 10. Consistent monolithicity was obtained with a high acid concentration (A=0.475). S* = 3.0, which will decrease for larger TMS volumes. A 1:1 mixture of CH_3OH/NH_2CHO was found to work best. Figure 6 shows silica gels dried to 60°C, with a processing time of 2 days. Table II shows some of physical properties of the silica gels after drying at 60°C.

FTIR SPECTROSCOPY LIQUID CELL CHARACTERIZATION

In order to understand more clearly the influence of HNO_3 and NH_2CHO on the sol gel conversion, a Nicolet MX-1 FTIR spectrometer was used to monitor the changes occurring in the sol. A drop of sol was pressed

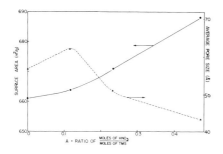

FIG. 3. Surface area and pore size of Sol 5 as a function of A.

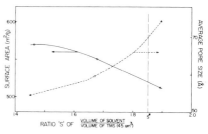

FIG. 4. Surface area and pore size of Sol 5 as a function of S.

15 cm^3 TMS (0.10008 Moles)

12.5 cm^3 CH$_3$OH

12.5 cm^3 NH$_2$CHO

2 cm^3 Conc. HNO$_3$ (0.0477 Moles)
↓
Stir Magnetically at 25°C
↓
Add 18.17 cm^3 (1.008M) of H$_2$O (R=10)
↓
Stir for 10 Minutes
↓
Cast into Sealed Polystyrene Container
↓
Gel at 60°C in 1 Hr.
↓
Age at 60°C for 12 Hrs.
↓
Dry Unidirectionally

FIG. 5. Flow diagram of optimized 100S gel manufacture.

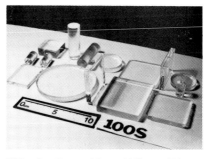

FIG. 6. Formamide modified silica gel monoliths dried at 60°C.

TABLE II. Physical properties of DCCA modified silica gel dried at 60°C.

Density	1.42 g/cm^3
Hardness	4 DPH (100g load)
Tensile Strength	200 psi
Index of Refraction	1.442
Surface Area	750 m^2/g
Pore Volume	1.0 cm^3/g
Average Pore Size	40 Å

between two cadmium telluride windows using a 0.015 mm liquid cell gasket. The change with time of the FTIR spectra of Sol #4 (see Table I) was monitored and compared to Sol #2 [5].

Figure 7 shows the IR spectra of TMS, CH$_3$OH and NH$_2$CHO, compared to Sol 5, after 1 minute of reaction. Each Sol 5 absorbance band can be assigned to a pure component band (Table III) except for the band at 948 cm^{-1}. This is the Si-OH peak due to the HNO$_3$ catalyzed hydrolysis of TMS, so none of the TMS bands are visible even after only 45 seconds of Fig. 7 reaction. The SiOH band also appears in Sol 4 (with no NH$_2$CHO) due to the presence of HNO$_3$, but doesn't appear in Sol 2 (with no acid), whereas all the TMS bands are visible (Fig. 8). With time Sol 5 loses its CH$_3$OH bands (1018 cm^{-1} and 1108 cm^{-1}) with the formation of Si-0-Si bridging bonds at 1082 cm^{-1} (Fig. 9). Sol 4 shows the same trend, only more slowly, confirming Fig. 2. The change with time of the CH$_3$OH (1018 cm^{-1}), molecular water (1663 cm^{-1}) and SiOH (948 cm^{-1}) bands intensity are shown in Fig. 10. CH_3OH and SiOH both decrease with time, and water initially increases before decreasing after 2 hours. It is hypothesized that this is due to the very rapid initial formation of SiOH bonds from TMS.

$$SiOCH_3 + H_2O \rightarrow SiOH + CH_3OH \qquad (1)$$

causing the formation of the large CH_3OH band intensity. The increase in H_2O intensity is caused by the formation of siloxane bonds by

$$SiOH + SiOH \rightarrow Si-O-Si + H_2O \qquad (2)$$

which forms the 1082 cm^{-1} absorbance band hidden until the CH_3OH and H_2O band intensities decrease due to their evaporation from the liquid cell. This is the opposite trend to Sol 2 [5] which showed an increase in CH_3OH and a decrease in H_2O band intensity, over 2 hours of reaction, due to a much slower rate of hydrolysis and the formation of CH_3OH i.e., rather than water, during siloxane bond formation.

$$SiOCH_3 + SiOH \rightarrow Si-O-Si + CH_3OH \qquad (3)$$

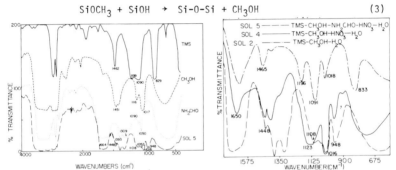

FIG. 7. FTIR spectra of TMS, CH_3OH, NH_2CHO and Sol 5 after 1 minute.

FIG. 8. FTIR spectra of Sol 2, 4 and 5 after 1 minutes.

TABLE III. Absorption Band Wavenumbers (cm^{-1}) Assignments.

TMS	NH_2CHO	CH_3OH	Sol 2	Sol 4	Sol 5
829			833		
		1018	1018	1014	1018
	1053				1056
1088	1087		1091		
		1116		1123	1108
1199			1199		
	1321				1321
	1388				1387
		1420		1413	
		1451		1448	1448
1465			1465		

DENSIFICATION

Small samples of dense silica have been produced by rapid heating with isothermal holds at critical temperatures [1] and by use of slow heating

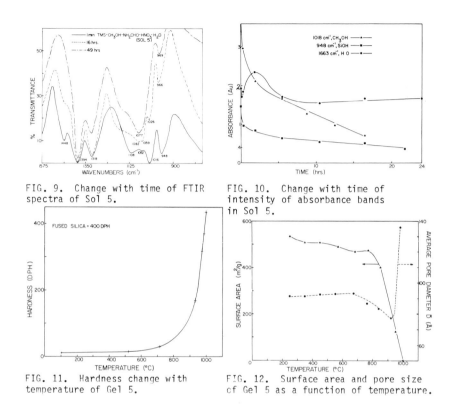

FIG. 9. Change with time of FTIR spectra of Sol 5.

FIG. 10. Change with time of intensity of absorbance bands in Sol 5.

FIG. 11. Hardness change with temperature of Gel 5.

FIG. 12. Surface area and pore size of Gel 5 as a function of temperature.

rates [6]. The critical temperatures for NH_2CHO modified silica gels were found using DSC and DTA. Samples were heated in a microprocessor controlled furnace with the facility for vacuum, air and argon atmospheres at these heating rates along with much slower rates. The hardness, surface area and average pore size was measured after densification at various temperatures. The molecular structure was monitored using x-ray diffraction and FTIR spectroscopy. The dried cylindrical samples used for densification were about 2.5 cm^3; and had a density of 1.40 cm^3/g. All samples fragmented by 300°C due to residual formamide (see Ref. 8), but densified to produce amorphous silica samples of 2-3 mm size. Viscous sintering started at about 800°C to produce dense silica with a hardness of 400 DPH at 1000°C (Fig. 11). Figure 12, showing the surface area, confirms that densification occurs at 1000°C. Figure 13 shows that the gel is still amorphous at 1000°C and that the NH_2CHO has not caused devitrification. Figure 14 shows the increase in intensity of the absorbance bands due to Si-O-Si at 1268 cm^{-1} and 1126 cm^{-1} between 250°C and 1000°C. Comparison with the amorphous silica spectra shows the presence of the SiOH 926 cm^{-1} band even at 1000°C similar to the results of Phalippou et al.[7].

CONCLUSIONS

Formamide is the first DCCA that shows the feasibility of the use of organic solvents as a means of improving the processing of metal organic derived gels. It reduces gelation, aging, drying times, the drying stress,

52

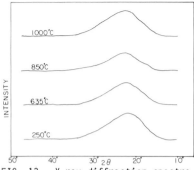

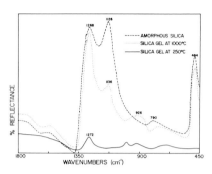

FIG. 13. X-ray diffraction spectra of Sol 5 at different temperatures.

FIG. 14. FTIR spectra of Sol 5 and amorphous silica.

and increases the size of gel monolith that can be made up to about 100 cm^3 after drying for 2 days. This is facilitated by acid catalysis and the correct solvent volume. The acid catalysis caused rapid hydrolysis allowing a stronger gel to be formed. FTIR spectroscopy liquid cell monitoring of the sol reactions confirms the influence of HNO_3 and NH_2CHO on the sol reaction rates. A DCCA modified silica gel can be densified by 1000°C without devitrification, but further work on a new DCCA is needed to densify large monoliths.

ACKNOWLEDGMENTS

The authors gratefully acknowledge the financial support of the Air Force Office of Scientific Research contract #F49620-83-C-0072 during the course of this work, and the technical assistance of Guy LaTorre.

REFERENCES

1. L.C. Klein and E. J. Garvey in: Ultrastructure Processing of Ceramics, Glasses and Composites, L.L. Hench and D. R. Ulrich, eds. (John Wiley and Sons 1984) in press.
2. J. Zarzycki in: Ultrastructure Processing of Ceramics, Glasses and Composites, L.L. Hench and D.R. Ulrich, eds. (John Wiley and Sons 1984) in press.
3. S. Wallace and L.L. Hench, "Metal Organic Derived 20L Gel Monoliths," to be published in the Proceedings of the 7th Annual Conf. on Composites and Advanced Ceramic Materials, Am. Ceram. Soc.
4. H. Schmidt, H. Scholze and A. Kaiser, "Principles of Hydrolysis and Condensation Reactions of Alkoxysilanes," Proceedings of 2nd International Workshop on Glasses and Glass Ceramics from Gels.
5. M. Prassas and L.L. Hench in: Ultrastructure Processing of Ceramics, Glasses and Composites, L.L. Hench and D. R. Ulrich, eds. (John Wiley and Sons 1984) in press.
6. M. Nogami and Y. Moruja, Yogyo-Kyokai-Shi 87, 34-42 (1979).
7. J. Phalippou, T. Woignier and J. Zarzycki in: Ultrastructure Processing of Ceramics, Glasses and Composites, L.L. Hench and D. R. Ulrich, eds. (John Wiley and Sons 1984) in press.
8. S.H. Wang and L.L. Hench, "Processing Variables of Sol-Gel Derived 20 Soda Silicates," to be published in the Proceedings of the 7th Annual Conf. on Composites and Advanced Ceramic Materials, Am. Ceram. Soc.

STRENGTH OF GEL-DERIVED SiO$_2$ FIBERS

WILLIAM C. LaCOURSE
Inst. of Glass Science and Engineering
N.Y.S. College of Ceramics, Alfred University
Alfred, N.Y. 14802

ABSTRACT

Strengths of as drawn gel-derived SiO$_2$ fibers are found to depend on H$_2$O and alcohol content of the sol, as well as drawing viscosity. Additional factors, including fiber diameter and drying and consolidation schedules are found to dramatically influence strengths of the consolidated glass fibers.

INTRODUCTION

Sakka and Kamiya [1-3] have provided considerable information regarding production of fibers from tetraethoxysilane (TEOS) sols. Acid catalysed TEOS with H$_2$O/TEOS ratios of less than 2 yields sols with viscosities which rise slowly through the fiberizing range (5-500N-s/m^2) and structures which permit continuous filaments to be drawn. Fibers can be consolidated at temperatures near 800 $^{\circ}$C to produce 20 μm diameter glass fibers with strengths near 600 MN/m^2. Strengths decrease rapidly with increasing diameter, and with increasing consolidation temperature above 800 $^{\circ}$C. [3]

A number of factors, including diameter and consolidation temperature are expected to influence the strength of gel-derived SiO$_2$ fibers. These include

A. SOL STRUCTURE
1. H$_2$O/TEOS ratio
2. Humidity-Temperature
3. Solvent (alcohol) content
4. pH and acid type
5. Sol viscosity

B. DRYING AND CONSOLIDATION
1. Sol structure
2. Ageing prior to consolidation
3. Atmosphere
4. Heating rate
5. Maximum temperature
6. Fiber diameter

In the present paper several of these factors are investigated to determine effects on the strength of "as drawn" (unconsolidated) and consolidated SiO$_2$ fibers.

PROCEDURES

Sols were prepared using the procedure of Dahar [4,5]. TEOS and anhydrous ethyl alcohol were premixed at room temperature for ½ hour. An aqueous solution of 0.1N HCl was then added to provide the desired water content. The sol was allowed to react further while covered, for an additional 2

Mat. Res. Soc. Symp. Proc. Vol. 32 (1984) Published by Elsevier Science Publishing Co., Inc.

hours. It was then uncovered and placed in an oven at 50 $^{\circ}C$ and 20% relative humidity until the viscosity reached desired levels. This required 1.5 to 5 days depending on the alcohol and water contents.

Sols having H_2O/TEOS molar ratios of 1.4, 1.5, 1.6, 1.7 and 2.4 were prepared with a volume ratio of alcohol to TEOS of 0.5. The ROH/TEOS ratio was varied from 0.25 to 0.75 at a H_2O/TEOS ratio of 1.6. The acid content, humidity and temperature were held constant.

Fibers could be drawn when the viscosity reached 5-500 N-s/m^2. Drawn fibers were aged at room temperature in air for various times and then fired in air using heating rates of either 2.0 or 4.0 $^{\circ}C$/min. Strengths were determined at various stages in processing using a tensile test at a strain rate of 10 mm/min, and a guage length of 75-150 mm. Fibers were first fixed between two pieces of paper, using either tape or glue, and the paper was then placed in the tensile apparatus for testing.

RESULTS AND DISCUSSION

1. As Drawn Fibers

a. Fiber Geometry: Fibers could not be drawn from sols with 1.4 H_2O/TEOS even though viscosities reached 100-500 N-s/m^2. Fibers from all other sols had a non-circular cross section, generally either elliptical (   ) or rectangular with rounded edges (   ). Diameters measured along the short axis ranged from 10 to 150 μm. Average aspect ratios varied with composition as shown in Table I.

Table I. Aspect ratios for fibers produced from sols having an alcohol /TEOS ratio of 0.5.

H_2O/TEOS	ASPECT RATIO
1.5	2.3
1.6	2.0
1.7	1.9
2.4	2.5

Reasons for this behavior are not clear. It may be related to the fact that sols transform slowly to the semi-solid gel after drawing, and transform first at the surface. Continued reaction and shrinkage of the interior would cause stresses on the surface. This, coupled with possible molecular orientation during drawing, and anisotropic relaxation of elongated silicate molecules may produce stresses that are relaxed most readily by formation of a non-circular cross section. Measurements of fiber geometry as a function of time after drawing will be required before a satisfactory explanation can be developed.

b. Effect of Sol Viscosity: Results shown in Table II indicate that unfired fiber strengths are determined in part by the viscosity of the sol during drawing. In the present experiments increasing viscosity is caused by increasing molecular weight of the silicate molecules, as well as decreasing alcohol content due to evaporation. Although not measured in the current experiments, alcohol contents are expected to be fairly constant. Previous work [4] indicated that under similar conditions most of the alcohol evaporated during the first 3 days of reaction, then remained constant. Therefore, for sols containing 1.5-1.7 H_2O/TEOS the alcohol content was relatively constant at the various drawing viscosities. For the 2.4 H_2O/TEOS sol drawing occured during the time that the alcohol content was decreasing. However, there was only a few hours between low and high viscosity draws, and changes in alcohol content are expected to be small.

Table II. Effect of sol viscosity on the strength of as drawn fiber. Fiber aged one week prior to testing.

SOL	VISCOSITY $N-s/m^2$	AVE. STRENGTH MN/m^2
1.6 H_2O/TEOS .25 ROH/TEOS	< 10	30
1.6 H_2O/TEOS .25 ROH/TEOS	≈ 50	34
1.6 H_2O/TEOS .25 ROH/TEOS	>100	37
1.6 H_2O/TEOS .75 ROH/TEOS	< 10	38
1.6 H_2O/TEOS .75 ROH/TEOS	>100	44
2.4 H_2O/TEOS .50 ROH/TEOS	10-50	75
2.4 H_2O/TEOS .50 ROH/TEOS	>100	110

If after drawing condensation reactions continue producing a structure that is independent of drawing viscosity, one would observe no effect on strength when the alcohol content is constant. Current results therefore suggest that the final fiber structure varies with drawing viscosity.

c. Effect of Fiber Diameter: Figure 1 shows the strength of unconsolidated fibers as a function of the longest fiber dimension, for fibers drawn at viscosities near 20-50 $N-s/m^2$. While there is considerable scatter it is clear that there is no strong dependence on diameter for the as drawn fibers. Similar results were obtained for all compositions, suggesting that the diameter effect observed by Sakka[3] arises only after extended drying at low temperatures, and/or upon firing to elevated temperatures.

d. Effect of H_2O and Alcohol: Effects of H_2O content on the strength of as drawn fibers are summarized in Table III. Based on an expected increased degree of polymerization with increased water content one might expect strengths to increase with the H_2O/TEOS ratio. The apparent decrease between ratios of 1.5 and 1.6 is therefore unexpected, and cannot be accounted for at present.
It might be noted that the strength of fibers prepared with 2.4 H_2O/TEOS is of the same order of magnitude as those

measured for commercial glass containers. Heating of these
fibers to 150 °C increased the average strength to more than
170 MN/m^2. Heated fibers did exhibit a dependence of strength
on diameter, with those having a maximum dimension of less than
70 μm having an average strength of 220 MN/m^2, and those with
larger maximum diameters having an average strength of 130
MN/m^2. A maximum strength of more than 300 MN/m^2 was measured.
While considerable work remains these results suggest the possibility of low temperature production of moderately high
strength, fine diameter SiO$_2$ fibers.

Table III. Effect of water and alcohol content on strength of
as drawn fibers.

H$_2$O/TEOS (molar)	STRENGTH MN/m^2	ROH/TEOS (volume)	STRENGTH MN/m^2
1.5	55	0.25	37
1.6	40	0.50	40
1.7	69	0.75	45
2.4	110		

Unfired strengths appear to increase almost linearly with
increasing alcohol content, over the range studied. Reasons
for this are not clear. In earlier studies it was shown that
the only effect of increasing the alcohol content from 0.5 to
1.0 ROH/TEOS was to shift the point at which the viscosity was
suitable for fiber drawing to longer times. Beyond that time
the viscosity-time relationships were almost identical, suggesting similar sol structures.

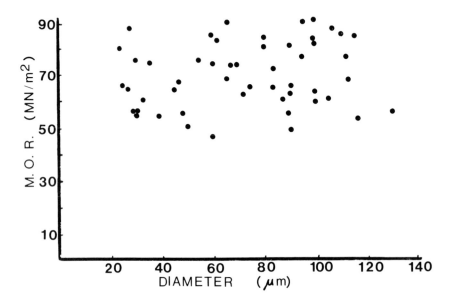

Fig. 1. Strength vs. large diameter for unfired 1.7 H$_2$O/TEOS
fibers.

On the other hand, with low alcohol contents there is a tendency for sols to be inhomogeneous. For example, if the H_2O is added too rapidly to a sol containing 0.25 ROH/TEOS the sols become slightly cloudy and remain so during subsequent processing. For the sols used in this study the rate of H_2O addition was low enough to produce clear sols, but it is likely that some structural differences exist, depending on the ROH/TEOS ratio.

Effects of firing on fiber strengths are summarized in Fig. 2 and Table IV. All fired samples exhibit a dependence of strength on diameter, and there is an indication of an inverse relationship between as drawn and consolidated strengths. Fibers prepared with 2.4 H_2O/TEOS attained strengths of only 200 MN/m^2 at 500 ^OC and 250 MN/m^2 at 700 ^OC. These samples appeared brown after firing. Fibers with 1.6 H_2O/TEOS fired clear and exhibited average strengths of 240 and 450 MN/m^2 after firing to 500 and 700 ^OC respectively. Ageing prior to firing also prevents one from obtaining optimum strengths.

Table IV. Strength of consolidated glass fiber.

SOL	FIRING SCHEDULE	AVE. STRENGTH MN/m^2
1.6 H_2O/TEOS .25 ROH/TEOS	Fired immediately after draw. 2 ^OC/min to 700 ^OC	450
1.6 H_2O/TEOS .25 ROH/TEOS	Fired immediately after draw. 500 ^OC	240
1.6 H_2O/TEOS .25 ROH/TEOS	Aged prior to fire. 2 ^OC/min 500 ^OC	140
1.6 H_2O/TEOS .75 ROH/TEOS	Fired immediately 2 ^OC/min 700 ^OC	410
2.4 H_2O/TEOS .50 ROH/TEOS	Fired immediately 2 ^OC/min 500 ^OC	200

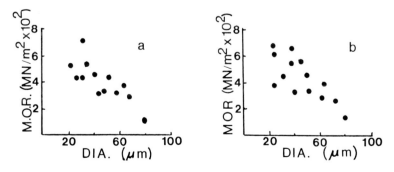

Fig. 2. Strength vs. diameter for fired fibers. a. 1.6 H_2O/TEOS---.75ROH/TEOS. b. 1.6 H_2O/TEOS---.25 ROH/TEOS.

58

One expects this behavior if flaws formed during firing
are related to the degree of crosslinking in the unfired fiber,
and to the porosity. Small diameter, weakly crosslinked, open
structures resulting from low water contents and immediate firing enhance the ability of the gel structure to relax stresses
produced by evaporation of residual alcohol and molecular water
at low temperatures, and to carbonization and dehydroxylation
at higher temperatures. A detailed study of the strength of
fired samples as a function of the properties and structure of
the as drawn fiber should permit one to design compositions,
preparation procedures and firing schedules which will produce
substantially higher strengths than those observed here.

REFERENCES

1. K. Kamiya and S. Sakka, J. Mater. Sci., 15, 2937-2939
    (1980).
2. S. Sakka, Treatise on Materials Science and Technology,
    M. Tomozawa and R. Doremus, eds. (Academic Press, New
    York 1983).
3. S. Sakka, Hoyomen, 19, 430-442 (1981).
4. S. Dahar, M.S. Thesis, N.Y.S. College of Ceramics,
    Alfred University, May 1983.
5. S. Dahar, W. C. LaCourse and M. Md. Akhtar, J. Amer.
    Ceram. Soc., in press.

THE CERAMIST AS CHEMIST - OPPORTUNITIES FOR NEW MATERIALS

D. R. UHLMANN, B.J.J. ZELINSKI AND G.E. WNEK, Department of
Materials Science and Engineering, Massachusetts Institute
of Technology, Cambridge, Mass. 02139

ABSTRACT

The use of sol-gel techniques to prepare glasses and
crystalline ceramics offers outstanding opportunity for
breakthroughs in technology. The areas of particular
promise include novel glasses; crystalline ceramics with
exceptional microstructures; coatings for modification of
electrical, optical, mechanical and chemical properties;
porous media with high surface area and tailored chemistry;
ceramic powders with high chemical homogeneity and narrow
distributions of particle size; matrix materials in ceramic-
ceramic composites; and a wide spectrum of specialty ceramic
materials, ranging from abrasives and fibers to glass ceramics
and films. Opportunities in each of these areas will be
discussed and related to the advances in understanding and
process technology required for their achievement. The
theses will be advanced that creative chemistry provides
the key to many of these advances, that ceramists simply
MUST learn more chemistry, but that we dare not rest from
our labors when the chemistry is done.

I. INTRODUCTION

The methods of wet chemistry offer outstanding promise for effecting
radical changes in ceramics technology. The attendance at the present
Symposium, like that at previous workshops held in Padua and Wurzburg,
provides an effective commentary on the interest and even excitement which
is characteristic of the field. Applications of the technology cover a
broad front; and significant advances are announced on almost a monthly basis
on developments ranging from varistors to optical waveguides.

With all the richly-deserved attention directed to applications of
sol-gel techniques, however, the underlying physics and chemistry has often
been sadly neglected. Fundamental kinetic data are almost always unavailable;
knowledge of structural features at various stages of processing is frag-
mentary at best; and the effects of process conditions on chemical species,
reaction sequences and intermediate products remain to be explored in
anything like satisfactory detail. The field thus seems to be constructing
an edifice of ever-increasing splendor and complexity, but on foundations
whose stilt-like character will soon compromise the overall structure.

With this perspective, the present paper will first review some of the
most attractive opportunities offered by sol-gel technology, as well as some
of the most striking recent advances. It will then direct attention to a
number of critical areas where our lack of knowledge of chemistry, structure
and kinetics is already limiting our ability to synthesize desired materials.
Finally, it will suggest the need of much greater interaction between
ceramists and chemists, and the importance of educational changes in both
disciplines to render easier such interaction.

Mat. Res. Soc. Symp. Proc. Vol. 32 (1984) Published by Elsevier Science Publishing Co., Inc.

II. APPLICATIONS OF SOL-GEL TECHNOLOGY IN CERAMICS

The application of wet chemical techniques to the formation of ceramic materials is based on a number of important characteristics. These include:

A. The ability to prepare powders with a narrow distribution of particle sizes and a small mean size. This permits sintering to high density, and in some cases to substantially full density, at temperatures lower by several hundred Centigrade degrees than those used in conventional processing. Examples of the effective use of this ability are provided by the preparation of transparent mullite (1) and of high density TiO_2 at remarkably low sintering temperatures (2).

B. The ability to prepare amorphous powders and dried gels which can be processed to full density before the onset of crystallization, and subsequently crystallized to form more refractory crystalline bodies. In some cases, significant residual amorphous phases may be retained, thereby producing a glass-ceramic body; while in others, substantially complete crystallization may be effected. This ability should permit substantially complete densification to be achieved at remarkably low temperatures (compared with those required to densify powders of the already-crystallized material). Such low-temperature densification can be critical in a wide range of applications, including matrix materials in ceramic-ceramic composites where degradation of the fibers imposes decisive limitations on processing temperatures.

C. The ability to prepare glasses with novel compositions and properties. This ability is based on the avoidance of limitations imposed by the critical cooling rate for glass formation, since glasses may be formed by a low temperature viscous sintering process rather than by cooling a melt. The availability of such low temperature processing routes should be most important for compositions which form fluid melts at their liquidus temperatures. Examples of the effective use of the ability to form novel glasses are provided by $Na_2O-ZrO_2-SiO_2$ glasses with unusually high ZrO_2 contents (3) and $CaO-SiO_2$ glasses within the miscibility gap (4).

D. The ability to fabricate ceramic bodies with novel microstructures and distributions of phases. This in turn is based on the capability of generating fine powders, often of unstable or metastable phases, and carrying out low temperature processing. Examples of the effective use of this ability include novel Al_2O_3-based abrasives (5) and the ZnO-based varistors with exceptionally fine grain size recently announced by workers at the Oak Ridge National Laboratory (6).

E. The ability to produce coatings with tailored chemistry and desired electrical, optical and transport properties. While much attention in the glass community has been directed to establishing the conditions required to form monoliths, it seems clear--to the present authors at least--that the most natural application of gel techniques is to the formation of thin layers (coatings). The low temperature processing made possible by these techniques affords the opportunity of developing novel coatings with minimal deleterious effect on the substrate. Effective uses of gel technology in coating are illustrated by passivation coatings on glass for liquid crystal displays (7), colored reflective coatings on glass for architectural or automotive applications (8), and antireflection coatings on glass for other optical applications (9).

F. The ability to produce shaped bodies and coatings with controlled levels of high porosity, and with the pore surfaces having tailored chemistry. Surface areas exceeding 100 m^2 gm^{-1} have already been obtained;

and a wide range of chemistries have been (are being) explored. While much attention has been directed to fabricating porous bodies of various shapes, the most effective application may again lie in the area of coatings. Effective examples are provided by the lovely work on porous fibers produced by unidirectional solidification of gels (10, 11), as well as a range of catalyst applications.

G. The ability to produce materials with mixed organic-inorganic functionality, having unique combinations of properties--in many cases, properties unobtainable with either organic polymers or conventional ceramics. This area remains largely unexplored at the present time; but initial results seem highly encouraging [e.g., the impressive work of Schmidt and Philipp and their co-workers (12) on organic-modified silicates].

H. The ability to modify the surfaces or ceramic or polymer bodies so as to modify their response to mechanical forces and promote their adhesion to other materials. With suitable coatings produced by sol-gel techniques, it is possible to increase appreciably the scratch resistance of polymers; and similar techniques (using different chemistries) can be used to provide engineered gradations in chemistry for improved bonding. As an example of the latter, consider the difficulty of achieving good adhesion between many polymers and 400 series stainless steels, which can be greatly ameliorated by depositing a coating which is tenaciously adherent to the steel and also provides an external surface to which the polymers can readily adhere.

I. The ability to protect underlying metals, ceramics or polymers from harsh chemical environments using adherent, chemically-resistant coatings. The expanded range of possibilities for glass-lined reaction vessels has already been noted. Other examples include the provision of water-impervious coatings on polymers and possibly even the protection of refractory metals against oxidation.

J. The ability to provide coatings over large areas and complex geometries (including the insides of tubes). Large area coatings are already a practical reality, as demonstrated by the architectural glass coatings developed by Dislich and his co-workers at Schott (8). Questions remain concerning the uniformity of coating over large areas, the speed at which coatings can be applied, and the extent to which pinholes can be eliminated; but the present authors believe strongly that this capability is destined to play an increasingly-important role in future developments.

K. The ability to provide materials with a high degree of chemical homogeneity. This is a widely-cited advantage of sol-gel processing techniques, with great potential pay-off for ceramic science and technology. There are remarkably few cases, however, for which such high chemical homogeneity has been demonstrated; and as shall be discussed below, there are good reasons for expecting microscopic inhomogeneity unless careful attention is directed to processing conditions and to the kinetics of the various rate processes.

With capabilities such as these provided by sol-gel technology, it is not surprising that the range of technological applications is impressively broad. Among the applications which seem particularly promising, the following may be noted:

1. Applications in electronics. Considered here are materials and devices where controlled and/or fine grain size and controlled grain and grain boundary chemistry are important or critical to performance. Examples include varistors, thermistors, ferrites and dielectrics, with materials ranging from ZnO to $BaTiO_3$ and PZT. Also considered are substrates with desired dielectric and thermal properties, thin glassy dielectric layers, and

a broad spectrum of materials with exceptional dielectric and electrical
properties. Examples of the last include crystalline and amorphous fast
ion conductors and materials with desired dielectric loss characteristics.

2. Optical applications. Considered here are preforms for optical
waveguides and joints between optical fibers, as well as a broad range of
electro-optic materials whose chemistry and homogeneity are critical to
performance. Work on waveguide preforms is being actively pursued at
several laboratories in this country and abroad. Also considered are novel
optical glasses with exceptional index-dispersion relations.

3. Mechanical applications. Considered here are matrix materials for
ceramic-ceramic composites and oxide materials quite generally with fine,
uniform microstructures. Also considered are novel glass-ceramic materials
with desired combinations of properties and abrasive products with exceptional grinding performance in selected applications.

4. Coating applications. Considered here are conductive coatings
(e.g., ITO or cadmium stannate), infrared reflective coatings, passivation
coatings (e.g., SiO_2 in CC displays), oxidation-resistant coatings,
corrosion-resistant coatings (e.g., glasses on metals), release coatings,
adhesion coatings (e.g., coatings with graded chemistry), scratch-resistant
coatings, porous coatings (e.g., catalytic coatings on solid supports),
impervious coatings, antireflection coatings (e.g., graded index coatings)
and refractory coatings. The potential applications in this area are too
numerous for brief citation.

5. Catalysis and membrane applications. Considered here are porous
materials with exceptionally high surface areas and tailored chemistries,
for use as supports for catalytically-active species as well as directly
active materials. Also considered are filtration media with high thruput
and high selectivity.

6. Shaped body applications. Considered here are fibers and bodies
with complex geometric shapes with unique properties. Examples include
fibers with outstanding resistance to alkali attack and molded parts with
fine, uniform microstructures. Fibers with attractive mechanical properties
prepared using sol-gel techniques are already articles of commerce.

7. Scientific applications. Considered here is the use of sol-gel
methods to prepare samples for scientific investigations of phenomena.
The unique chemistries and microstructures obtainable with these methods
should expand greatly the range of conditions accessible for study.

The range of potential applications of sol-gel techniques is sufficient
to whet the appetite of even a jaded ceramist. The realization of this
potential depends critically, however, upon the ability to understand and
control the often-complex chemistry involved in processing. The ceramist
operating without a chemist at his side is therefore asking for intellectual
discomfort and practical inefficiency. To appreciate the criticality of
chemical insight, it seems appropriate to review briefly some of the chemical
considerations which are important in the sol-gel formation of ceramics.

III. SELECTED CHEMICAL CONSIDERATIONS

Chemical insight is vitally important in all stages of sol-gel
processing, at least up to the point where a dried powder or gel is prepared.
Among the areas where such insight seems critical, the following are
deserving of particular note:

A. <u>Selection of starting materials</u>. The materials used in the synthesis can critically influence the course of the reactions which take place during processing. Selection of metal salts--e.g., metal chlorides or hydroxides--can lead to quite different approaches and considerations than the use of corresponding metal alkoxides. Metal salts are often attractive from the perspective of cost, but present problems with respect to elimination of undesired species (e.g., HCl when using $AlCl_3$ as a starting material). Even when using alkoxides, which are attractive with respect to elimination of undesired species, the nature of the alkoxide can greatly affect the sequence of events.

For reasons of brevity and the papers presented at this meeting, the present paper will focus attention on alkoxide precursors; but the reader is reminded of the alternative approach, which is attractive for many applications.

With the exception of Si and Ge, the alkoxides of Group III-Group V and transition metals which are important to ceramics generally do not occur as isolated species but as complexes. The origin of such complex formation is the phenomenon of coordinate unsaturation. This may be illustrated by the case of titanium, where treatment e.g., of $TiCl_4$ with THF yields a solid of stoichiometry $TiCl_4$ $(THF)_2$. The Ti atom uses its empty d-orbitals to coordinate with oxygen lone pairs of THF, thus maximizing its coordination number (six). It may therefore be said that the Ti in $TiCl_4$ itself is coordinately unsaturated. The high viscosity of titanium alkoxides is due to a similar phenomenon, in this case inter-molecular coordination with alkoxide oxygens as "bridging" ligands, viz.

(1)

Structure of $Ti_3(OEt)_{12}$ (ref. 13)

The contrast in reactivity between $SiCl_4$ and CCl_4 is also enlightening in this regard. CCl_4 is indefinitely stable in water, whereas $SiCl_4$ is rapidly hydrolyzed. The primary reason for this dramatic difference is that CCl_4, having no available low-energy d-orbitals, is coordinately saturated, whereas Si can expand its coordination shell and allow access to the nucleophile H_2O. Thus, while both CCl_4 and $SiCl_4$ are thermodynamical-ly unstable towards hydrolysis, $SiCl_4$ is kinetically unstable.

$$Cl_{\prime\prime\prime}\overset{\underset{\textstyle Cl}{|}}{\underset{\textstyle Cl}{Si}}{-}Cl \quad \xrightarrow{H_2O} \quad \left[Cl_{\prime\prime\prime}\overset{\underset{\textstyle Cl}{|}}{\underset{\textstyle \underset{\textstyle |}{\overset{\textstyle \oplus}{O}}{-}H}{\overset{\ominus}{Si}}}{-}Cl \right] \quad \longrightarrow \quad SiCl_3OH + HCl \qquad (2)$$

"sp^3d" intermediate-not
possible for carbon

In general, the degree of complex formation (oligomerization) depends on the nature of the metal atom and the alkoxy group. The complexity usually increases with the size of the metal atom, and decreases with the size and degree of branching of the alkoxide group (for steric reasons). The character of oligomer formation can also depend on temperature, con-

concentration and time, as well as on the nature of the solvent. As we shall see, the nature of the metal ion and alkoxide group, and their attendant effects on association and oligomer formation, can have a profound effect on subsequent hydrolysis and polymerization behavior.

B. Selection of solvent. The nature of the solvent can likewise play a critical role in the synthesis. The two most widely-used solvents water and alcohols, are not inert with respect to the reaction sequence. Water plays a direct role in driving the forward reaction in hydrolysis and the back reaction in condensation:

$$\equiv MOR + H_2O \rightleftarrows \equiv MOH + HOR \tag{3a}$$

$$\equiv MOH + HOM \equiv \rightleftarrows \equiv M\text{-}O\text{-}M \equiv + H_2O \tag{3b}$$

Alcohol solvents play a direct role through the process of alcohol exchange:

$$\equiv MOR + R'OH \rightleftarrows \equiv MOR' + ROH \tag{4a}$$

they can also drive the reverse reaction in condensation by alcoholysis:

$$\equiv MOH + ROM \equiv \rightleftarrows \equiv M\text{-}O\text{-}M \equiv + ROH \tag{4b}$$

For these reasons, it is rather surprising that more attention has not been directed to non-alcohol organic solvents as reaction media--at least for purposes of generating scientific insight.

C. Hydrolysis behavior. For single alkoxides, the rate of hydrolysis depends on both the nature of the metal ion and the alkoxy group. The hydrolysis rate generally decreases (14) with increasing length of the n-alkyl chain and with increasing branching of the alkyl group, presumably for steric reasons. It is also expected that the hydrolysis rate should increase as the metal ion is more electropositive, and hence as one goes down a column or from right to left across a row of the Periodic Table. It should be pointed out, however, that detailed kinetic data on hydrolysis kinetics of alkoxides are quite sparse.

In general, both acids and bases catalyze hydrolysis. Taking silicon as an example, in each case the reaction may be viewed as a nucleophilic attack on the electrophilic silicon atom. Acids increase the electrophilicity of Si and transform -OR into the better-leaving group ROH, whereas bases increase the nucleophilicity of H_2O by producing OH^-. Possible mechanisms (shown for "flank" attack) may be the following:

Acid catalyzed:

$$RO\text{-}Si(OR)(OR)\text{-}OR \xrightarrow[H_2O]{H^\oplus} \cdots \longrightarrow (RO)_3SiOH + HOR + H^\oplus \tag{5a}$$

Base catalyzed:

$$RO\text{-}Si(OR)(OR)\text{-}OR \xrightarrow[H_2O]{OH^\ominus} \cdots \longrightarrow (RO)_3SiOH + HOR + OH^\ominus \tag{5b}$$

The use of alcohol as a solvent can have important implications for the extent of completion of hydrolysis. Consider, for example, the hydrolysis reaction:

$$M(OR)_n + xH_2O \overset{\rightarrow}{\leftarrow} M(OH)_x(OR)_{n-x} + xROH \tag{6}$$

With alcohol as a solvent, the system will reach an equilibrium condition with free water remaining in the system and hydrolysis remaining incomplete.

D. **Multicomponent systems.** When two alkoxides are mixed in solution (prior to adding water), binary alkoxide complexes--termed double alkoxides-- can be formed. As examples, Meerwein and Bersin (15) titrated strongly electropositive metal alkoxides with other alkoxides in the parent alcohol and obtained double alkoxides such as $KZn(OEt)_3$, $NaSn(OEt)_6$ and $Mg[Al(OEt)_4]_2$. A wide range of double alkoxides can be obtained by mixing "acidic" and "basic" alkoxides, or by combinations of two strong Lewis acids [as $Al[U(OEt)_6]_3$.

In many cases, the solubility of the double alkoxide is higher than that of either single alkoxide. For example, the ethoxides of Al and Mg are only little soluble in cool ethanol, whereas the double alkoxide $Mg[Al(OEt)_4]_2$ is highly soluble (16). This behavior very likely reflects the fact that double alkoxides are generally much less associated in alcohols or other organic solvents than are the constituent alkoxides.

In many systems, more than one double alkoxide can be formed; and hence the nature of the species present in solution will depend critically on the molar ratios of the constituents. Impressive work in this area has been carried out by Novoselova et al. (17), who determined solubility diagrams for a number of alkoxide systems, e.g., $Ti(OEt)_4$-$Ba(OEt)_2$-$EtOH$ and $Ti(OEt)_4$-$Ba(OEt)_2$-C_6H_6.

It is also noteworthy that little information is available on the hydrolysis behavior of double alkoxides.

E. **Polymerization in Single-Alkoxide Systems.** Condensation polymerization between metal hydroxides or hydroxides and alkoxides is a complex, poorly understood process. However, there appear to be a few "rules of thumb" based on realistic chemical principles which may be useful in predicting which alkoxides will polymerize upon hydrolysis. An important factor is the electropositivity of the metal atom in question, since this governs whether a stable oxide can be formed at all.

In aqueous systems, alkali metal oxides such as Na_2O are quite readily hydrolyzed owing to the stability of Na^+ in such solutions. Therefore, it is not surprising that upon hydrolysis of $NaOCH_3$, which yields NaOH and methanol, condensation does not occur to form $Na_2O + H_2O$.

Aluminum alkoxides apparently exhibit similarities and differences compared with Group I systems. Hydrolysis of aluminum alkoxides under highly acidic conditions affords aquo complexes of Al^{3+} [e.g., $Al^{3+}(H_2O)_6$]. These can "polymerize", especially upon increasing the pH, by the following mechanism (18, 19):

$$Al^{3+}(H_2O)_6 \xrightarrow{-H^{\oplus}} Al^{2+}(H_2O)_5OH \xrightarrow{-H^{\oplus}} Al^+(H_2O)_4(OH)_2 \text{ etc.} \tag{7}$$

$$2 \underset{\diagdown}{\overset{\diagup}{Al}}\underset{OH_2}{\overset{OH}{<}} \xrightarrow{-2H_2O} \underset{\diagdown}{\overset{\diagup}{Al}}\underset{H}{\overset{H}{\underset{O}{<\;>}}}\underset{\diagdown}{\overset{\diagup}{Al}} \tag{1}$$

Polymer 1 can be considered as a "highly hydrous" aluminum oxide. This type of mechanism may be generally operative for low valent metals (e.g., Cr^{2+}, Al^{3+}, Fe^{3+}, which form stable, aquated ions at low pH values.

Elements such as Si (and several transition metals), do not form aquated tetravalent ions, and it is interesting that hydrolysis of alkoxides of these elements leads to extensive condensation polymerization over a wide range of pH values:

$$\equiv Ti-OR \overset{H_2O}{\underset{-H_2O}{\rightleftarrows}} \equiv Ti-OH + ROH \tag{8a}$$

$$2 \equiv Ti-OH \overset{-H_2O}{\underset{\leftarrow}{\rightarrow}} \equiv Ti-O-Ti \equiv \tag{8b}$$

In contrast, alkoxides of Al, when hydrolyzed above pH 5, form $Al(OH)_3/H_2O$ and, at high pH, $Al(OH)_4^-$. Condensation apparently occurs only above 80°C. Thus, a fundamental difference exists between polymerization reactions of rather "basic" hydroxides [such as $Al(OH)_3$] and "acidic" species [such as $Si(OH)_4$]. These differences may be important when attempting to prepare copolymers by co-hydrolysis of two metal alkoxides.

The presence of catalysts and the ratio of water to alkoxide can have a profound effect on the course of polymerization. For example, under conditions of a large excess of water and base catalysis, $Si(OEt)_4$ polymerizes to form fully condensed colloidal particles. At low water contents, however, weakly crosslinked clusters appear to be formed under basic conditions and short 'linear' polymers under acidic conditions (20).

Far less is known about the polymerization behavior of other alkoxides. Perhaps the most striking feature of results obtained to date, based on experiments carried out with low water additions at the boiling points of the parent alcohols, is (14) that the groups which hydrolyze and condense are those present in solution prior to the addition of water [e.g., trimers in the case of $Ti(OEt)_4$]. Also noteworthy is the general paucity of data on the effects of acid or base catalysis on polymerization kinetics for metals other than Si. It should also be observed that almost nothing is known about the kinetics of polymerization, nor about the structural features present at various stages of polymerization for materials other than $Si(OEt)_4$.

Polymerization of hydrolyzed metal alkoxides is undoubtedly a step-growth condensation reaction. The term step-growth implying a slow buildup of molecular weight as monomers form dimers, dimers and monomers form trimers, dimers for tetramers, etc. Organic reactions which fall under this category, such as esterification, from difunctional monomers, are typically amenable to simple kinetic treatments by the experimentally justified assumption that functional group reactivity is independent of the size of the oligomer (or polymer) to which it is attached. Such an assumption is unlikely to have any merit in sol-gel chemistry since the functionality of most "monomers" is 3 or 4, making steric (and electronic) factors important in determining the step-wise growth kinetics as has been shown for multi-functional organic systems such as epoxies (21). Moreover, the issue is further complicated by the fact that alkoxides themselves are not "monomers"; the amount of water and pH will determine the "monomer" production and subsequent polymer growth.

F. Polymerization in Multicomponent Systems. The situation in multicomponent systems is more complicated and even less understood than in systems with a single type of metal ion. Issues of liquid-liquid immiscibility and solubility problems can arise frequently particularly when salt solutions are added to alkoxide solutions [e.g., the work of Mukherjee (22) on the sodium borosilicate system].

When mixtures of alkoxides are hydrolyzed and polymerized, complications associated with differences in rates of the kinetic processes can
lead to microscopic inhomogeneities. If the rate of polymerization of one
species is much faster than that of another, the first species can grow to
a considerable extent--in some cases, even precipitating out of solution--
while the second species remains substantially unreacted. The character
and extent of reaction between the species will depend critically on the
respective rates of hydrolysis and condensation of the two species, and on
the structure of the polymerized species produced (e.g., open networks vs.
condensed colloidal particles). Almost nothing is known about this area,
which is critical for the production of compositionally homogeneous materials.
In one noteworthy study, Schmidt, Scholze and Kaiser (23) suggest the
occurrence of Si-O-B condensation products in the system $Si(OMe)_4$-$B(OEt)_3$-
EtOH.

Other investigators have suggested that slow hydrolysis-polymerization,
as produced by exposure to humidity in the air, will yield highly homogeneous products; but this suggestion remains to be documented in detail.
The approach appears subject to problems of a high residual organic content
and relatively low network connectivity at the point of gelation.
Preliminary partial hydrolysis of the slower-reacting species has also been
suggested as promoting homogeneity; but this approach is likewise not free
of problems. When carrying out preliminary partial hydrolysis, the order
in which alkoxides are added to the solution can significantly affect the
results [see Yoldas (24)]. This work also illustrated the role of one
species influencing the solubility of the hydrated species of a second
component; and considerations of solubility suggest that preliminary partial
hydrolysis should generally be effective only in regions of a multicomponent system.

Of the various approaches to preparing homogeneous multicomponent
materials, the results of Yamani et al. (25) on a Na_2O-TiO_2-SiO_2 composition
suggest that simultaneous hydrolysis and polymerization of alkoxides with
similar reactivities yields the most homogeneous glass (more homogeneous
than slow hydrolysis or preliminary partial hydrolysis of alkoxides with
different reactivities).

Finally, it should be noted that the addition of water to a solution
of alkoxides can generate species which can act as catalysts in subsequent
polymerization reactions. It should also be noted that the sequence of
reactions and structures involved in carrying out polymerization through
to the gel state, or to a fully developed sol, remain to be characterized
in anything like satisfactory detail for almost all systems of interest to
ceramists.

G. Post-Precipitation/Gelation Phenomena. A range of post-polymer-
ization phenomena involving significant chemical considerations are also
important in the use of sol-gel techniques to produce ceramics. Included
here are the stabilization of colloidal sols and particulates, the collapse
of gels, the drying of gels, and the elimination of residual organic
functionalities. In each case, particularly the last two, our present
state of knowledge leaves much to be desired.

IV. CRITICAL NEEDS INVOLVING CHEMISTRY

The discussion of the preceding section has uncovered a broad range
of areas where chemical insight is vital, but where our present knowledge
is abysmal. These include:

1. Solubility diagrams for a broader range of alkoxides and solvents, including more complex systems.

2. Kinetic data on the hydrolysis of single alkoxides, to provide quantitative insight in place of qualitative trends. Variables of importance include the type of metal ion, the alkoxy moiety, the nature and polarity of the solvent, the concentrations of species including water, and temperature. Specific attention should be directed to issues of stepwise hydrolysis.

3. Characterization of the species present at various stages of hydrolysis and their relation to those present before hydrolysis. It seems important to establish the conditions and materials for which the character of association (oligomerization) changes on hydrolysis.

4. Kinetics of polymerization of single alkoxide systems, exploiting the variables cited above.

5. Characterization of the species present at various stages of polymerization. This insight is almost totally lacking for systems other than SiO_2; and even in that case, the suggestions regarding structure are not cast in concrete.

6. Effects of partial condensation on kinetics of hydrolysis, and effects of the extent of hydrolysis on condensation.

7. Catalysis effects on kinetic behavior and on species produced, for non-aqueous as well as aqueous solvents.

8. Kinetics of hydrolysis and polymerization of double alkoxides, and characterization of the species present at various stages of processing. Double alkoxides may provide the key in many cases to the production of microscopically homogeneous ceramics, but they are relatively underutilized and their behavior relatively unexplored at the present time.

9. Kinetics of hydrolysis and polymerization of mixtures of alkoxides, and characterization of the species present at various stages of processing. Specific attention should be directed to the nature of the copolymers formed and to the determination of reactivity ratios.

10. Development of models of the type used successfully to describe thermosetting organic polymers, with the objective of predicting structural development in systems of interest to ceramics.

11. Growth of particulates; variation and control of particle size. Detailed attention here should yield great benefits for the production of tailor-made powders.

12. Stabilization and destabilization behavior of sols.

13. Conditions of gelation; phase transitions with structural collapse, exploring the principles developed by Tanaka and his co-workers (26, e.g.) for polyacrylamide gels and their application to ceramic systems; effects of process conditions on microstructural features of gels.

14. Polymerization behavior after precipitation or gelation (in the case of organic polymers, precipitation severely limits the molecular weight); processes involved in elimination of hydroxyl groups and organic residues with heat and/or vacuum.

15. Investigation of a broader range of systems, including (a) non-aqueous, non-alcohol solvents; (b) use of tailored alkoxides, such as those with moieties having different reactivities; more complex systems, especially non-silicate systems; organic-modified inorganic materials; and use of non-alkoxide precursors, such as metal salts and modified silanes.

V. CRITICAL AREAS WHERE NON-CHEMICAL ISSUES ARE IMPORTANT

Insights of the type outlined here seem important if the field is ever going to advance beyond the stage of cookbook recipes. Even if such information were available, however, the job of the materials engineer would by no means be complete. There would remain a number of critical issues once the synthetic chemistry is done. These include:

1. Drying of gels: kinetics, mechanistics, properties as a function of time, stress development (including relaxation), modeling, hypercritical drying, effects of exposure to atmosphere, orientation effects.

2. Densification of gels: kinetics, mechanistics, properties and chemistry as a function of time (including effect of hydroxyl groups), modeling.

3. Crystallization: kinetics as a function of structure, chemistry, temperature, time and thermal history. Kinetics here should include a separate evaluation of nucleation and growth processes. Impurities can have a large effect [see the results of Mukherjee and Zarzycki on the effect of hydroxyl groups on the crystallization kinetics of lanthanum silicates (27)], and should be investigated with care.

4. Effects of process history on structure and properties of final ceramic bodies, including residual effects even after subsequent melting at temperatures well above the glass transition (T_g). See, e.g., the critical questions raised by Neilson and Weinberg (28, 29) and by Yoldas (30) who showed important differences between melted glasses and gel-derived glasses even after melting both at temperatures hundreds of Centigrade degrees above T_g. A broad range of properties needs to be investigated and understood, from corrosion behavior to mechanical properties. The work of Hench and his colleagues (31) provides a worthwhile example of such studies.

VI. CONCLUDING REMARKS

In the present paper, we have attempted to outline some of the attractive opportunities offered by wet chemical techniques to ceramics. We have focussed on chemical issues and areas where chemical insight seems critical, since we believe that the key to the next generation of ceramic materials lies in a creative fusion of ceramics and chemistry. It is unlikely that detailed attention will be directed to each of the areas discussed above; but a considerable expansion of activity in such areas seems essential for the effective development of new materials and technologies.

The task of effecting the desired fusion between ceramics and chemistry is not an easy one, and may require innovative approaches to education as well as to research organization. The opportunities offered by the field seem, however, to merit extraordinary efforts in this regard.

VII. ACKNOWLEDGEMENTS

Financial support for the present work was provided by the Air Force
Office of Scientific Research. This support is gratefully acknowledged.

REFERENCES

1. S. Prochazka and F.J. Klug, J. Am. Ceram. Soc., 66, 874 (1983).
2. E.A. Barringer and H.K. Bowen, J. Am. Ceram. Soc., 65, C199 (1982).
3. K. Kamiya, S. Sakka and Y. Tatemichi, J. Mater. Sci., 15, 1765 (1980).
4. T. Hayashi and H. Saito, J. Mater. Sci., 15, 1971 (1980).
5. H.G. Sowman, and M.A. Leitheiser: U.S. Patent, 4,314,827 (1982).
6. Oak Ridge National Lab. development reported in Am. Ceram. Soc.
    Bull., 62, 1089 (1983).
7. F. de Monterey, "Passivation Coatings on Glass for Liquid Crystal
    Displays," private communication.
8. H. Dislich, "Coatings on Glass," Treatise on Glass, 2, D.R. Uhlmann
    and N.J. Kreidl, eds. (Academic Press Inc., 1984).
9. C.J. Brinker and M.S. Harrington, Solar Energy Materials, 5, 159 (1981).
10. W. Mahler and M.F. Bechtold, Nature, 285, 27 (1980).
11. S. Sakka, "Formation of Glass and Amorphous Oxide Fibers from Solution,"
    the Proceedings.
12. G. Philipp and H. Schmidt, "New Materials for Contact Lenses Prepared
    from Si and Ti Alkoxides by the Sol-Gel Process," to be published
    in J. Non-Cryst. Solids; H. Schmidt "Organically Modified Silicates by
    the Sol-Gel Process," this Proceedings.
13. W.R. Russo and W.H. Nelson, J. Amer. Chem. Soc., 92, 152 (1970).
14. D.C. Bradley, R.C. Mehrotra and D.P. Gaur, Metal Alkoxides (Academic
    Press Inc., London, 1978).
15. H. Meerwein and T. Bersin, Ann., 476, 113 (1929).
16. S. Govil and R.C. Mehrotra, Syn. React. Inorg. Metal-Org. Chem.,
    5, 267 (1975).
17. A.V. Novaselova, N. Ya. Turova, E.P. Turevskaya, M.I. Yanovskaya,
    V.A. Kozunov, and N.I. Kozlova, Izv. Akad. Nauk SSSR, Neorgan.
    Materialv. 15, 1055 (1979).
18. H.J. Emeléus and A.G. Sharpe, Modern Aspects of Inorganic Chemistry,
    4th ed., Chp. 10 (Wiley, New York 1973).
19. F.G.R. Gimblatt, Inorganic Polymer Chemistry, Chaps. 3 and 4
    (Butterworths, London, 1963).
20. D.W. Schaefer, "Structure of Soluble Silicates," this Proceedings.
21. G.V. Di Filippo, Ph.D. Thesis, Massachusetts Institute of Technology,
    Cambridge, Mass. (1983).
22. S.P. Mukherjee, "Homogeneity of Gels and Gel-Derived Glasses," to be
    published in J. Non-Cryst. Solids.
23. H. Schmidt, M. Scholze and A. Kaiser, J. Non-Cryst. Solids, 48, 65 (1982).
24. B.E. Yoldas, "Modification of Polymer-Gel Structures", to be published
    in J. Non-Cryst. Solids.
25. M. Yamane, S. Inoue and K. Nakazawa, J. Non-Cryst. Solids, 48, 153 (1982).
26. T. Tanaka, Sci. Am., 244, 124 (1981).
27. S.P. Mukherjee and J. Zarzycki, J. Non-Cryst. Solids, 20, 455 (1976).
28. M.C. Weinberg and G.F. Neilson, J. Am. Ceram. Soc., 66, 132 (1983).
29. M.C. Weinberg and G.F. Neilson, J. Mater. Sci., 13, 1206 (1978).
30. B.E. Yoldas, Ceramics and Glasses, 105 (1981).
31. L.L. Hench, "Environmental Effects in Gel-Derived Materials,"
    this Proceedings.

PROCESSING AND PROPERTIES OF SOL-GEL DERIVED 20 MOL% Na$_2$O-80 MOL% SiO$_2$(20N) MATERIALS

S. H. Wang and L. L. Hench
Materials Science and Engineering, University of Florida
Gainesville, Florida 32611 USA

INTRODUCTION

A number of investigators [1-6] have studied the sol-gel-glass trans-formation and the aging, drying and densification processes involved. However, monolithic, amorphous, large scale xerogels are still difficult to produce because of insufficient understanding of basic changes in micros-tructure during the sol-gel-glass transformation and the chemical reactions of the precursors, solvents and catalysts used. Some papers suggest ways of obtaining monolithic gels by (1) increasing the mechanical strength of gel by aging [7]; (2) diminishing the magnitude of capillary forces by enlarging the pore size and/or decreasing the surface energy by using sur-factants [8]; (3) reducing the rate of evaporation of the solvent from the pores by using a semipermeable membrane during drying [9] (4); and elimi-nating the pore liquid-solid interface by hypercritical evacuation [10].

The object of this investigation is the understanding of the principal mechanisms of sol-gel processing that uses a new reliable method of forming monolithic xerogels. In this paper the (20N) 20 mole % Na$_2$O-80 mole % SiO$_2$ system was chosen for study using a drying control chemical additive (DCCA) in the sol to prevent fracture during drying. Properties of gels made in this way are compared with properties of gels made without use of a DCCA.

EXPERIMENTAL PROCEDURE: SOL-GEL PREPARATION AND DRYING

Two kinds of samples were made, either with or without a DCCA. Sodium hydroxide (40 mole percent) was first added to 85 total volume percent of methanol while mixing; a silica precursor tetramethylortho-silicate (80 mole percent) was then added followed by water to hydrolyze the TMS thereby producing a 20N gel. The second type of sample used formamide as a DCCA while making the sol. The difference between the first and second processes was only in the amount of methanol, half of which was replaced by formamide in the second case.

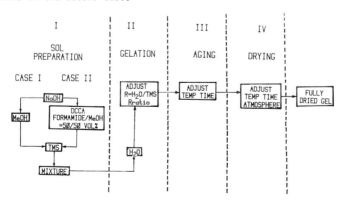

Fig. 1. Flow chart of 20N gel processing.

Mat. Res. Soc. Symp. Proc. Vol. 32 (1984) Published by Elsevier Science Publishing Co., Inc.

Figure 1 shows the flow chart of the 20N gel processing which is divided into four stages, (1) sol preparation (2) gelation (3) aging and (4) drying. During the second stage, the gelation time, which varied with water content, was measured both with and without formamide. The results are shown in Fig. 2. The gelation time was recorded from the time that the final drop of water was added until the sol became a gel. All the water was added within 4 minutes and the amount of water was calculated as the molar ratio to TMS content.

In the third stage the two types of samples were cast into polystyrene vials. The caps were closed and sealed with vacuum tape and placed into a 85°C oven for thermal aging. They were removed from the oven at intervals during the aging process. Figure 3 shows the weight loss due to shrinkage which varied with the aging time.

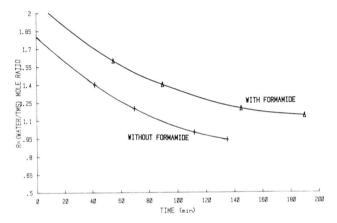

Fig. 2. Gelation time vs. water content with and without formamide. Formamide/methanol was 50/50 mixed by volume at 25°C.

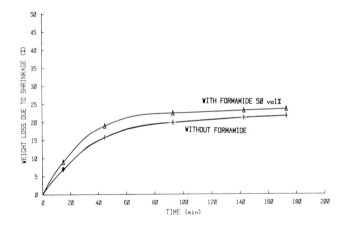

Fig. 3. Weight loss vs. time in the aging stage. Samples were aged at 85°C

In the fourth stage both types of samples were aged for 2 1/2 hours at 85°C, then the vials were then opened and the exhausted pore liquid poured out. They were then weighed to provide the baseline for further weight loss measurements, then replaced in the oven for drying. They were weighed at intervals during drying with the results shown in Fig. 4.

In the fourth stage the evaporation rates were estimated at different drying temperatures and under different drying conditions. One set of samples was opened to air and the other set was kept under a well closed polystyrene wrap. The evaporation rate was estimated by measuring the average surface area exposed to air and under the wrap. The average surface area was calculated by averaging the initial and the final dimensions of the sample during a fixed period of time at that drying temperature. The weight loss due to evaporation and shrinkage was measured in g/cm^2 as a function of time as shown in Fig. 5. Microhardness was measured after

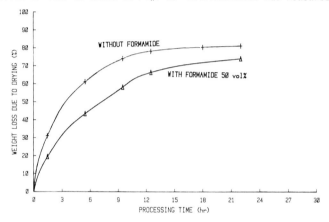

Fig. 4. Free weight loss vs. time in drying stage. Samples were preaged 2.5 hr. and then dried at same 85°C.

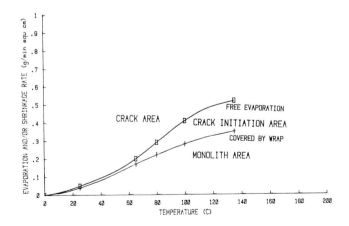

Fig. 5. Free evaporation and shrinkage rates vs. temp. in drying stage.
(I) Above line is crack area. (II) Below line is monolith area.

aging for 2 hours at 85°C for the gels made with an without the DCCA. Shrinkage was also measured as a percent of the original total gel volume. The results are shown in Fig. 6.

Further samples with and without DCCA were aged two and half hours then opened and allowed to dry at different temperatures, 25°, 65°, 80° and 125° centigrade, for 20 hours followed by microhardness tests. The results are shown in Fig. 7.

The processing conditions for stages III (aging) and IV (drying) required to produce reliable monoliths vs cracked samples are summarized in Fig. 8. This curve was obtained in the following manner. All the samples were aged at 65°C for a fixed time, then placed in a oven with variable temperature control. The temperature was slowly raised by 5°C intervals until the gels cracked. The combination of aging times and drying temperatures which lead to large monolithic dried 20N gels are above the curve drawn in Fig. 8.

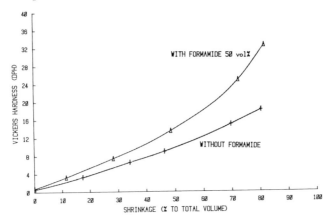

Fig. 6. Microhardness vs. percent shrinkage of wet gel. Samples were aged 2 hrs. at same 85°C drying temperature.

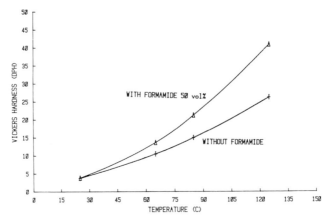

Fig. 7. Microhardness vs. drying temperature. Samples were aged 2.5 hrs. then dried at same temp. 20 hrs.

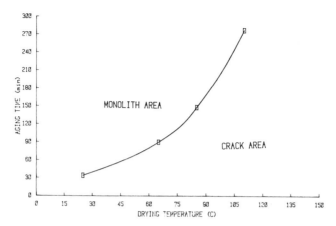

Fig. 8. Monolith and crack areas between the aging time vs. temp. curve
samples were aged 65°C x mins. then opened and placed in drying
oven.

Density was measured on samples aged in a 85°C oven 2.5 hours and
dried 22 hours. Results are shown in Fig. 9. The surface area variations
of the 20N gels made with formamide were measured by the BET method. The
samples prepared contained formamide at varying levels. Results are shown
in Fig. 10.

DISCUSSION OF RESULTS

The experimental results show that the addition of formamide to the
alkoxide 20N system prolonged the gelation time (Fig. 2), increased weight
loss during the aging stage (Fig. 3), decreased the weight loss in the
drying stage (Fig. 4) and increased the Vickers hardness (Figs. 6 and 7).
The data indicates that formamide added to the 20N solution hindered the
reaction between TMS and water. This is believed to be due to the strong
amine group on the side of the formamide molecule. Once the sol was transformed to a gel, formamide helped the gel body to shrink more quickly and
to increase in stength. This is proposed to be due to hydrogen bonds which
form between the formamide amine groups and the hydroxide groups on the
silica particle surface, thus raising the gel strength. Figure 8 shows
that aging improved the strength of the gel, so that at a fixed temperature
there is a critical aging time beyond which the gel will not crack. Figure
5 shows the critical evaporation and shrinkage rate for the 20N gels with
formamide. Below the rate-temperature curve shown the gels do not initiate
fracture and remain uncracked. If drying exceeds the rates shown, cracking
will result. The upper of the two curves corresponds to drying via evaporation of the liquid as well as shrinkage. This condition permits a more
rapid rate of drying since the extent of particle rearrangement and pore
changes in the gel are minimized.

Using the results of Fig. 5, a critical closed cell shrinkage rate of
0.22 g/min/cm^2 was established for the 20N gel with formamide. Figure 9
shows the change in density of the gels due to the 85°C aging and drying at
this critical rate. The density increased by approximately 50% from 1.08
g/cc to 1.51 g/cc during the 22 hour schedule. Control over the density
during this stage of processing is very important because it will control

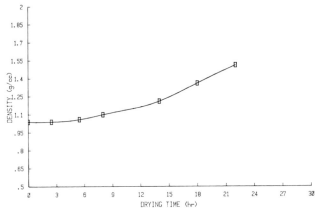

Fig. 9. The variation of density vs. drying time. Samples were aged and dried at 85°C for 2.5 hrs. and up to 22 hrs.

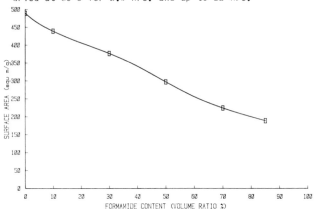

Fig. 10. The variation of surface area vs. formamide content as DCCA. Samples were aged and dried at 85°C for 2.5 hrs. and 22 hrs. Then 150°C for 20 hrs. for bet.

the eventual gel-glass transformation behavior [11] and also the gel durability [12]. Both acids and bases catalyze the silicic sol to form high molecular weight units, but they perform differently during drying. With basic catalysis (in these experiments, sodium hydroxide) the monomer particles grow much bigger than those catalyzed by acid before the three dimensional gel network is established. This is why the surface area changes proportionally to the amount of formamide added as shown in Fig. 10. In other words, as formamide is added the solution becomes even more basic, the sol particle growth is enhanced and the surface area is decreased. In this case the gel structure is no longer fibrillar, and thus even very small internal impurities can destroy a monolithic dried gel, as indicated in Ref. 7 (pg. 230). Consequently even control over the aging and drying process with formamide may be insufficient to prevent cracking if dust particles are incorporated in the gel. This is due to local stresses from the differential shrinkage rate of the dirt particles and the gel being higher than the gel strength will endure.

CONCLUSION

Essentially a optimized aging process can improve gel strength so that it will withstand a critical rate of evaporation at a fixed low temperature. If the evaporation rate exceeds the rate of gel shrinkage, cracks will be initiated. Evaporation rate is a function of temperature, with a decrease in drying temperature decreasing the evaporation rate. However this results in increased drying times.

The drying time can be minimized by controlling the evaporation rate such that it is always equal to or just slightly less than the shrinkage rate of the gel. A computerized vapor phase controller can be used to optimize and control the rates and thereby lead to reliable production of dried monolithic xerogels with a 24 hour processing schedule.

A primary advantage of the DCCA in the gel is that the critical shrinkage rate is substantially higher. Consequently, the evaporation rate can be higher and the drying time shortened severalfold. The increase in critical shrinkage rate appears to be due to the formamide increasing the gel strength through amine group bonding of the gel particles. A substantially higher temperature is required to remove the formamide compared with only methanol which enables interparticle necks in the gel to develop sufficiently to resist drying stresses.

ACKNOWLEDGMENTS

The authors gratefully acknowledge the partial financial support of AFOSR Contract # F49620-83-C-0072.

REFERENCES

1. C.J. Brinker, K.D. Keefer, D.W. Schaefer and C.S. Ashley, J. Non-Cryst. Solids, 48, 47-64. (1982)
2. H. Schmidt, H. Scholze and A. Kaiser, J. Non-Cryst. Solids, 48, 65-77 (1982).
3. J.D. Mackenzie, J. Non-Cryst. Solids, 48, 1-10 (1982).
4. Masayuki Nogami and Yoshiyo Moriya, J. Non-Cryst. Solids, 37, 191 (1980).
5. M. Prassas, Ph.D. Thesis, Montpellier, France 1981.
6. L.C. Klein and G.J. Garvey, "Monolithic Dried Gels", J. Non-Cryst. Solids (1982).
7. R.K. Iler in: The Chemistry of Silica, (Wiley-Interscience, New York, 1979) Chapters 1-6.
8. J. Zarzycki, "Monolithic Xero-and Aerogels for Gel-Glass Processes", in Ultrastructure Processing of Ceramics, Glasses and Composites, L.L. Hench and D.R. Ulrich, eds.,(John Wiley & Sons, New York, 1984).
9. M. Prassas, J. Phalippou, L.L. Hench and J. Zarzycki, J. Non-Cryst. Solids, 48 (1982), 79.
10. J. Phalippou, T. Woignier, and J. Zarzycki, "Behavior of Monolithic Silica Aerogels at Temperatures Above 1000°C" in Ultrastructure Processing of Ceramics, Glasses and Composites, L.L. Hench and D.R. Ulrich, eds. (J. Wiley & Sons, New York, 1984).
11. M. Prassas and L.L. Hench, "Physical Chemical Factors in Sol-Gel Processing in Ultrastructure Processing of Glasses, Ceramics and Composites, L.L. Hench and D.R. Ulrich, eds. (J. Wiley & Sons, Inc. New York, 1984).
12. L.L. Hench, "Environmental Effects in Gel Derived Materials", in this Conference Proceedings.

PHYSICAL-CHEMICAL VARIABLES IN PROCESSING Na_2O - B_2O_3 - SiO_2 GEL MONOLITHS

Gerard Orcel and Larry L. Hench
Materials Science and Engineering, University of Florida,
Gainesville, Florida 32611 USA

ABSTRACT

The addition of a drying control chemical additive
(DCCA) to the solvent allows short drying times for
monolithic gel samples. The variables involved in the use
of a DCCA in preparing gels from the Na_2O-SiO_2-B_2O_3 system
are described. Due to its higher reactivity, a DCCA
influences the chemical behavior of the material at each
step of the processing. The effects of formamide as a DCCA
on the gelation process, surface charge, and surface area
are presented. The variations of the physical properties of
the gel, due to the addition of the DCCA, are related to the
reactions of the formamide with the sol. Although the DCCA
helps in producing monolithic dried gels, the problem of
densification is not solved as yet.

INTRODUCTION

It has been recently described that the addition of a Drying Control
Chemical Additive (DCCA) to the solvent allows short drying procedures
during the preparation of gels in different chemical systems such as:
SiO_2, SiO_2 - Li_2O, SiO_2 - Na_2O, SiO_2 - B_2O_3 - Na_2O and Al_2O_3 [1-4]. However, fully dried monolithic gels are more difficult to obtain when the
number of constituents of the gel increases. In addition, the mechanisms
responsible for the general influence of a DCCA is not fully understood as
yet. Our purpose in this paper is to increase the understanding of the
DCCA approach by studying the effect of formamide as a drying control
chemical additive on the chemical and physical properties of a gel in the
ternary system SiO_2 - B_2O_3 - Na_2O.

EXPERIMENTAL PROCEDURE

The composition studied was: 42 mole % SiO_2 - 30 mole % B_2O_3 - 28 mole
% Na_2O. The procedure followed for the manufacture of the gel is shown
schematically in Fig. 1. The total amount of water added corresponds to
the stoichiometric quantity for the hydrolysis of the alkoxides. The gels
were cast in Pyrex® glass vials, and aged at 60°C for 24 hours in closed
containers. The drying process involved a combination of heating treatments under regular air atmosphere and in approximately 1 Torr of vacuum.

RESULTS AND DISCUSSION

The chemical reactions which occur during the gelation process can be
written as follows:

$$Si(OC_2H_5)_4 + 4\ H_2O \rightarrow Si(OH)_x(OC_2H_5)_y + xC_2H_5OH + y\ H_2O \qquad (1)$$

$$B(OCH_3)_3 + 3\ H_2O \rightarrow B(OH)_u(OCH_3)_v + u\ CH_3OH + v\ H_2O \qquad (2) \quad \text{Hydrolysis}$$

$$NaOCH_3 + H_2O \rightarrow Na^+ + OH^- + CH_3OH \qquad (3)$$

Mat. Res. Soc. Symp. Proc. Vol. 32 (1984) Published by Elsevier Science Publishing Co., Inc.

$$\equiv Si-OR' + R''O-Si\equiv \rightarrow \equiv Si-O-Si\equiv + R'OR'' \qquad (4)$$

$$\equiv Si-OR' + R'''O-B= \rightarrow \equiv Si-O-B= + R'OR''' \qquad (5)\ \text{Polycondensation}$$

$$=B-OR''' + R''''O-B= \rightarrow =B-O-B= + R'''OR'''' \qquad (6)$$

with $x + y = 4$ and $u + v = 3$

The rate of these reactions (1-6), and subsequently the structural characteristics of the resulting gel, are influenced by the experimental conditions such as: temperature, pH, nature and concentration of the reactants [6-9]. The amount of DCCA in the solvent is also a very important variable. The influence of these parameters can be visualized by recording the gelation time for different processing conditions. In Fig. 2 we report the gelation time as a function of the proportion of formamide in the solvent, for various gelling temperatures. Another variable was the acidity of the water added at the end of the process (see Fig. 1).

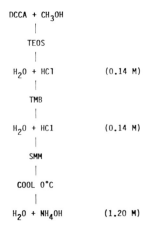

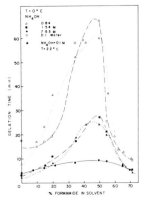

FIG. 1 Schematic diagram of sol-gel process for 42S-30B-28N gel derived monoliths.

FIG. 2 Effect of the molarity of the NH_4OH solution (see Fig. 1) on gelation time versus concentration of formamide.

The general trend of the curves in Fig. 2 is in good agreement with what has been published: the gelation time increases when the temperature decreases and/or when the pH decreases. The shape of the curves obtained suggests that there is a competition between several mechanisms during the gelation process. A sol can be destabilized by increasing the brownian motion of the sol particles (i.e. by increasing the temperature) and/or inducing electrostatic interactions between them (usually by adding an electrolyte). With the procedure we followed, the final "pH" was about 10.5, slightly depending on the amount of formamide in the solvent, and the acidity of the water used. Thus, the second mechanism should be predominant for this particular system. We already have proposed that the formamide is forming a complex with the sodium ions in the tetrahedral boron sites [3]. This complexing reaction can explain the rising portion of the curves. Due to the high pH, the apparent particle charge is negative [11]. By reacting with these sodium ions, the formamide reduces the number of positive sites on the sol particles, making the surface charge even more negative. Less electrostatic attractions are possible which leads to an increase of the gelation time with increasing formamide content.

However, when more formamide is supplied to the system, several reactions become more apparent: the scissioning of stressed bridging oxygen bonds, as proposed by Michalske et al. [5] the hydrolysis of formamide and the neutralization of negative charge centers through ionic attractions (Fig. 3). In a basic medium, the formamide can be hydrolysed to produce formic acid and ammonia. The reaction can be written as follows:

$$HCONH_2 + H_2O \rightarrow NH_3 + HCOOH \tag{7}$$

This reaction explains two facts which seem contradictory; although the formamide is a base weaker than water (the formamide belongs to the acid amide group) the "pH" of the solution before gelation was slightly more basic when the formamide concentration was higher. The formic acid can easily react with an alcohol to produce an ester [12] and thus does not significantly displace the pH, compared to the ammonia. NH_3 is a relatively strong base and can poison the negative sites, when combined with an hydrogen ion. The surface charge of the sol particles becomes less negative and the repulsive forces are diminished in proportion. Thus the gelation time decreases, as shown by the curves in Fig. 2 for concentrations in DCCA higher than 47 vol %.

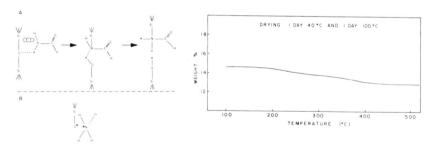

FIG. 3 Schematic representation of a) scissioning of stressed bridging oxygen bond by formamide, b) charge neutralization through ionic attraction between negative and positive centers.

FIG. 4 TGA thermogram of a gel prepared with 50 vol % of formamide in solvent and dried in air at 40°C for 24 hours and at 100°C for 24 hours.

The maxima of each of the curves in Fig. 2 are located at about 47 vol % of formamide. The data show that the maxima do not depend on the pH of the water added at the end of the process. This suggests that the influence of the electrolyte is on the hydrolysis and condensation rates rather than on the surface charge of the sol particles. Although the quantity of the basic water added at the end of the process was very large (7 vol %), the "pH" of the sol before the gelation was around 10.5, independent of the molarity of the solution of ammonium hydroxide used. Thus, the sol behaves as a buffer. Because of the high ionizing power of the formamide, i.e. with a dielectric constant $\varepsilon = 110$, the electrolyte tends to combine with the solvent instead of fixing on the sol particles. So the amount of electrostatic interactions between the sol particles remains the same, regardless of the pH of the water added. The number of moles of formamide which interact with the sol particles and affect their apparent surface charge is independent of the molarity of the solution of NH_4OH used. This may explain the relative constancy of the location of the maxima of the curves.

Due to the complexation of the sodium ions in the tetrahedral sites with the formamide, the surface charge of the sol particles is affected. If the formamide linked to the sodium ions is not removed during the heat treatment, then the surface charge of the gel should be a function of the composition of the solvent. The apparent surface charge should be more negative as the proportion of DCCA in the solvent increases. In order to show that formamide was still chemically bonded to the gel after heating, the following experiment was conducted. Freshly prepared gels were heated in air at 40°C for 24 hours and then at 100°C for another day. All the solvent was evacuated, as shown by the TGA curve in Fig. 4. The gels were crushed to a powder. A zeta potential measurement was performed on the powdered gels that had been produced with different concentrations of formamide in the solvent. The resulting curve (Fig. 5) clearly shows that the surface charge of the gel is function of the amount of DCCA used. The apparent surface charge of the particles is more negative as the concentration of the drying control chemical additive in the solvent increases, i.e. as more sodium ions are combined with the formamide. The composition 42S-30B-28N, corresponds to the eutectic point. Since all the gels display approximately the same melting temperature (580°C), the variation in the apparent surface charge of the samples does not correspond to the extraction of the sodium ions by the solvent, but to the screening effect by the formamide. The quantity of positive centers which are covered by the DCCA increases with the concentration of the drying control chemical agent in the solvent.

The presence of adsorbed molecules on the gel surface will influence the densification process. Due to steric hindrances some hydroxyl groups will be prevented from reacting together and less bridging oxygen bonds will be formed. This is confirmed by the fact that the true density of the gels decreases when the concentration of formamide in the solvent increases. It is well known that the silanol groups, which are acidic, can hydrogen bond with different families of molecules [6]. In the case of amides, two different types of hydrogen bond are possible:

$$\equiv Si-OH \cdots\cdots N - C \overset{\displaystyle R \quad\; O}{\underset{\displaystyle R' \quad R''}{}} \qquad or \qquad \equiv Si-OH \cdots\cdots O = C \overset{\displaystyle NRR'}{\underset{\displaystyle R''}{}}$$

It has been shown that for the dimethyl acetamide

$$O = C \overset{\displaystyle CH_3}{-} N(CH_3)_2$$

that a hydrogen bond occurs to a silica surface through the oxygen in the compound [11]. It is difficult to extrapolate from this compound to formamide because the steric hindrance of the nitrogen group is very different for the two amides. When the number of hydrogen bonds with the silanol and/or the boroxol groups increases, the densification process through a dehydration reaction:

$$\equiv Si-OH + HO-Si \equiv \;\rightarrow\; \equiv Si-O-Si \equiv + H_2O \qquad\qquad (8)$$

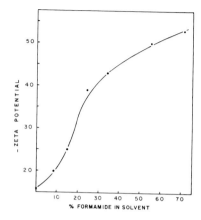

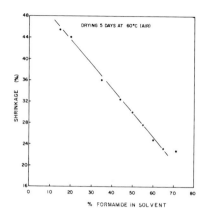

FIG. 5 Effect of the concentration formamide on the Zeta potential of 42S-30B-28N gel particles.

FIG. 6 Radial shrinkage of cylindrical gels versus formamide concentration after 5 days drying at 60°C.

is hindered. This is illustrated in Fig. 6 where the radial shrinkage of a cylindrical gel is plotted as a function of the DCCA proportion in the solvent, for a 5 day drying at 60°C in air. The gels contract less when the quantity of formamide is increased.

This behavior is less marked when the drying temperature is higher (Fig. 7). In this figure we plotted the bulk density of the gels, measured by pycnometry with mercury as liquid, as a function of the DCCA content in the solvent. The curves show that for an amount less than 20 vol % the formamide helps the densification of the gels. Above 20 vol % of DCCA, the density does not vary significantly.

Since the hydrolysis and polycondensation rates are affected by the addition of a DCCA, the surface area of the gels should vary with the solvent composition. The plot obtained for various heat treatments (Fig. 8), does not exhibit an obvious relationship between the two properties (surface area and formamide concentration). The relative low value corresponding to a pure methanol solvent and a 200°C heat treatment, suggests that some of the pores have been closed, probably trapping organic molecules. When the temperature is increased, the internal pressure of the pores rises and leads to the bloating of the material. That is probably why we obtained higher surface area measurements when the samples have been directly heated to higher temperature.

CONCLUSION

The addition of a Drying Control Chemical Agent (DCCA) to the solvent helps in shortening drying procedures. It influences the gelation process, due to the bonding of formamide with sol particles. The physical properties of the gels such as the density and the surface area are greatly influenced by the concentration of DCCA in the solvent. More work needs to be done in order to produce pure monolithic gel derived glasses.

84

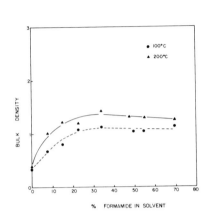

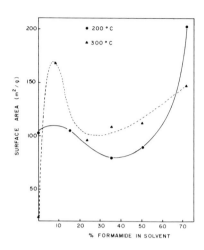

FIG. 7 Effect of the drying temperature on the bulk density versus
formamide concentration.

FIG. 8 Effect of the drying temperature on the surface area versus
formamide concentration.

ACKNOWLEDGEMENTS

The authors gratefully acknowledge the partial financial support of
AFOSR Contract # F 49620-83-C-0072.

REFERENCES

1. S. Wallace and L. Hench, 8th Annual Conference on Composites and
   Advanced Ceramic Materials, Jan. 15-18, 1984, Cocoa Beach, FL.
2. S.H. Wang and L. Hench, 8th Annual Conference on Composites and
   Advanced Ceramic Materials, Jan. 15-18, 1984, Cocoa Beach, FL.
3. G. Orcel and L. Hench, 8th Annual Conference on Composites and Advanced
   Ceramic Materials, Jan. 15-18, 1984, Cocoa Beach, FL.
4. J. Lannutti and D. Clark, 8th Annual Conference on Composites and
   Advanced Ceramic Materials, Jan. 15-18, 1984, Cocoa Beach, FL.
5. T. Michalske and S. Freiman, J. Am. Ceram. Soc., 66 [4], 4 (1981).
6. C.J. Brinker in: Ultrastructure Processing of Ceramics, Glasses and
   Composites, L.L. Hench and D.R. Ulrich, eds., (John Wiley & Sons, New
   York, NY 1984), Chp. 5.
7. L.C. Klein, and G.J. Garvey, Effect of Water on Acid and Base Catalyzed
   Hydrolysis of Tetraethylorthosilicate (TEOS), presented at Materials
   Research Society (1984), Spring meeting, Albuquerque, N. M.
8. D.P. Partlow and B.E. Yoldas, J. N. C. S., 46, 153 (1981).
9. H. Schmidt, H. Scholze, A. Kaiser, J. N. C. S., 68, 65 (1982).
10. M. Voronkov, V. Mileshkevich and Y. Yuzhelevskii, The Siloxane Bond,
    Consultants Bureau, New York, 1978.
11. R.K. Iler, The Colloid Chemistry of Silica and Silicate, Cornell
    University Press, 1955.
12. J.A. Riddick and W.B. Bunger in: Techniques of Chemistry - Vol. 2,
    Organic Solvents, 3rd Ed. (Wiley-Interscience, New York, 1970).

THE ROLE OF WATER IN DENSIFICATION OF GELS

T. A. GALLO,[+] C. J. BRINKER,* L. C. KLEIN,[+] AND G. W.
SCHERER**, Rutgers University, Ceramics Dept. Piscataway,
NJ, *Sandia National Laboratories, Albuquerque, NM,
**Corning Glass Works, Corning, NY

ABSTRACT

The densification behavior of a gel-derived
borosilicate glass was studied. Surface area, thermal
gravimetric and infrared measurements were used to calculate
the surface hydroxyl coverage as a function of time at
several temperatures. Application of a viscous sintering
model to the isothermal shrinkage data showed viscosity to
increase isothermally by almost two orders of magnitude at
the lowest investigated temperature. Although most of this
increase is attributable to increased crosslinking
accompanying dehydroxylation, structural relaxation was also
shown to significantly affect the densification kinetics.

INTRODUCTION

Glasses produced from metal alkoxides via the sol-gel process are known
to have higher purity and lower processing temperatures than comparable
glasses produced by melting [1,2,3]. The higher purity of these glasses
results from using high purity reagents and from the method of densifying
gels by sintering rather than melting, so that metallic impurities are not
introduced from a crucible [4]. The higher purity of gel-derived glasses
generally refers to metallic impurities. It is equally important to
consider the residual hydroxyl concentration, [OH]. Because gel glasses are
densified at low temperatures without melting, they may contain over 20
times more [OH] than the corresponding melted composition [5]. Therefore,
bloating may occur when the sintered gel is heated near its softening
temperature. Bloating results when water vapor generated in the dense or,
at least, closed-pore gel expands as the temperature is raised. This
problem increases with thickness of monolithic gels [3,4,6].
In two previous investigations of a borosilicate gel it was shown that,
during gel densification, the isothermal viscosity can increase by over two
orders of magnitude [7,8]. Condensation reactions:

$$\equiv Si-OH + HO-Si \longrightarrow \equiv Si-O-Si\equiv \ + H_2O$$

which serve to polymerize the network and thus increase the viscosity[9] are
thought to make the larger contribution to the viscosity increase, but
because gels sinter at very high viscosities, 10^{13} to 10^{16} poises [7,8],
structural relaxation may also make a contribution. The primary purpose of
this investigation, therefore, is to investigate the effect of [OH] on
viscosity, in order to elucidate the contributions of both condensation
reactions and structural relaxation to the observed isothermal increases in
viscosity.

EXPERIMENTAL

The gel chosen for study had nominal composition: 83% SiO_2, 15% B_2O_3,
1.2% Na_2O, and 0.8% Al_2O_3 by weight. The preparation technique[10]

Mat. Res. Soc. Symp. Proc. Vol. 32 (1984) © Elsevier Science Publishing Co., Inc.

consisted of partially hydrolyzing TEOS and sequentially adding Al-sec-butoxide, trimethylborate (TMB), and NaCOOCH$_2$ (2M). The final pH and water-to-alkoxide ratio was modified by addition of an aqueous solution of NH$_4$OH. The solution was cast into polypropylene molds, allowed to gel at room temperature and dried at 50°C for 7 days to produce cylindrical specimens (diameter = 0.8 cm, length = 10 cm).

Weight loss and linear shrinkage were measured using a Theta Dilatronic dual push rod dilatometer and a DuPont 1090 Thermal Analyzer, respectively. All heat treatments were performed in ultra high purity oxygen (135 ppm H$_2$O) at a flow rate of 30 cm^3/min. BET surface area measurments were made using a Micromeritics Digisorb 2500 pore analysis system. Water produced during gel densification (a by-product of condensation) was measured by titrating the gas evolved during the dilatometry experiments with Karl Fischer reagent using a Mitsubishi Coulometric Moisture Meter. Water contents of dense specimens were measured by FTIR assuming an extinction coefficient of 56 l$\cdot$mol$^{-1}\cdot$cm^{-1} [11].

RESULTS

The experiments consisted of first heating the gel at 2°C/min to full density while titrating the evolved gas. A companion TGA experiment was performed and the weight loss measured above 300°C was found to equal the amount of water titrated above 300°C. On the basis of these results weight loss above 300°C for a heating rate of 2°C/min was assumed to be all water. The hydroxyl content of this fully dense gel was determined by FTIR to be 640 ppm. Using this value it was possible to establish from the weight loss curve the hydroxyl content at any temperature between 300 and 640°C for a heating rate of 2°C/min.

A separate set of experiments was performed in which the gel samples were heated at 2°C/min in both the dilatometer and the thermal analyzer to 440, 490, 540 and 595°C and either quenched to room temperature or held isothermally for 18 hours. Combining the weight loss measured isothermally with the initial hydroxyl content established by the previous experiment, it was possible to determine the change in hydroxyl concentrations during each isotherm. BET surface areas were measured on both quenched samples and samples held isothermally. For temperatures less than about 300°C the BET measurements were made only on samples held isothermally for 1 hour.

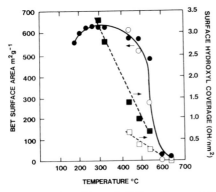

FIG. 1. Variation of BET surface area (circles) and hydroxyl coverage (boxes) with temperature for quenched samples (filled) and 18 hour isothermal heat treatments (open).

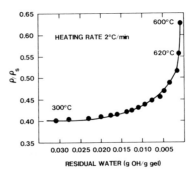

FIG. 2. Relative density vs. residual OH content for a constant heating rate of 2°C/min. Data taken at 20 degree increments.

Figure 1 shows the BET results. The surface area initially increases with temperature. Iler defines this initial heat treatment of a gel, up to about 300°C, as an activation stage[12] and the temperature of maximum surface area as the activated temperature, T_a. T_a corresponds to the temperature at which physically adsorbed water and residual organics are removed uncovering all accessible surface area, so that the gel surface structure approaches that of a porous, hydrated glass. Surface area decreases after the 18 hour hold for all temperatures except the 440°C sample, where the measured increase is 6%. The rapid decrease in surface area above 500°C corresponds to a rapid increase in density.

For T > 300°C, the surface hydroxyl coverage was calculated by dividing the hydroxyl concentration by the surface area. This assumes that there are not bulk hydroxyls above 300°C. We haven't verified this, but at high temperatures as the skeletal structure approaches that of an anhydrous glass, we expect that OH groups reside predominantly on the pore surfaces [13]. Surface hydroxyl coverage before and after 18 hour holds is shown in Figure 1. The hydroxyl coverage always decreases during the isotherm. The values of hydroxyl coverage are somewhat lower over this temperature range than those measured for pure silica gels of equivalent pore size[12]. This result however depends on whether or not the extinction coefficient choice is appropriate in our case.

The densification rate of a gel above T_a is dependent on temperature and firing atmosphere; therefore, the firing atmosphere was standardized using ultra high purity oxygen, and the temperature was varied. Changes in density, surface area and [OH] were measured as the gel was heated above T_a. A plot of relative density versus calculated residual [OH] is shown in Figure 2 for a heating rate of 2°C/min. Up to about 500°C there is a large reduction in [OH] with only a modest increase in density (∿5%). Most of this reduction is attributable to surface dehydroxylation which is not expected to contribute significantly to densification. Densification in this temperature range is attributable to skeletal densification which may occur by either structural relaxation or additional crosslinking within the bulk[7,8]. Above 500°C, there is a smaller change in [OH] whereas there is a sharp increase in density accompanied by a rapid reduction in surface area. This is consistent with viscous sintering being the predominant shrinkage mechanism above 500°C. This temperature boundary is not well-defined however and, as evidenced by the reduction in [OH] above 500°C, the skeletal composition appears to continually change with temperature.

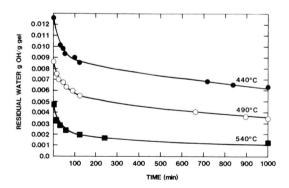

FIG. 3. Residual [OH] vs. time during isothermal heat treatments.

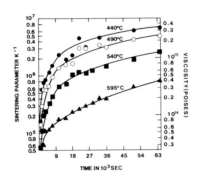

FIG. 4. Sintering parameter K^{-1} (and viscosity) vs. time during 18 hour isothermal heat treatments.

In previous studies of the sintering of gels, we have postulated that both structural changes due to loss of free volume and continued crosslinking, and compositional changes due to dehydration serve to increase the apparent viscosity during isothermal treatments[7,8]. In an attempt to separate these effects, isothermal densification data were analyzed using a viscous sintering model, from which the isothermal viscosity was derived. Changes in viscosity could, therefore, be directly related to dehydroxylation.

Isothermal changes in [OH] are shown in Figure 3 for the eighteen hour treatments at 440, 490, and 540°C. In each case, [OH] decreases rapidly during the first 100 minutes followed by a slower decrease thereafter. At the highest temperature, the rate of [OH] loss approaches zero at 1000 minutes. These results suggest that at each temperature an equilibrium [OH] is approached.

Isothermal densification data were analyzed using a sintering model based on a geometry consisting of a cubic array of intersecting cylinders[14]. To test the applicability of this model, surface areas calculated from relative density and pore size, according to the model, were compared to the values measured by BET. The calculated surface areas were 20-25% lower than the measured values, but this discrepancy does not seriously affect the predictions of the sintering model[14].

According to the model, the quantity K is calculated using:

$$K = (\gamma/\eta \ell_o) (\rho/\rho_s)^{1/3}$$

where γ is the surface tension, η is the viscosity, ℓ_o is the initial cylinder length, ρ_s is the skeletal density assumed to be 2.146 gm/cm^3 and ρ_o is the initial bulk density. The reciprocal of K is proportional to viscosity and for this gel over the densification range studied, η is between 1×10^{12} and 5×10^{15} poise, given ℓ_o = 77A and γ = 250 ergs/cm^2. A plot of K^{-1} versus time is shown in Figure 4. The viscosity of the gel increases with time; larger increases are observed at lower temperatures. Even after 18 hours at 595°C, a constant value of η is not attained.

Both [OH] and excess free volume affect the viscosity of a gel or glass. Over a narrow range of temperature, the equilibrium viscosity of a glass can be expressed by:

$$\eta = \eta_o \exp (E/RT).$$

The constant η_0 and the activation energy, E , both decrease as [OH] increases. If the glass is not in structural equilibrium, i.e. if it contains "excess" free volume, η also depends on the fictive temperature, T_f, as follows [15]

$$\eta_{T,t} = \eta_0 \exp\left(\frac{xE}{RT} + \frac{(1-x)E}{RT_{f_t}}\right) \qquad \text{where}$$

$0 \leq x \leq 1$. At equilibrium, $T = T_f$ and this reduces to the Arrhenius equation; otherwise there is a quantity of excess free volume proportional to $T-T_f$. As T_f decreases toward T, the apparent activation energy varies between E and xE; typically x = 1/2.

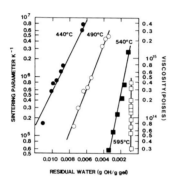

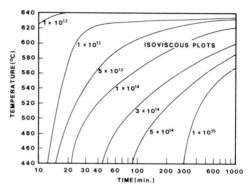

FIG. 5. Sintering parameter K^{-1} (and viscosity) vs. [OH] during 18 hour isothermal heat treatments.

FIG. 6. Time vs. temperature curves corresponding to constant viscosity as calculated from the cylinder model [15].

To assess the relative effects of dehydration and skeletal relaxation, a plot of log (K^{-1}) versus residual [OH] is shown in Figure 5. Although the linear relationship suggests that the increase in viscosity is related to dehydration, the viscosity is clearly dependent on factors other than [OH] at higher temperatures. At 595°C, [OH] loss is no longer measurable, but the viscosity still increases, by a factor of ∿20, most likely because of slight skeletal changes in the gel. Although it could be argued that some of the η increase at 595°C results from dehydroxylation in a closed pore system (i.e. water is generated which cannot escape), there was no evidence of bloating, and near infrared spectroscopy showed no evidence of molecular water. Thus, we are led to the conclusion that structural relaxation, which reduces excess free volume and thus reduces T_f, is significantly increasing the viscosity and that it becomes more important as the sintering temperature is increased toward the measured (using DSC) T_g of the fully densified gel (620°C).

Finally, when the apparent viscosity calculated from the model is plotted on a time-temperature field, curves of approximately constant viscosity can be drawn in a manner similar to a time-temperature-transformation plot [3]. Thus, for the investigated composition, the time required to reach a given viscosity at a particular temperature can be predicted, which will aid in planning a densification schedule. Because of the role of [OH] and free volume in gels, the viscosity increases with time. The combined effects of [OH], free volume, and temperature lead to the curves in Figure 6.

90

CONCLUSIONS

In order to determine the role of water in the densification of gels, the bulk density, surface area and hydroxyl content were measured during heat treatments above the activation temperature, T_a. At low temperatures, dehydration occurred with little associated densification, indicative of surface dehydroxylation. At the highest temperatures rapid densification occurred with little dehydration (indicative of viscous sintering).

Values of apparent viscosity, derived from a viscous sintering model, were observed to increase isothermally during treatments ranging from 440 to 595°C. Some of this increase is attributable to increased crosslinking of the oxide skeleton due to condensation reactions; however, a substantial viscosity increase occurred at 595°C with no apparent dehydration (as determined by TGA). This suggests that a portion of the viscosity increase is due to structural relaxation which serves to reduce T_f and hence increase the viscosity.

ACKNOWLEDGEMENT

This work was conducted during T. A. Gallo's summer employment at Sandia National National Laboratories. The technical assistance of C. S. Ashley, M. S. Harrington, S. M. Lappin, R. Z. Lawson, D. R. Salmi, M. C. Oborny, and K. L. Higgins, all of Sandia, is greatly appreciated.

REFERENCES

1. M. Yamane, S. Aso, S. Okano, T. Sakaino. J. Mat. Sci. 14 (1979) 607-611.
2. C. J. Brinker, K. D. Keefer, D. W. Schaefer, and C. S. Ashley, J-Non-Crystal. Solids 48 (1982) 47-64.
3. J. Zarzycki. J. Non-Crystal. Solid 48 (1982) 105-116.
4. L. C. Klein, T. A. Gallo and G. J. Garvey. J. Non-Crystal. Solids 63 (1984).
5. C. J. Brinker and D. M. Haaland to be presented 1984 Annual Meeting of the Am. Ceram. Soc.
6. K. Kamiya, S. Sakka, I. Yamanaka. Proc. 10th Internat. Cong. Glass 13 (1974) 44-47.
7. C. J. Brinker and G. W. Scherer, Proceedings of Ultrastructure Processing of Ceramics, Glasses and Composites, L. L. Hench, D. R. Ulrich Eds., John Wiley & Son, NY (1984).
8. G. W. Scherer, C. J. Brinker, E. P. Roth. Submitted to J. Non-Cryst. Solids.
9. G. Hetherington, K. H. Jack and J. C. Kennedy. Phys. Chem. Glasses 5 (1964) 139-136.
10. C. J. Brinker and G. W. Scherer, submitted to J. Non-Crystal. Solids.
11. J. P. Williams et al. Am. Ceram. Soc. Bull. 55 (1976) 524-527.
12. R. K. Iler, The Chemistry of Silica, John Wiley and Sons Inc., NY 1979.
13. D. M. Krol and J. G. Van Lierop. J. Non-Crystal. Solids 63 (1984)(1983).
14. G. W. Scherer. J. Am. Ceram. Soc. 60 (1977) 236-46.
15. O. S. Narayanaswany. J. Am. Ceram. Soc. 54 [10] 491-498 (1975).

FORMATION OF GLASS AND AMORPHOUS OXIDE FIBERS FROM SOLUTION

SUMIO SAKKA
Institute for Chemical Research, Kyoto University,
Uji, Kyoto-Fu 611, Japan

ABSTRACT

A review has been made of our works on two types of
sol-gel methods for obtaining oxide fibers through low
temperature process: (1) the alkoxide method and (2) the
freezing-of-gel method. In the alkoxide method, metal
alkoxides in an alcoholic solution are hydrolyzed and poly-
condensed into chain-like polymers, drawn into gel fibers
near room temperature and heated to several hundred $^{\circ}$C.
Various problems encountered in the process are discussed
with SiO_2-based fibers. In the freezing-of-gel method, a
hydrogel is unidirectionally frozen by lowering the gel
cylinder into a cold bath. Thawing of unidirectionally
grown ice crystals leaves a bundle of hydroxide fibers.
Factors affecting the formation of fibers are discussed with
titanium and zirconium hydroxide fibers as examples.

INTRODUCTION

Conventionally, glass and ceramic fibers have been prepared by drawing
from the high temperature melts of corresponding compositions. This
method needs conversion of raw materials to a homogeneous melt. This makes
it difficult to prepare fibers of the compositions which are high melting,
immiscible in the liquid state or easily crystallizable during cooling.
The sol-gel technique of fiber preparation in which fibrous gels drawn from
the solution near room temperature are converted to glass or ceramic fibers
at several hundred degrees makes it possible to avoid these difficulties.
Recently, various kinds of glass and ceramic fibers such as SiO_2, Al_2O_3,
TiO_2, ZrO_2 and SiC fibers have been prepared by the sol-gel technique.[3] It
can be said that the chemistry at low temperatures is important in this
technique in contrast with the conventional method involving melting.
There are several variations in fiber preparation based on the sol-gel
method, as shown in Table I. In this paper the preparation of fibers using

TABLE I. Variations of fiber preparations through sol-gel processes

Method	Fiber formation	Starting materials	Forming temperature	Fibers
1.	Low temp., Drawing during hydrolysis.	Metal alkoxide	Near room temp.	SiO_2, SiO_2-ZrO_2, SiO_2-TiO_2
2.	Low temp., Drawing by adding visco- sity increasing reagent.	Metal alkoxide, Inorganic com- pounds	Near room temp.	SiO_2, Al_2O_3, TiO_2, ZrO_2
3.	High temp., Draw- ing from glass rods	Metal alkoxide, Colloidal silica	Higher than 1250°C	SiO_2
4.	Unidirectional freezing of gel	Inorganic com- pounds	near 0°C	SiO_2, TiO_2, ZrO_2

Mat. Res. Soc. Symp. Proc. Vol. 32 (1984) © Elsevier Science Publishing Co., Inc.

metal alkoxides as starting materials (called as the alkoxide method in this paper) and the preparation of fibers by unidirectional freezing of gels (called as the freezing-of-gel method) will be reviewed mainly on the basis of experiences in our laboratories of Mie and Kyoto Universities.

PREPARATION OF FIBERS BY THE ALKOXIDE METHOD

Hydrolysis and accompanying polycondensation of metal alkoxides in solution produce alkoxide polymers. This converts the solution to sol and subsequently to gel. On heating, the gel is converted to glass or ceramics. This method gives bulk materials, fibers and coating films of glasses and ceramics [1-4].

The preparation of fibers needs the appearance of spinnability of the solution in the course of hydrolysis-polycondensation and the quick solidification as the fiber. The spinnability may appear when some kind of viscosity-increasing reagent is added to the solution. In this case, however, drawn fibers have to be immediately fixed by some special means, for example, by passing the fibers through the setting solution. Otherwise, fibers once formed may be cut, becoming a round drop of the solution. Therefore, we looked for the condition in which the solution becomes spinnable as a result of progress of hydrolysis-polycondensation, that is, the condition for the appearance of spinnability without addition of viscosity-increasing reagent. In this case, the fibers drawn from the viscous, spinnable solutions are quickly solidified during drawing due to the progress of hydrolysis-polycondensation caused by the moisture in air, because the degree of hydrolysis-polycondensation reaction progressing before drawing is very large and only a small degree of extra reaction is required for fixing the fibers.

In the followings, the results obtained on the formation of silica glass fibers from silicon alkoxide will be described. The condition required for fiber drawing will be stressed.

Pertinent Compositions of the Starting Solutions for Fiber Drawing

First of all, it should be noted that all starting solutions do not necessarily become drawable. The composition of the solution must be appropriate, in order for the fiber drawing to be possible.

Generally, it has been shown [5] that the alkoxide solution becomes spinnable when the content of water used for hydrolysis is low and the catalyst is acid. When the water content is high or an alkali like ammonia is chosen as the catalyst, the solution is non-Newtonian, becoming an elastic gel without exhibiting spinnability.

Figure 1 is a diagram [6] showing the fiber drawing behavior versus composition relationship for the formation of SiO_2 glass fibers from the $Si(OC_2H_5)_4-H_2O-C_2H_5OH$ solutions. The ratio $[HCl]/[Si(OC_2H_5)_4]$ was kept at 0.01. The hydrolysis reaction has been carried out at $80°C$. The triangular diagram can be divided into four areas.

The components of the solution are not miscible with each other in area I, where the C_2H_5OH content is low. In area II, where the $[H_2O]/[Si(OC_2H_5)_4]$ ratio r is over 5, the solutions become of elastic nature just before and after solidification to a gel, exhibiting no spinnability. Too much time is required for the solution to become viscous and solidify in area III, in which r is less than 1.5. In area IV, where r ranges from 1.5 to 4, the solutions exhibit spinnability and therefore the fiber drawing is possible. On heating up to 500-900°C the drawn gel fibers become SiO_2 glass fibers. Besides the possibility of fiber drawing, there is a problem that the cross section of the fibers thus prepared are not necessarily circular unlike the ordinary glass fibers drawn from the high temperature melt. A circular cross-section can be obtained in some limited compositions of area IV. This problem has been discussed elsewhere [6].

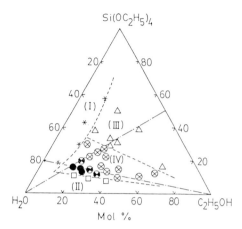

FIG. 1. Relation between fiber drawing behavior and composition of $Si(OC_2H_5)_4-H_2O-C_2H_5OH$ solution with $HCl/Si(OC_2H_5)_4$ =0.03 hydrolyzed at $80°C$[6]. ✲ immiscible(area I), ◻ not spinnable(area II), △ not-geling(area III), ○ circular cross-section(area IV), ⊗ non-circular cross-section(area IV), ⊖ circular and non-circular cross-section (area IV).

Time Period Required for the Reaction Leading to Appearance of Spinnability

It has been shown [5] that fibers can be drawn from the solutions of pertinent compositions when their viscosity reaches about 10 poises as a result of progress of the hydrolysis-polycondensation reaction. The time required for a solution to reach the drawable state is a function of ambient temperature. It is more than several days for room temperature, while it is two or three hours for $80°C$. Two or three hours may be reasonable for the practical application of the procedure.

Another requirement for the practical application is that the length of time during which fibers can be drawn continually should be long. This possibility has been found [7] by examining the time dependence of viscosity for the alkoxide solutions of different water contents shown in Figure 2. The reaction temperature has been fixed at $80°C$, while the measurement of viscosity has been made at $25°C$. It is seen in Figure 2, that in the solution with the $[H_2O]/[Si(OC_2H_5)_4]$=1.7, the increasing rate of the viscosity markedly decreases after the viscosity reaches about 10 poises. This is interpreted to be caused by the exhaustion of the hydrolysis-reaction water in the solution. Accordingly, it is possible to continue drawing for a prolonged time.

Process of Hydrolysis-Polycondensation

In the previous section, it has been shown that the lower water content of the alkoxide solution is favorable for fiber drawing, while no spinnable state appears when the water content is high.

We assumed that linear polymers are formed in the low water content solutions which show spinnability in the course of progress of hydrolysis-polycondensation reaction. It was also assumed that three-dimensional net-

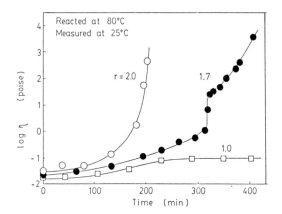

FIG. 2. Change of viscosity of $Si(OC_2H_5)_4-H_2O-$
C_2H_5OH solution at $80°C$ [7]. $r = [H_2O]/[Si(OC_2H_5)_4]$.

works or colloidal particles may be formed in the high water content solutions which show an elastic nature before gelation and no spinnability.

In order to confirm these, the molecular weights and intrinsic viscosities of the solutions taken in the course of gelation of the tetraethyl
orthosilane (TEOS) have been measured and discussed in terms of the shape of
the polymers produced in the solutions [8].

The compositions of the solutions used in the measurement are shown in
Table II. Solutions 1 and 2 are characterized by the lower water contents,
solution 3 by the intermediate content and solution 4 by the higher water
content. The solutions have been kept at $30°C$ for the hydrolysis-poly-con-
densation reaction. A portion of the solution has been taken at various
times in the course of the reaction. The silicon alkoxide polymers in the
solution have been trimethylsilylized [9] for stabilization, dissolved in
benzene and subjected to the measurement of molecular weight and viscosity.

Figure 3 shows the variation of number average molecular weight M_n with
the reduced reaction time t/t_g, where t_g is the gelling time. Figure 4
shows the η_{sp}/C - C plots, where η_{sp} is the reduced viscosity and C is the
concentration, for the alkoxide polymers taken from solution 1 at various
reaction times. The numbers attached to the lines in the figure denote the
measured value of M_n. It is seen that the slopes of the straight lines and
the values of η are larger for larger M_n values. The occurrence of the
slope in the lines indicates that linear polymers are found in the solution

TABLE II. Compositions and properties of $Si(OC_2H_5)_4$ solutions.

Solution[a)	$Si(OC_2H_5)_4$ (g)	H_2O r[b)	C_2H_5OH (ml)	Concentration of SiO_2 (wt%)	Spinn- ability	Time for geling (h)
1	169.5	1.0	324	33.3	Yes	233
2	178.6	2.0	280	42.3	Yes	240
3	280.0	5.0	79	61.0	No	64
4	169.5	20.0	47	33.5	No	138

a) The ratio $[HCl]/[Si(OC_2H_5)_4]$ is 0.01 for all glasses.
b) r represents the ratio $[H_2O]/[Si(OC_2H_5)_4]$.

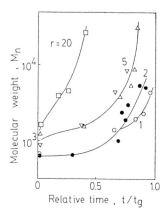

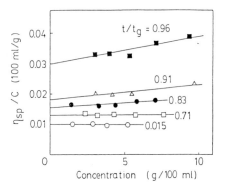

FIG. 3. Change of number-average molecular weight M_n of the trimethylsilylated alkoxide polymers with relative time t/t_g (t_g is the geling time) for $Si(OC_2H_5)_4$ solutions with different r's. The marks $\triangle$ and $\triangledown$ for the solution with r of 5.0 correspond to the polymers trimethylsilylated with trimethylchlorosilane (TMC) and hexamethyldisiloxane (HMDS), respectively.

FIG. 4. Relations between reduced viscosity η_{sp}/c and concentration of the trimethylsilylated alkoxide polymers for solution 1 with r of 1.0.

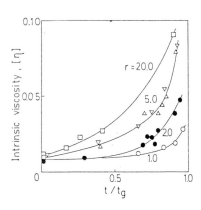

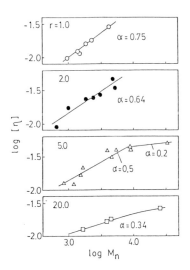

FIG. 5. Change of intrinsic viscosity $[\eta]$ with relative time t/t_g for $Si(OC_2H_5)_4$ solutions with different r's.

FIG. 6. Relation between $[\eta]$ and M_n of the trimethylsilylated alkoxide polymers for $Si(OC_2H_5)_4$ solutions with different r's.

[5]. Figure 5 shows the variation of intrinsic viscosity $[\eta]$ with t/t_g.
Figure 6 shows the log M_n versus log$[\eta]$ plots. The slope of the plot
is larger than 0.5, that is, 0.75 and 0.64 respectively for solutions 1 and
2 of which the water content is low. It is smaller than 0.5, that is, 0.34
for solution 4 of which the water content is high. The slope for solution
3 which has the intermediate water content is about 0.5 at the early stage
of the hydrolysis-polycondensation reaction and about 0.2 at the later stage.
   It is known that for the polymer solutions $[\eta]$ is related to M_n by the
expression [10, 11].

$$[\eta] = kM_n^{\alpha} \qquad \dots\dots\dots\dots(1)$$

where k is a constant depending on the kind of the polymer, solvent and temperature. The exponent α, that is, the slope of the log $[\eta]$ - log M_n plot
takes a value between 0-2.0. The value depends on the shape of polymers: α
=0 for rigid spheric particles, α=0.5-1.0 for flexible, chainlike or linear
polymers and α=1.0-2.0 for non-flexible or rigid, rod-like polymers [11].
It is reported for high polymers containing siloxane bondings Si-O-Si that
α=0.5 for linear polydimethylsiloxanes, α=0.21-0.28 for branched or crosslinked polymethylsiloxane and α=0.3 for spheric polysilicates 12 .
   Referring to these, the present experimental results on the $[\eta]$-M_n
relationships can be interpreted to show the followings. In solution 1 with
r= $[H_2O]/[Si(OC_2H_5)]$ equaling 1.0, the alkoxide polymers grow as linear
polymers with increasing molecular weight, that is, linear polymers are
formed in the solution. Linear polymers are also found in solution 2 with
r=2.0. Solution 3 with r=5.0 contains linear polymers at the early stage
of the reaction but three-dimensional or spheric growth of polymers occurs
at the later reaction stage close to the onset of gelation. In solution 4
containing a much amount of water expressed by r=20, three-dimensional or
spheric growth of alkoxide polymers are predominant.
   The above results are summarized in Table III. Linear polymers are

TABLE III. The exponent α's for the alkoxide polymers and properties
of $Si(OC_2H_5)_4$ solution.

Solution	$H_2O(r)$	α	Type of polymer	Spinnabilyty
1	1.0	0.75	Linear	Yes
2	2.0	0.64	Linear	Yes
3	5.0	0.5	Branched	No
		0.2	Three-dimensional	
4	20.0	0.34	Three-dimensional, Spherical	No

main reaction products in solution 1 and 2 which become drawable. Three-
dimensional polymers or particles are dominant in solution 4 which does not
show spinnability but forms a large bulk gel and glass. Solution 3 is intermediate between the above two cases, resulting in no drawability and no
bulk gel formation. This systematic conclusion may prove the author's prediction that the fiber drawing is only possible for the alkoxide solutions
containing linear polymers [2, 3].

PREPARATION OF FIBERS BY THE FREEZING-OF-GEL METHOD

   Mahler and Bechtold [13] made bundles of silica gel fibers of 11 cm in
length and 40 to 260 μm in diameter having the composition $Si_3O_5(OH)_2$ by
unidirectional freezing of hydrated silica gel in 1980. The gel fibers were
converted to high strength SiO_2 glass fibers by heating at $925°C$. In this
paper, our experience on the formation of zirconia and titania gel fibers by
unidirectional freezing will be described and the freezing process will be
discussed in terms of formation of long fibers and change of fiber diameter.

Preparation of Gels

For successful formation of gel fibers by this method, preparation of pertinent hydrogels is important, in which there are sufficient metal-oxygen-metal bondings and accordingly the oxide network structure is partially constructed before freezing. This is achieved by geling the sol by removing electrolytes in the process of dialysis in distilled water.

In order to prepare zirconia and titania hydrogels, $ZrOCl_2.8H_2O$ and $TiCl_4$ [14], respectively, were dissolved in water, partially neutralized by adding KOH solution, placed in cellulose tube of 6mm in diameter and 120mm in length and dialyzed in distilled water for 96hr. Thus translucent, fairly hard gel masses were formed.

Undirectional Freezing

For unidirectional freezing, the cellulose tube containing the hydrogel was put into a polyethylene cylinder, which was lined with polyurethane foam for thermal insulation in the radial direction of the cylinder, and the cylinder was lowered into a $-78^{\circ}C$ cold bath of dry ice-ethanol mixture at a rate of 2-8 cm/h. After the whole cylinder was immersed in the cold bath the frozen hydrogel columm was allowed to thaw on a glass dish at room temperature. This resulted in the formation of various forms of oxide hydrogels such as long fibers nearly corresponding to the length of the cylinder, shorter fibers and granules, depending on the lowering rate of the cylinder in the unidirectional freezing process. It should be noticed, however, that other conditions of preparation of the starting hydrogels such as the oxide concentration of the solution, degree of the partial neutralization, time of dialysis and size of cylinder used in freezing of gel are also important.

Freezing Condition for Long Fiber Formation

The lowering rate of the cylinder markedly affects the formation of fibers and the fiber diameter. This effect is described below for the case of zirconia fibers [15]. The ZrO_2 concentration of the starting solution was fixed at 1 M. The polyethylene cylinder for freezing was 10 cm long and 1 cm in diameter. Six lowering rates of cylinder were adopted: 2, 2.5, 4, 6, 7 and 8 cm/hr.

Long fibers corresponding to the full length of the cylinder (10cm) were produced at intermediate lowering rates of 4 and 6 cm/hr. Fibers were continuous up to about 5 cm from the bottom of the cylinder for lowering rates of 2.5 and 7 cm/hr. Only granules and very short fibers were formed for lower and higher lowering rates of 2 and 8 cm/hr.

It is assumed that the formation of concentrated oxide gel products in unidirectional freezing may be caused by the concentration of oxide components in regions between ice crystals on their freezing. The shapes of concentrated oxide gels may be defined by the shape of frozen ice crystals. Accordingly, the formation of fibers and fiber diameters may be discussed in terms of the freezing rate R and temperature gradient at the freezing front G, which may depend on the lowering rate of the cylinder.

Tiller et al. [16] expresses the condition for the plane front growth of the solvent in the unidirectional solidification of the solution as follows.

$$G/R \geqq \frac{m.Co(1-k)}{k \cdot D} \qquad \ldots\ldots\ldots(2)$$

Here, m is the slope of liquidus line, Co is the initial concentration of the solution, k is the partition ratio and D is the diffusion coefficient. The similar reasoning of solidification is assumed to hold in the unidirectional feezing of ice, which is considered in this experiment. In freezing of the same gels, the value of the right-hand side of formula (2) is identical.

When G/R is large owing to the small lowering rate, the ice crystals grow by the plane front growth mechanism. Then, zirconia component is expelled to the front area of the freezing ice crystals and no zirconia fibers may be formed. This explains why no long fibers are formed for lowering rates smaller than 2 cm/hr. For high lowering rates causing small G/R values, ice crystals grow by the cellular growth mechanism as ice fibers. The ZrO_2 component is concentrated in the region between ice fibers, forming zirconia gel fibers on freezing. The formation of long fibers for larger lowering rates is explained by this reasoning.

For too high lowering rates causing very low G/R values, however, the occurrence of large constitutional supercooling causes formation of discrete, particulate ice crystals and no long fibers are formed. This explains why no long fibers are formed for lowering rates higher than 8 cm/hr.

Figure 7 shows G/R as a function of the distance from the bottom of the cylinder for various lowering rates [17]. In the figure, two dashed lines indicate the G/R range where continuous fibers can be obtained. It is indicated that continuous fibers can be obtained for G/R range between 2 and $7.5°C.h/cm^2$.

Diameter of Fibers

The diameter of fibers thus obtained increases with increasing distance from the bottom. For instance, the diameter was 30μ m at the bottom and 50 μm at the top of the cylinder for the lowering rate of 4 cm/hr. According to Flemings [16], the intercellular distance, that is, the fiber diameter linearly increases with increasing 1/RG in the cellular growth of unidirectional solidification. The above fiber diameter means the diameter of ice crystals. In the present case, it is assumed that the diameter of zirconia fibers may be proportional to the diameter of ice crystals, because the diameter of long spaces between ice fibers increases with that of ice fibers at the same ice content. Accordingly, the diameter of oxide gel fibers linearly increases with 1/RG. Figure 8 shows that the diameter of zirconia gel fibers follow the above relation.

The similar linear relation between the diameter of the oxide fiber and

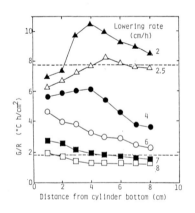

FIG. 7. Change of G/R with distance from cylinder bottom in unidirectional freezing of zirconia hydrogel. G/R values ranging between dashed lines give continuous fibers. G: Temperature gradient. R: Freezing rate.

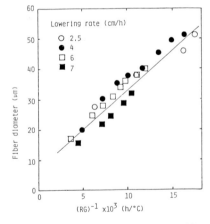

FIG. 8. Change of diameter with 1/RG for zirconia fibers produced by unidirectional freezing of gel.

1/RG has been also obtained for titania fibers [18].

Properties of Freeze-formed Fibers

Oxide gel fibers have polygonal cross-sections. They are porous hydrate gels after thawing and drying at low temperatures. Silica fibers prepared by Mahler [13] were of $Si_3O_5(OH)_2$ composition with surface area of 900 m^2/g after drying at $150°C$. Zirconia fibers prepared in our laboratory had a composition $ZrO_2.H_2O$ after drying at $25°C$ and then heating at $290°C$. Titania fibers prepared in our laboratory had a composition $TiO_2.1.4H_2O$ with surface area of 350 m^2/g. The zirconia fibers were amorphous up to $290°C$, where the cubic and monoclinic ZrO_2 crystals started to precipitate. The titania fibers as prepared contained a small amount of anatase crystals.

CONCLUSION

Two types of methods in which sols or gels are formed into fibers at low temperatures near room temperature and oxide glass and ceramic fibers can be obtained by heating those fibers to several hundred $°C$ have been reviewed. It should be stressed that the chemistry in preparing sols leading to fibers are very important.

REFERENCES

1. H. Dislich, Angew. Chem. Int. ed. 10, 363 (1971).
2. S. Sakka, J. Non-Crystal. Solids 42, 403 (1982).
3. S. Sakka in: Treatise on Materials Science and Technology Vol.22, Glass III, M. Tomozawa and R. Doremus, eds. (Academic Press, New York 1982) pp. 129-167
4. H. Dislich, J. Non-Crystal. Solids 57, 371 (1983)
5. S. Sakka and K. Kamiya, J. Non-Crystal. Solids 48, 31 (1982)
6. S. Sakka and K. Kamiya, The 19th University Conference on Ceramics, University of North Carolina, Raleigh, North Carolina, U. S. A., November 8-11, 1982
7. S. Sakka, K. Kamiya and T. Kato, Yogyo-Kyokai-Shi 90, 555 (1980)
8. S. Sakka, K. Kamiya, K. Makita and Y. Yamamoto, 2nd International Workshop on Glasses and Glass-Ceramics from Gels, Wurzburg, Germany, July 1-2, 1983
9. C. R. Masson, J. Non-Crystal. Solids 25, 3 (1977); C. W. Lentz, Inorg. Chem. 3, 574 (1964)
10. W. J. Budley and H. Mark in: High Molecular Weight Organic Compounds, R. E. Burk and O. Grunmitt, eds (Interscience Publisher, New York 1949) pp. 7-112
11. H. Tsuchida, Science of Polymers [in Japanese], Baihukan Publishing Company, 1975, pp. 85-87
12. Y. Abe and T. Misono, J. Polymer Sci. Polymer Chem. 21, 41 (1983)
13. W. Mahler and M. F. Bechtold, Nature 285, 27 (1980)
14. L. Kruczynski, H. D. Gesser, C. W. Turner and E. A. Spears, Nature 291, 399 (1981)
15. T. Kokubo, Y. Teranishi and T. Maki, J. Non-Crystal. Solids 56, 411 (1983)
16. W. A. Tiller, K. A. Jackson, J. W. Rutter and B. Chalmers, Acta Met. 1, 428 (1953); M. C. Flemings, Solidification Processing, McGraw-Hall, New York (1974) p.84
17. T. Kokubo and Y. Teranishi, private communication
18. T. Maki and T. Kokubo, private communication

ENVIRONMENTAL EFFECTS IN GEL DERIVED SILICATES

L.L. HENCH
Ceramics Division, Department of Materials Science and Engineering,
University of Florida, Gainesville, Florida 32611

INTRODUCTION

Various methods have been developed for producing crystals, powders, coatings and monoliths from gels [1]. The scientific basis for understanding compositional effects, gelation, aging, drying and densification is also advancing rapidly [2]. However, there is as yet relatively little information on the durability, weathering, or corrosion resistance of glasses, glass-ceramics, ceramics, or composites made via the gel route. Data is also sparse on the effects of vacuum, thermal exposure, or mechanical stress on the stability of gel-derived solids. Relationships between sol-gel processing variables and environmental stability are especially lacking at the present time. Since many end-use applications of gel derived materials involve exposure to severe environments, it is essential that the durability of these materials be established during their development.

Consequently, the goal of this paper is to review the current status of understanding: (1) the mechanisms of weathering, corrosion and atmospheric attack of silicate glasses and glass-ceramics; (2) the effects of atmosphere on pre-densified gels, the phases developed, and the consequences of those phases on densification; (3) the rates of aqueous corrosion of gel-derived vs. melt-derived glasses; and (4) the weathering resistance of gel-derived coatings.

Corrosion Mechanisms

Studies of the interaction of many glass compositions with a variety of environments have led to the identification of at least ten types of corrosion phenomena which are summarized in Table I. A glass or a glass-ceramic may encounter one or more of the corrosion processes simultaneously or sequentially. The durability of gel-derived materials is controlled by these ten processes.

TABLE I. Corrosion Processes

1. Ion Exchange or Selective Leaching
   Involves the exchange of mobile species from the glass with protons or hydronium ions from the solution. Results in surface film formation.
2. Network Dissolution (Congruent and Surface Film Dissolution)
   Involves breaking down of structural bonds in the glass or surface film. May occur uniformly or locally.
3. Pitting
   Localized network dissolution due to surface heterogeneities, stresses or defects.
4. Solution Concentration
   Involves the concentration of the solution with respect to species from the glass. May result in a reduction of the corrosion rate.
5. Precipitation
   Involves the formation of insoluble compounds on the glass surface by reaction of the dissolved constituents from the glass with species already present in the solution. May be influenced by solution pH.

Mat. Res. Soc. Symp. Proc. Vol. 32 (1984) Published by Elsevier Science Publishing Co., Inc.

6. Stable Film Formation
   Involves the alteration of the glass surface composition by either diffusion processes or interfacially controlled reactions.
7. Surface Layer Exfoliation
   Involves the flaking off of surface layers formed by ion exchange. Usually occurs after the glass has been removed from the solution due to dehydration and accompanying stresses.
8. Weathering
   Involves the interaction of humidity and reactive gases from the atmosphere with the surface. Usually results in the accumulation of precipitates (both soluble and insoluble) on the surface.
9. Stress Corrosion
   Interaction of tensile stresses and chemical reactions leading to accelerated attack.
10. Erosion Corrosion
    Due to the mechanical abrasion and/or chemical reactions on the surface. Typical examples are high velocity wind or water carrying sand over glass surfaces.

Illustrations of each of these 10 types of corrosion processes are available in ref. #3 and a more extensive discussion of them is present in ref. #4.

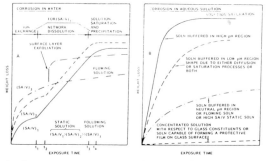

Fig. 1. Weight loss versus exposure time for glasses exposed to A) pure water, and b) other types of solutions.

A practical understanding of glass corrosion can be obtained by plotting weight loss (based on either weighing the glass sample before and after exposure, or solution analysis) versus time. Figure 1 illustrates the general behavior in terms of mechanism, kinetics and solubility limits for static and flowing pure water (A), and for other types of solution (B). During the early stage of corrosion in pure water, ion exchange is the rate controlling mechanism and the weight loss is root-time dependent. After a critical time, t_c, in static solutions ion exchange produces a high pH (i.e. high OH^- concentration) which causes network dissolution to become the rate controlling mechanism of corrosion. The extent of weight loss due to this mechanism is linearly dependent on time. The time, t_c, corresponding to a changeover in the rate controlling kinetics may not be so abrupt as indicated in Fig. 1. It is, however, dependent on surface area/volume (SA/V); as SA/V is increased, t_c is decreased (see t_1 and t_2). Furthermore, as SA/V is increased, solution concentration effects reduce the rate of ion exchange so that infinitely dilute or flowing solutions result in the fastest rates of ion exchange and the longest t_c (t_c may never be reached in a flowing solution if the flow rate is high).

In static solutions, the constantly changing pH causes a corresponding change in solubility limits of constituents in the glass. When an equilibrium pH is achieved, solution saturation may occur for some or all of the constituents from the glass. When saturation is achieved with respect to all of the glass constituents the sample weight will remain constant (t_3 and t_4). This does not mean that the corrosion reactions have ceased but merely that the rate of precipitation from the solution is equal to the rates of ion exchange and network dissolution. Thus, the surface of the glass may continue to change even after saturation has been achieved. However, the equilibrium rates are usually much lower than non-equilibrium rates and surface alterations may be imperceptible over long periods.

Stable surface films form when the extent of a glass structural constituent required to achieve saturation is small. Thus, glasses containing a sufficient quantity of Al_2O_3 form stable films in static solutions when the pH is between 3 and 10.7. A small SA/V requires more weight loss for saturation than does a large SA/V. Extensive removal of the passivating species may prevent stable film formation under conditions of small SA/V or flowing solutions. If surface layer exfoliation occurs, an abrupt increase in the weight loss will be observed as shown in Fig. 1 and the usual kinetics equations will not be obeyed. In the case of flowing solutions a new surface layer will be formed but the rate of ion exchange will be more rapid immediately after exfoliation.

Atmosphere Effects on Pre-Densified Gels

A broad compositional range of Na_2O - SiO_2 gels has been studied by Prassas, et al [5,6]. Tetramethoxysilane (TMS) was used as the source of SiO_2 and sodium methylate diluted in methanol was used for Na_2O. After 0°C gelation, the samples were aged at 55°C for 5 hrs. followed by 10 hrs. of drying at 130°C under a methanolic atmosphere.

X-ray diffraction and IR transmission and reflection spectroscopy of the Na_2O - SiO_2 gels all showed spectral shifts characteristic of the increasing number of non-bridging oxygen bonds as the Na_2O content was increased from 3.7 mole % to 40 mole % [5]. The changes in the S,D, and R bonds of the IRRS spectra with increasing Na_2O content were nearly equivalent to melt glasses of the same composition [7] [8]. Consequently, Prassas et al [5] concluded there was "structural similarity between the gels and glasses". A subsequent thermal densification study on the same gels showed that full density glasses could be obtained from the Na_2O - SiO_2 gels by heating in the range of 440 - 640°C [6] [9].

Although there is structural similarity between Na_2O - SiO_2 gels and glasses, there are major differences in atmospheric sensitivity. Weathering studies of binary Na_2O - SiO_2 melt derived glasses by Chao and Clark [10] show that the primary reaction product of weathering for many weeks in air is Na_3H $(CO_3)_2$. $2H_2O$, called trona mixed carbonate. The same reaction product also forms on some of the Na_2O - SiO_2 gels [5]. However, an intermediate Na_2CO_3 . H_2O phase generally appears on the Na_2O - SiO_2 gels within hours after their exposure to ambient atmosphere. The weathering process begins very quickly at the outer surface of the gel and the concentration of carbonate is less on the inside of the porous gel. The content of the Na^+ in the internal layers decreases as they are replaced by H^+ ions and physically adsorbed water.

Increasing the alkali content of the gel significantly increases the rate of the weathering process, as shown in Fig. 2 based in part upon Prassas et al [5]. After 20 hours weathering the 15 mole% Na_2O gel is just beginning to show a carbonate peak whereas substantial crystallization has developed in the gels with higher Na content.

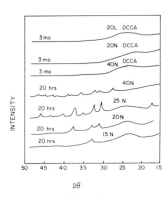

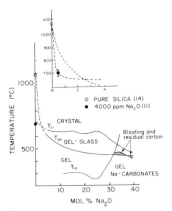

Fig. 2. X-ray diffraction spectra of gels after weathering in ambient air for times indicated.

Fig. 3. Gel thermal processing diagram for the Na_2O-SiO_2 system.

Processing variations in preparing a gel can have a major effect on weathering susceptibility. Prassas and Hench [11] showed that a range of initial (pre-densification) surface areas from $9m^2$ to $355m^2/g$ could be obtained for 20 mole% Na_2O - SiO_2 gels. The gels with the higher surface specific area showed the higher proportion of carbonate crystals.

Densification of the Na_2O - SiO_2 gels is strongly affected by the carbonate weathering history. Figure 3 summarizes the gel processing diagram for the Na_2O - SiO_2 system [11]. T_{ce} on the diagram corresponds to the temperature where the carbonate was eliminated. T_{ci} corresponds to the temperature for initiation of crystallization and T_{db} for beginning of densification. Figure 3 shows that conversion of carbonate weathered gels can be difficult. This is because sintering can begin before the Na carbonates have not been completely eliminated. The T_{ce} and T_{db} curves can be so close to each other that bloating or retention of residual carbon due to incomplete carbonate decomposition may be nearly impossible to prevent. An important observation is that there is no apparent effect of immiscibility on T_{db}, T_{ce}, T_{ci} or weathering of gels in the Na_2O-SiO_2 system.

The best method for reducing the effect of the accelerated carbonate weathering on gel densification is to prevent its occurance. Recent findings of the use of drying control chemical additives (DCCA's) to control the rate of pore liquid evaporation and drying stresses have also led to a solution of the problem of accelerated weathering. Figure 2 includes X-ray diffraction spectra of DCCA alkali-silicate gels exposed to an ambient air atmosphere for three months without appreciable development of carbonate weathering products. Infrared reflection spectra a 20N, 20L, and 40N alkali silicate gels made with DCCA's, Fig. 4, show no evidence of carbonates and excellent reflection intensity. In contrast, the earlier soda-silica gels made without a DCCA showed a loss of the Si-O-Na bridging oxygen vibration at $850cm^{-1}$ after only a few hours of weathering [5].

It is hypothesized that the DCCA molecules, such as formamide, are chemically bonded to the surface of the gel pores and thereby prevent the chemical reactions with CO_2 and H_2O. TGA, DSC, and DTA data indicate that temperatures of >180°C are necessary to dissociate the DCCA's from the porous gels. When densification follows directly after such dissociation there is no problem with carbonate formation or elimination.

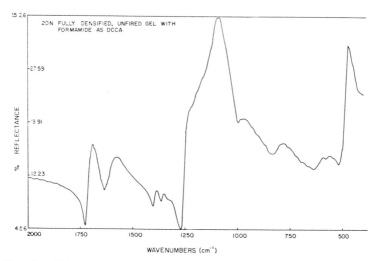

Fig. 4. FTIR spectrum of 20N gel made with formamide DCCA.

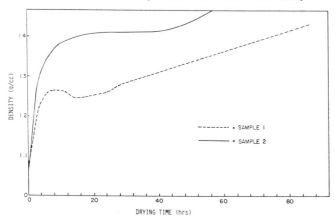

Fig. 5. Time dependent change in density of two SiO$_2$ gels made with
formamide DCCA.

The gels made with DCCA's can show variations in drying behavior,
however [16]. Figure 5 shows the continued increase in density of a 100%
SiO$_2$ gel made with TMS using formamide as a DCCA [15]. Both samples
increased in density by nearly 40% during ambient drying. For reasons as
yet unknown, the first sample, made with the same procedure took 5X as long
to reach a density of 1.4 g/cc.

One of the important features of monolithic gels made with DCCA's are
very high surface areas; e.g. as much as 1100m^2/g. A possible concern is
whether such high surface area monoliths can be exposed to a vacuum and
thermal cycles without cracking. Figure 6 shows the partly favorable
results of a simple experiment to answer that question. Three 100% SiO$_2$
samples, one (A) made with formamide as a DCCA and fully dried at 85°C, the
second (B) made with an organic acid DCCA and dried at 85°C, and a third

(C) made the same as B but densified to 90% theoretical at 720°C were allowed to absorb moisture under ambient conditions for several weeks. All three were then evacuated at 30°C causing an initial weight loss of ~2%. The effects of progressively higher temperatures under vacuum are shown in Fig. 6. Sample A fractured during exposure to 130°C under vacuum. However, samples B and C remained intact. Samples B and C were subsequently allowed to equilibrate with the atmosphere and then rapidly exposed to 10^{-6} Torr. No cracking was observed. Figure 7 is a composite photograph of several 100% SiO_2 gel samples made with DCCA's after exposure to vacuum and thermal cycles.

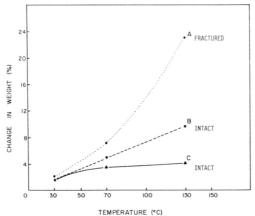

Fig. 6. Weight changes of SiO_2 exposed in-vacuo to temperatures. indicated.

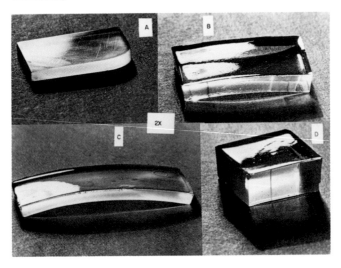

Fig. 7. Monolithic SiO_2 samples made with DCCA's exposed to vacuum.

Certain DCCA chemicals however can lead to environmental degradation if precautions are not undertaken. As an example, a recent study [12] of formamide as a DCCA in the 2ON system showed that conversion of the dried gels to glass was hampered by moisture absorption from the air. The absorption occurs preferentially on the surface of the gel and results in uneven stresses between the gel surface and the bulk thereby producing cracks. The following equation describes the absorption mechanism:

This surface reaction is more serious in a fully dried gel than in a partially containing dried gel. The stresses developed in almost fully drived gels containing residual are sufficiently great formamide that they crack explosively when exposed to air. Reducing the quantity of formamide is only a partial solution. Only complete removal by thermal or chemical treatment eliminates the hydration cracking. Another alternative is to use a DCCA which does not produce a reaction with moisture. Glycerol ($C_3H_8O_3$) is a possible alternative. However, glycerol with its -C-O-H bond will decompose and react with the sodium ions in high surface area 20N gels bodies to form a sodium carbonate crystalline phase at a higher temperature. The equations shown below indicate the glycerol decomposition reactions which should occur during drying in air.

$$C_3H_8O_3 + 3\tfrac{1}{2}O_2 \quad \text{-------->} \quad 3CO_2 + 4H_2O$$

$$CO_2 + H_2O + Na^+ \quad \text{------>} \quad NaHCO_3 + H^+$$

$$\text{or} \quad CO_2 + H_2O + 2Na^+ \quad \text{----->} \quad Na_2CO_3 + 2H^+$$

Thus, the primary advantage offered by use of the DCCA is control of the large shrinkage stresses during drying. However, the residual DCCA left in the pores can lead to subsequent environmental instability and therefore careful removal of such chemicals must be achieved if they are used in gel processes.

Phalippou, et al. [21] have shown that vapor phase chlorination of silicate gels is very effective in removing Si-OH bonds. The - Si - Cl bond is not stable and densification must occur immediately after chlorination to avoid rehydration. The chlorination procedure, using CCl_4, occurs within 60 minutes at 500°C. Residual water contents of <10 ppm in fully densified silica can be achieved in this manner.

Pantano, et al. [22] have also shown that vapor phase ammonia treatment of chlorinated silica gels is an effective means of incorporating nitrogen in a sol-gel derived silica film. An increase in index of refraction of the film from n = 1.42 to n = 1.60 was obtained with the combined chlorine-ammonia treatment prior to densification

Relative Corrosion Rates of Gel-Derived Silicates

Two studies have addressed the question as to the relative durability of gel-derived glasses. Compositions of 33 Mole% Na_2O - 67 Mole% SiO_2 (33N) and 20 Mole% Na_2O - 80 Mole% SiO_2 (20N) were made in the same manner as described by Prassas et al [5]. The gels were heated through the gel-glass transition range of 490°C - 590°C, as previously described [9,17], to produce fully dense glass samples. Melt glasses of the same composition were made by casting the molten batch from 1350°C into graphite molds.

Comparative studies of weathering, ion exchange, and network dissolution attack (see Table I) of the gel-derived and melt derived glasses were made. IRRS profiles after 3 mo. of ambient weathering at 70-90% relative humidity showed a very different surface structural sequence on the 33N glasses (Fig. 8). The gel-derived glass showed a weathering layer approximately 2X deeper than the melt glass. The outer weathered zone of the gel-derived glass was typically soda depleted (see Fig. 8), the same as the melt glass. However, an equivalently thick intermediate transitional layer had developed during weathering that had a structure similar to the gel prior to complete gel-glass transformation. In fact, some 33N gel-derived glasses reverted back to a gelatinous state during weathering, became soft and pliable [1] and completely lost their rigid glassy character.

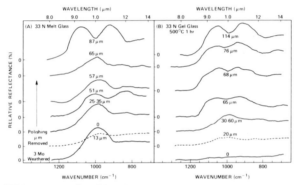

Fig. 8. IRRS spectra for 33N glasses after polishing to remove various thickness of weathered layers, (A) melt glass, (b) gel-derived glass.

Aqueous corrosion of the same series of 33N glasses showed that dealkalization (Mechanism #1 of Table I) was somewhat slower at first for the gel-derived glass [17]. However, the SiO_2-rich film that developed on the gel-derived glass was less dense and network dissolution (Mechanism #2) proceeded much more rapidly leading to extensive pitting (Mechanism #3). Consequently, the gel derived 33N glasses made with the Prassas et al process [5], are considerably less durable than melt glasses of equivalent composition.

Another study [18] showed that density and aqueous corrosion resistance of 20N alkoxide-derived glasses increased with increasing aging temperature and time. A 20N sample aged at 50°C went through alkali exchange and network dissolution (see Table I) within just 60 minutes in 100°C water whereas a sample aged at 65°C resisted several hours of 100°C corrosion. Studies to determine whether gel-derived glasses made with the DCCA method [12,14] are more or less durable than melt derived glasses are still underway.

Weathering Resistance of Gel-Derived Coatings

Sol-gel dip coatings of titania [19] and indium-tin oxide [20] surface layers on large float glass planes are successful examples of sol-gel chemistry. In order for these processes to be commercially feasible it is essential that the coatings have long term stability and weathering resistance. Two year exposure in outdoor weathering stands at Mainz, Ibiza (southern Mediterranean) with high UV radiation, and laboratory tests all show no change in appearance or degradation. This result indicates that sol-gel processing can produce materials with corrosion resistance equivalent to standard processing methods.

CONCLUSION

At present, it is still to early to generalize as to which of the many processing variables are important in controlling corrosion rates of gel-derived materials. Final density is obviously important but the concentration of bridging oxygen bonds may be even more important. Figure 9 indicates the potential difference in bond energy distributions that may be present in gel-derived silicate glasses in comparison with melt derived glasses of the same composition. Depending upon processing history, a substantially broader distribution of bond distances and bond angles, and therefore bond energies, may be present. The low energy tail of the bond distribution will be much more susceptible to attack by H_2O, CO_2, Cl_2, etc. Thus, the various environmental interactions described above occur at a rate proportional to the breadth of the bond distribution curves shown in Fig. 9.

ACKNOWLEDGMENTS

The author gratefully acknowledges support of AFOSR Contract #F49620-83-C-0072.

REFERENCES

1. L.L. Hench and D.R. Ulrich, eds., Ultrastructure Processing of Ceramics, Glasses and Composites, (J. Wiley & Sons, Inc., New York, 1984).
2. L.L. Hench and D.R. Ulrich, eds., Ultrastructure Processing of Ceramics, Glasses and Composites, (J. Wiley & Sons, Inc., New York, 1984) Chapters 4-14.
3. D.E. Clark and L.L. Hench, Mat. Res. Soc. Symp. Proc. 15 113-124 (1983).
4. D.E. Clark, C.G. Pantano and L.L. Hench in: Corrosion of Glass, (The Glass Industry, New York, 1979).
5. M. Prassas, L.L. Hench, J. Phalippou, and J. Zarzycki, J. Non-Cryst. Solids 48, 79-95 (1982).
6. M. Prassas, J. Phalippou, L.L. Hench, J. Non-Cryst. Solids, 63, 375-389 (1984).
7. J.R. Ferraro and M.H. Manghnam, J. Appl. Phys. 43, 4595 (1972).
8. D.M. Sanders, W.B. Person and L.L. Hench, Appl. Spectra., 28 [3] 247-255 (1974).
9. L.L. Hench, M. Prassas, and J. Phalippou, J. Non Cryst. Solids 53, 183-193 (1982).
10. Y. Chao and D.E. Clark, "Weathering of Binary Alkali Silicate Glasses and Glass-Ceramics, to be published in Glass Technology.
11. M. Prassas and L.L. Hench, "Physical Chemical Factors in Sol-Gel Processing" in: Ultrastructure Processing of Ceramics, Glasses and Composites, L.L. Hench and D.R. Ulrich, eds. (J. Wiley & Sons, New York. 1984).
12. S.H. Wang and L.L. Hench, "Processing Variables of Sol-Gel Derived Soda Silicates", Proceedings 8th Annual Conf. on Composites and Advanced Ceramic Materials, Am. Ceram. Soc., to be published.

13. S. Wallace and L.L. Hench, "Organometallic-Derived 20 mol% Li_2O-80 mol% (20L) Gel Monoliths", Proceedings 8th Annual Conf. on Composites and Advanced Ceramic Materials, Am. Ceram. Soc., to be published.
14. G. Orcel and L.L. Hench, "Effect of the Use of a Drying-Control Chemical Agent (DCCA) on the Crystallization and Thermal Behavior of Soda Silicate and Soda Borosilicate", Proceedings 8th Annual Conf. on Composites and Advanced Ceramic Materials, Am. Ceram. Soc., to be published.
15. S. Wallace and L.L. Hench, "The Processing and Characterization of DCCA Modified Gel-Derived Silica", Proceedings MRS Spring Mtg. Albuquerque, NM, 1984.
16. S. Wallace and L.L. Hench, to be published.
17. L.L. Hench, M. Prassas and J. Phalippou, Ceramic Engineering and Science Proceeding, Am. Ceram. Soc., 3, No. 9-10, 447-483 (1982).
18. L.L. Hench, S. Wallace, S. Wang and M. Prassas, Ceramic Engineering and Science Proceeding, Am. Ceram. Soc., 4 No. 9-10, 732-739 (1983).
19. H. Dislich and P. Hinz, J. Non Cryst., 48, pp. 11-16 (1982).
20. N. Arfsten and H. Dislich in: Ultrastructure Processing of Ceramics, Glasses and Composites, L.L. Hench and D.R. Ulrich, eds. (John Wiley & Sons, New York, 1984).
21. J. Phalippou, T. Woignier, and J. Zarzycki, "Behavior of Monolithic Silica Aerogels at Temperatures Above 1000°C," in Ultrastructure Processing of Ceramics, Glasses and Composites, L.L. Hench and D.R. Ulrich, eds., (J. Wiley & Sons, New York 1984).
22. C.G. Pantano, P.M. Glaser, and D.M. Armhurst, "Nitridation of Silica Sol-Gel Thin Films," in Ultrastructure Processing of Ceramics, Glasses and Composites, L.L. Hench and D.R. Ulrich, eds. (J. Wiley & Sons, New York, 1984).

GELS AND GEL-DERIVED GLASSES IN THE SiO_2 GeO_2 SYSTEM

SHYAMA P. MUKHERJEE
Applied Mechanics Technology Section

Jet Propulsion Laboratory; California Institute of Technology
Pasadena, CA 91109, U.S.A.

ABSTRACT

Gels and gel-monoliths in the SiO_2-GeO_2 system were prepared by different procedures using metal alkoxides as starting materials. Several gel compositions with increasing GeO_2 were prepared. Crystallization behavior of gels in relation to the preparation procedures, homogeneity, and molecular structure of gels is discussed. Supercritically dried porous gel-monoliths of high silica compositions were densified into glass by sintering at 1280°C.

INTRODUCTION

One important aspect of the sol-gel process is the possibility of the molecular scale distribution of a second component (network former or network modifier) in the silicate network matrix [1-3]. However, in order to take advantage of this possibility, it is important to know how the gel-processing parameters influence the distribution, structure, and chemical composition of a second component. Because of the structural similarity of GeO_4 and SiO_4 tetrahedra, the SiO_2-GeO_2 system could be a good model system for the scientific understanding of the influence of gel-processing parameters on the distribution and structure of GeO_4 tetrahedra in silica networks. If the second network-forming oxide gel (e.g., GeO_2 gel) tends to crystallize faster than the network forming matrix gel, the crystallization kinetics of the binary gel-monoliths prepared by different procedures would be different during gel-to-glass transformation. Moreover, the nature and concentration of hydroxyl groups and the rate of dehydroxylation during subsequent thermal treatment will play significant roles in the kinetics of crystallization [4-5]. The conventional technique of making GeO_2-containing glasses tend to form non-stoichiometric germania glasses at high temperatures during melting [6]. The non-stoichiometry of GeO_2 and SiO_2 influences the kinetics of crystallization [6] and optical properties [7] of glasses. The sol-gel processing might eliminate the problem of non-stoichiometry. Presuming that a heterogeneous nucleation mechanism [8] is active in the crystallization of a gel system like the SiO_2-GeO_2, it might be possible to get an understanding of the structural effects of different gels on the crystallization rate, after factoring out the effect of hydroxyl groups.

Hence, the objectives of the present research are (a) to develop different gel preparation procedures that might significantly influence the homogeneity (chemical and structural) of gel-monoliths in the SiO_2-GeO_2 system, (b) to investigate the crystallization kinetics of gels prepared by different procedures, (c) to investigate the effect of increasing GeO_2 concentration on the crystallization, and (d) to transform gel-monoliths into glass by sintering at temperatures $\leq 1300°C$. This work gives a qualitative picture of the influence of gel preparation on the crystallization kinetics of gels. Attempts have been made to explain the results in terms of the structure of the GeO_4 tetrahedra-rich phase of the binary gels.

Mat. Res. Soc. Symp. Proc. Vol. 32 (1984) Published by Elsevier Science Publishing Co., Inc.

EXPERIMENTAL WORK AND RESULTS

Compositions of Gels

The starting-batch compositions of the gels investigated are given in Table I.

Gel Preparation Procedures

The following two preparation approaches were developed for preparing the gels and gel-monoliths.

Approach A. The starting compounds were tetraethoxysilane or tetra-methoxysilane and germanium ethoxide. The general procedure consisted of the following steps: (1) Mix alkoxysilane with three times its volume of anhydrous ethanol at 40°C. (2) Partial hydrolysis at 40°C with 1 mol water per mol of $S(OR)_4$, acidified with HCl (0.003 mol per mol $Si(OR)_4$). (3) Stir for 2 hours at 40°C, pH ~ 2. (4) Cool to room temperature and add dilute ethanolic solution of $Ge(OC_2H_5)_4$. (4) Stir for 1/2 hour to 1 hour at room temperature. (6) Add 4 mols of H_2O (diluted with ethanol) per mol of $M(OR)_4$ (7) Stir at room temperature for 1/2 hr to 1 hr and cast for gel-monolith formation.

In addition to the use of HCl as the only catalyst, some experiments were done in which a small amount of HF acid was added at the final step. The objective was to decrease the gelling time and to see the influence of HF on the pore structure of gels.

Approach B. The starting compounds were tetramethoxysilane and ger-manium ethoxide. The general procedure consisted of the following steps: (1) Mix $Si(OCH_3)_4$ with CH_3OH at room temperature. (2) Add aqueous NH_4OH solution 0.0002 mol NH_4OH per mol $Si(OCH_3)_4$ and 1 mol H_2O per mol $Si(OCH_3)_4$. (3) Stir for 3/4 hr at room temperature; pH ~ 8. (4) Add ethanolic solution of $Ge(OC_2H_5)_4$ and stir for 1/4 hr to 1/2 hr. (5) Add H_2O diluted with C_2H_5OH, 4 mols H_2O per mol $M(OR)_4$, mixed with 0.0001 mol NH_4OH per mol alkoxide. (6) Stir for 1/4 hr to 1/2 hr at pH ~ 7. (7) Cast for gelation. (The details of the gel preparation param-eters for different procedures are given in Table II.)

Drying of Gel-Monoliths

Air Drying: The sol was cast at room temperature into teflon molds, which were covered with plastic wrap sheet. The gel-monolith formed was allowed to age at the room temperature in an alcohol atmosphere for approximately one month. The radial shrinkage was about 55%. Then, it was slowly heated in an air oven up to 70°C.

Supercritical Drying or Critical Point Drying: In order to avoid crack formation caused by the stresses as due to capillary forces, the supercritical drying technique of gel-monoliths was applied [9,10]. The gel-monoliths were supercritically dried at about 250°C and at a pressure of around 1150 psi [9]. No significant volume changes occurred during supercritical drying.

Differential Thermal Analysis and Crystallization of Gels

Crystallization of gels was monitored by the differential thermal analysis (DTA) and x-ray powder diffraction techniques. The results of x-ray diffraction studies show that all binary compositions were non-crystalline after air drying at 70°C. The compositions having GeO_2 up to 56% did not show any crystallization when heated up to 600°C at the rate

Table I. Compositions of gels

Composition No.	Weight Percent	
	SiO$_2$	GeO$_2$
SG1	95	5
SG2	90	10
SG3	80	20
SG4	44	56
SG5	20	80

Table II. Gel Preparation Parameters

Composition No.	Starting Compounds	Gel Preparation Approach	Molar Ratios of H$_2$O/M(OR)$_4$ Initial	Molar Ratios of H$_2$O/M(OR)$_4$ Final	Catalyst Used	pH Initial (before adding Ge(OR)$_4$)	pH Final (before casting)	Solution Conc. (g/l)	Gelling Time
SG1	a) Si(OC$_2$H$_5$)$_4$ Ge(OC$_2$H$_5$)$_4$	A	1	4	HCl	2	2 to 2.5	100	72 hrs
SG1	b) "	A	1	4	HCl plus HF	2	≤ 2	100	5 to 6 hrs.
SG2	a) Si(OC$_2$H$_5$)$_4$ Ge(OC$_2$H$_5$)$_4$	A	1	4	HCl	2	2.6	100	72 hrs
SG2	b) Si(OCH$_3$)$_4$ Ge(OC$_2$H$_5$)$_4$	B	1	4	NH$_4$OH	8	7	100	1 to 2 hrs.
SG3	a) Si(OC$_2$H$_5$)$_4$ Ge(OC$_2$H$_5$)$_4$	A	1	4	HCl	2	2.5	100	72 hrs
SG3	b) Si(OCH$_3$)$_4$ Ge(OC$_2$H$_5$)$_4$	A	1	4	HCl	2	2.5	100	20 hrs
SG4	Si(OCH$_3$)$_4$ Ge(OC$_2$H$_5$)$_4$	A	1	4	HCl	3 to 4	2	100	2 min.
SG5	a) Si(OCH$_3$)$_4$ Ge(OC$_2$H$_5$)$_4$	A	1	4	HCl	3 to 4	2	100	2 min.
SG5	b) Si(OCH$_3$)$_4$ Ge(OC$_2$H$_5$)$_4$	A	1	1	HCl	≤ 2	2 to 3	100	5-6 days

of 10°C/min. The DTA results showing the peak positions and nature of the peaks are given in Tables III and IV.

The phase crystallized with the high silica composition (SG2) was α-cristobalite. The precipitation of hexagonal GeO_2 was observed with the high GeO_2 composition, e.g., SG4. No systematic investigation has yet been made on the nature of crystalline phases precipitating with different compositions in relation to the time and temperature of heat treatment. Hexagonal GeO_2 crystallized on heat treatment of noncrystalline GeO_2 gel.

Infrared Spectroscopic Studies

The infrared absorption spectra of gels were taken in the wave number region 1800 cm^{-1} to 400 cm^{-1} using the KBr pellet technique. The IR ab-

Table III. DTA results of gels in the SiO_2-GeO_2 system (Heating rate: 10°C/min; Atmosphere: oxygen flowing)

COMP. NO.	STARTING COMPOUNDS	GEL PREPARATION PROCEDURE	THERMAL HISTORY	PEAK POSITION (°C)	NATURE OF PEAK
SG1	$Si(OC_2H_5)_4$ $Si(OC_2H_5)_4$	Approach A Catalyst HCl only pH ∿ 2	Air Dried at 70°C	80 180-260 400 500	Endo* Sharp Large Exo.** Exo, Small Exo, Small
SG1		Approach A Catalyst HCl plus HF pH ∿ 2		300-440	Broad Exo
SG2	$Si(OC_2H_5)_4$ $Ge(OC_2H_5)_4$	Approach A Catalyst HCl pH ∿ 2.6	Super- critically Dried	220-230 400	Sharp Exo Small Exo
SG2	$Si(OCH_3)_4$ $Ge(OC_2H_5)_4$	Approach A Catalyst HCl pH ∿ 2.6	"	200 240 400	Sharp Exo Small Exo Small Exo
	"	Approach B Catalyst NH_4OH pH ∿ 7.0	"	200 1060 1180 1210	Sharp Exo Small Exo Small Exo Small Exo
SG2	"	"	Air Dried at 70°C	300 1190 1220	Broad Exo Exo Small Exo
SG3	"	Approach A Catalyst HCl pH ∿ 2.5	Air Dried at 70°C	160 280 540	Endo Exo Exo
SG4	"	Approach A Catalyst HCl pH ∿ 2 to 3	Air Dried at 70°C	140 340 800 1050	Endo, Br Exo, Sharp Small Exo Small Endo
GeO_2 Gel	$Ge(OC_2H_5)_4$ - Noncrystalline	Air Dried at 70°C	340 780	Exo Small Exo

*Endo = Endothermic **Exo = Exothermic

Table IV. Exothermic DTA peaks due to crystallization of different gels

Composition No.	Preparation Procedure	Thermal History	Maximum Temp: 1300°C Heating Rate: 10°C/min Peak Position (°C)
SG2	Approach A	Air dried at 70°C	No peak
SG2	Approach A	Supercritically dried	No peak
SG2	Approach B	Air dried at 70°C	1190 1220
SG2	Approach B	Supercritically dried	1060 1180 1210
SG3	Approach A	Air dried at 70°C	No peak
SG4	Approach A	Air dried at 70°C	800 V. Small
GeO$_2$ gel	—	Air dried at 70°C	780 V. Small

Table V. IR Absorption Bands in Wave Numbers (cm^{-1}) of Gels of Different Compositions after Drying at 70°C for Several Days

SG1	SG2	SG3	SG4	SG5	GeO$_2$-SiO$_2$ Glass (Composition: SG4) Observed [11]	Designation [11]	Noncrystalline GeO$_2$ Gel (present work)
470 m	470 m	460 s	460 bs	460 m			
570 m	550 w 580 w.sh	550 m	550 m 580 m	520 m 550 m		Si-O-Ge	570 mb
800 m	800 m	790 m	780 m	590 m 760 w	650	Si-O-Ge Stretch	780 m
930 w	880 m		900 bs	890 vs	925-880	Si-O-Ge Stretch	880 vs
	970 vw	1000 msh	1020 sh	970 w	925-880	Ge-O-Ge Stretch	1040 w
1090 vs	1090 vs	1080 vs	1090 vs	1090 s	1020	Si-O-Ge Stretch	1100 vw
1180 sh	1170 sh	1180 sh	1180 sh	1085 sh	1085	Si-O-Si Stretch	

sorption band positions and intensities are listed in Table V [12]. The spectra of gels with high silica compositions prepared by Approach B were not significantly different from that of gels prepared by Approach A.

Gel-to-Glass Transformation

Based on the thermal dilatometric analysis of a supercritically dried SG2 gel-monolith (Approach A), the densification was done in an oxygen atmosphere by adopting a heat-treatment schedule consisting of step-wise heating to 1280°C and holding for 15 min. A holding at 600°C for 11 hours was also scheduled to remove hydroxyl groups. No crystallization occurred with the SG2 gel-monolith (Approach A), but the SG2 gel-monolith prepared by Approach B showed appreciable crystallization when sintered under identical conditions.

DISCUSSION

A critical analysis of the results of gel preparation (see Table II) shows that the parameters that strongly influence the gelling rate when the concentration of water added is kept constant are: (1) the pH of the polymeric solution, (2) the presence of HF acid catalyst, (3) the nature of the alkoxy group in the alkoxysilane, and (4) the concentration of $Ge(OC_2H_5)_4$.

The gelling rate of the solution at pH ~ 7 is much higher than that of the solution at pH ~ 2. This effect is primarily due to the higher rate of hydrolytic polycondensation reaction by alkoxysilane [12] and presumably of $Ge(OC_2H_5)_4$ at the higher pH. It may be anticipated that the pronounced polycondensation reaction might influence the distribution of GeO_4 tetrahedra in silica gel matrix. In other words, the homopolymerization or clustering of each network former might be more pronounced when the polycondensation rate is high. The role of HF acid in enhancing the gelling rate of silicic acid was reported by Iler [13]. He postulated [13] that at low pH ≤ 2, the F^- ion causes a temporary increase in the coordination number of silicon, forming a highly reactive intermediate. The intermediate complex containing the F^- ion may then regenerate the fluoride ion. One important aspect of F^- ion catalysis in this context is that HF promotes only the formation of Si-O-Si bonds and is not effective as OH^- ion for promoting rearrangement of the siloxane polymers once they are formed [14]. It appears that the catalytic effect of HF acid on the hydrolytic polycondensation of alkoxysilane in ethanol is similar to that of aqueous silicic acid. It is relevant to note that the high rate of gelation at low pH might play an important role in the homogeneous distribution of hydrolyzed alkoxysilane and germanium alkoxide species which are expected to be of low molecular weight at low pH. Finally, the higher gelling rate of methoxysilane as compared to ethoxysilane is primarily due to less steric hindrance of C_2H_5 groups [12]. The higher rate of gelling with the increase of $Ge(OC_2H_5)_4$ concentration indicates that the hydrolysis and polycondensation reaction of $Ge(OC_2H_5)_4$ are much faster than that of alkoxysilane.

Results of the DTA studies (Table IV) indicate that the crystallization rates of gel-monoliths prepared from the solution at pH ~ 2 are less than those of gel-monoliths prepared from the solution at pH ~ 7. Note that in order to avoid contamination of Na^+ ions which increase crystallization of silica gel [14], the gel was prepared in teflon containers. Moreover, the difference in specific surface areas of gel-monoliths did not significantly influence the crystallization rate. The air-dried SG2 gel (Approach B) showed crystallization during the differential thermal analysis, in spite of its lower specific surface areas as compared to that of supercritically dried SG2 gel (Approach A) having higher surface areas.

Several factors might contribute to the crystallization of gel-monoliths prepared at pH $\sim$7. At higher pH, the polycondensation rate is more pronounced than the hydrolysis [12]; consequently, at the gel point, the possibility of separation of a GeO_4-rich phase in the silica gel matrix is more likely; moreover, at higher pH, the primary gel particles undergo structural rearrangements during aging when coalescing or ripening of primary particles occurs. If there are two types of metal oxygen bonds, the rearrangement might be preferential because of the difference in the chemical reactivity of bonds like hydroxlated Si-O-Si, SiO-Ge, or Ge-O-Ge, and of the solubilities of the polymeric species in the residual solvents and thus a separation of GeO_4-rich phase might occur.

At pH $\sim$2, when the hydrolysis reaction is more pronounced and the polymeric species produced are of low molecular weights [15,16], the distribution of GeO_4 groups might be more uniform and random in the polysilicic acid network; in other words, the co-polymerization leading to the formation of the SiO-Ge bond is more likely at lower pH. The DTA results of GeO_2 gel and GeO_2-rich gels indicate that GeO_4 gel network tends to crystallize faster than SiO_4 gel network.

However, the contribution of the hydroxyl content to the crystallization of gels might be different in these two types of gels because of the difference in the dehydroxylation rate. A quantitative study of the crystallization rates after the removal of hydroxyl groups is necessary before making a definite conclusion. When the effects of hydroxyl content and stoichiometry are factored out, the effect of gel structures or other structural parameters can be determined quantitatively.

Results of the infrared spectroscopic studies show that the IR absorption bands of gels are similar to those of glasses made by melting [11,17]. No spectroscopic evidence of changing the coordination state from 4 to 6 was observed. Hence, it may be concluded that the Ge^{+4} ions are tetrahedrally coordinated in the gel structure. The IR spectrum of noncrystalline GeO_2 is similar to GeO_2 glass where tetrahedrally coordinated GeO_2 units are present [17].

CONCLUSIONS

A controlled change in the gel preparation procedures of a system containing two network formers, e.g., SiO_2-GeO_2, can produce gels of two types which might crystallize at different rates. The reasons might be related to the structure and chemical composition of GeO_4-network rich "clusters" which tend to crystallize faster. The stability toward crystallization decreases with increase of GeO_2 content. Supercritically dried gel-monoliths of high silica compositions prepared at pH $\sim$ 2 can be densified to glass by sintering at 1280°C.

ACKNOWLEDGEMENTS

Publication support was provided by the Jet Propulsion Laboratory, California Institute of Technology, under a contract with the National Aeronautics and Space Administration. The author acnowledges the laboratory assistance of T. Beam and J. C. Debsikdar of Battelle's Columbus Laboratories, where a part of the work was performed.

REFERENCES

1. S.P. Mukherjee, J. Zarzycki, and J.P. Traverse, J. Mater. Sc. 11, 341, (1976).
2. M. Yamane, S. Inoue, and K. Nakazawa, J. Non-Crystalline Solids 48 153 (1982).
3. S.P. Mukherjee: "Homogeneity of Gels and Gel-Derived Glasses" presented at the 2nd International Workshop on "Glasses and Glass-Ceramics from Gels" July 1-2, 1983, Würzburg, West Germany.
4. S.P. Mukherjee, J. Zarzycki, J.M. Badie, and J.P. Traverse, J. Non-crystalline Solids 20, 455-58 (1976).
5. S.P. Mukherjee, J. de Physique, Colloque C9 Supplement au No. 12, 43 Dec. 1982.
6. P.J. Vergano and D.R. Uhlman, Phys. Chem. Glasses 11 (2) 30-38 (1970).
7. A.J. Cohen and H.L. Smith, J. Phys. Chem. Solids, 7, 301-306 (1958).
8. J. Zarzycki, J. Non-Crystalline Solids, 48, 105-116 (1982).
9. S.P. Mukherjee and J.C. Debsikdar, Ceram. Bul. 62 (3), 413 (1983).
10. M. Prassas, J. Phalippou, and J. Zarzycki, J. de Physique, Colloque C9 Suppl. au n° 12 43, Rec. (1982).
11. N.F. Borrelli, Phys. Chem. Glasses 10 (2) 43-45 (1969).
12. E.C. Rochow, The Chemistry of Silicon, 1427 Comprehensive Inorganic Chemistry. Chap. 15 p. 1427, Pergamon Press. 1975.
13. R.K. Iler. Colloid Chemistry of Silica and Silicates, p. 29, Cornell Univ. Press. Ithaca, N. York 1955.
14. S.P. Mukherjee, "Inorganic Oxide Gels and Gel-Monoliths: Their Crystallization Behavior", presented at the 19th Univ. Conf. of Ceramic Science, "Emergent Process Methods for High Technology Ceramics", held at North Carolina State Univ. Nov. 8-10 1982 (to be published by Plenum Publ.)
15. R. Schwartz and K.G. Knauff, Z. Anorg. Allg. Chem. 275 176 (1954).
16. R. Aelion, A. Loebel, and F. Eirich, J. Am. Chem. Soc. 72 5705-5712 (1950).
17. G.E. Walrafen, J. Chem. Phys. 42 485 (1965).

SUPER-AMORPHOUS ALUMINA GELS

A. C. PIERRE[*] AND D. R. UHLMANN
Department of Materials Science and Engineering, Massachusetts
Institute of Technology, Cambridge, Mass. 02139; *On leave from
S.N.I.A.S. Company, St. Medard-en-Jalles, France

ABSTRACT

The structures of alumina gels made from aluminum sec butoxide have been studied by X-ray diffraction and scanning electron microscopy as a function of pH as well as of the temperature of peptization and gelling. The structures of these gels appear to be very sensitive to the conditions of preparation, and lead to a variety of gelled products and transformation behaviors on firing. Specific attention is directed to preferential orientation and to structures which are X-ray super-amorphous.

INTRODUCTION

The potential applicability of sol-gel technology for a wide range of applications is now well recognized. In the case of Al_2O_3 and Al_2O_3-based systems, one of the attractive applications involves their use as the matrices of ceramic-ceramic composites. The technology permits the impregnation of continuous tows of fibers, and leads naturally to benefit from processing techniques developed for resin-matrix composites. In the present case, degradation of fiber properties at firing temperatures used to densify the matrix becomes a matter of considerable concern. This in turn directs attention to the densification behavior and to its dependence on the structure and texture of the low-temperature precursors.

The present paper reports an investigation of the structure and textural features of some alumina gels prepared from alkoxides.

BACKGROUND

Gelation involves the development of chemical bonding between either atoms or molecules (polymeric gels) or solid particles (colloidal gels). It represents a cooperative phenomenon, with the gel point being the percolation threshold where the bonds are in sufficient number to realize a 3-dimensional infinite network.

While the organic gels have been the most extensively studied, an extensive literature exists for oxide gels as well. With SiO_2, e.g., siloxane bonds $\equiv Si-O-Si\equiv$ lead to gels termed polymeric [1], depending on whether they connect particles smaller or larger than 5 nm. In the case of alumina, the situation is much less clear, and the name "gel" has very often been given to a variety of different low temperature forms of alumina and its hydroxides.

Alumina hydrates with a fibrous texture which are describable qualitatively as gelatinous have been investigated for many decades, and can be obtained under a wide range of conditions. In 1870, Cossa [2] described a type obtained by direct reaction of Al with water. The diverse studies summarized by Gitzen [3] deal with gels obtained from Al salts (sulfates, nitrates, chlorides) and mineral bases. At least some such gels have been shown by Turkerich and Hillier [4] to be composed of tiny spheres arranged in fiber-like structures 100-500 Å in diameter. Use of these materials to obtain monoliths about 1 mil in thickness was reported by Bugosh [5] and was the subject of a patent [6]. Electrolysis as a technique of forming alumina gels has been employed since 1948 [7]; and alkoxide precursors have been used since 1953 [8]. The products which can most readily be classified as gels are usually obtained under acidic or neutral conditions. They typically yield X-ray diffraction patterns which indicate structures related to the monohydrate boehmite, γ-AlO(OH). Some of these gels appear rather well crystallized, while others yield diffraction patterns so devoid of features that the materials have been termed super-amorphous [9]. The better crystallized gels are generally obtained by aging in contact with the mother liquor. For some gels prepared under acidic conditions, the structures correspond to a boehmite, with all of the boehmite diffraction lines being present. For others, termed b boehmite gels, the main (020) diffraction peak corresponding to the interlayer distance in boehmite is absent. Strong preferred orientation in some of these gels has been reported [10]. Under basic conditions, particles of crystalline Bayerite, $Al(OH)_3$ rather than gels are formed.

One major complication of alumina gels obtained from salts is the presence of large amounts of anionic groups. Such complications can be avoided through the use of Al alkoxides which have been employed effectively by several workers [11-13, e.g.]. The pioneering work of Yoldas on alumina gels seems deserving of particular note. Of specific concern to the present work are his demonstration of a minimum gelling volume at about 0.07 mole of acid per mole of Al sec-butoxide for gels obtained at 80°C in excess H_2O, and his report of an amorphous gel obtained at temperatures about 25°C. He also mentions some conversion to Al_2O_3 at 1200°C, with no clear diffraction peak for d spacings between 2.16 and 2.72 Å.

STRUCTURES AND TEXTURES OF ALUMINA GELS DERIVED FROM ALKOXIDES

Experimental Procedures

Alumina gels were obtained following procedures similar to those of Yoldas [12]. The gels were prepared from Al sec-butoxide hydrolyzed in excess water [100 mole H_2O per mole $Al(OBu)_3$] using 0.035 to 1.12 mole of HNO_3 and, at room temperature NH_4OH up to 0.07 mole. Three gelation-drying procedures have been followed. One was at room temperature in neutral or slightly basic conditions (around pH8) using rapid centrifugation followed by 10^{-2} mm Hg vacuum. The aim of this procedure was to reproduce from alkoxides the super-amorphous product reported [9] as precipitated from $Al(NO_3)_3$ by NH_4OH. The two other procedures were derived from the Yoldas method. In one, hydrolysis and peptization were carried out in a closed erlenmeyer flask, with evaporation and gelation subsequently effected at 50°C in a lightly covered Pyrex dish. These gels are known to be boehmite-like. The last procedure was carried out entirely at 25°C, with the aim of obtaining amorphous gels from alkoxides. After hydrolyzing for 30 min, peptization and mixing were carried out for 3 days in a closed erlenmeyer flask. No mass loss was observed during this period. Gelation was then effected by evaporation in an open Pyrex dish. The mass of each sample was measured as a function of time until it remained constant to 0.1 g on a time scale of a few days. When extensive cracking occurred during drying, as with low acid additions, monoliths could easily be obtained by gelling over a layer of mercury [10] and cutting all around with a sharp knife to avoid nucleation of cracks from the sides.

X-ray diffraction patterns of the gels were taken using both thin solid layers and powders obtained therefrom. The microstructural features were

observed using a scanning electron microscope at magnifications of 20Kx to
50Kx. The SEM observations and diffraction analyses were carried out on
freshly prepared samples as well as samples heat treated at 600°C, 900°C
and 1150°C for various periods of time.

Results

Gelation Behavior. For both conditions of gelation investigated, the
phenomenological gel point was not marked by a change in the rate of drying.
This is illustrated by the data in Fig. 1 for gelation at 25°C by evaporation
in open dishes of samples containing various amounts of HNO_3. The mass of
high temperature gels (peptized at 90°C, gelled at 50°C) exhibited a
pronounced minimum as a function of acid concentration, as reported previously by Yoldas, while the low temperature gels (peptized and gelled at
25°C) displayed a monotonic decrease in mass with increasing concentration
of acid. These behaviors are illustrated in Fig. 2, where it is also seen
that the gelling mass is larger for the high temperature gels over most of
the range of acid concentration. In both cases, the final mass of the rigid
gel increases with the initial acid concentration (Fig. 3). It has also
been found that the pH buffers about 3.5 for a molar ratio $HNO_3/Al(OBu)_3$
between 0.14 and 0.56 for the high temperature gels, and about 4 for molar
ratios between 0.28 and 0.56 for the low T gels. Finally, in neutral or
basic conditions with NH_4OH, a white deposit occurs and we can no longer
speak of a gel point. During drying, this white powder shrinks like the
acidic gels to produce a solid monolith.

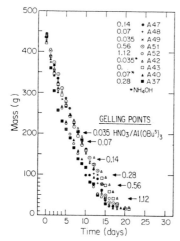

Fig. 1 Gelation at 20°C of
acidic gels

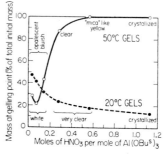

Fig. 2 Gelling mass of alumina gels
as a function of acid concentration

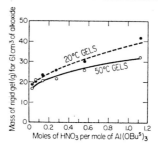

Fig. 3 Final mass of alumina gels
as a function of acid concentration
for 61 cm^3 of initial alkoxide

Structure of High Temperature Gels. The high temperature gels prepared at $HNO_3/Al(OBu^s)$ ratios of 0.035 are opalescent blue when observed in a dry dessicator. Viewed in the SEM at 20Kx, a fine fiber-like texture was observed. The fiber size was about 200 Å. This was confirmed by light scattering. At the minimum gel mass gelled monoliths were white, thin sections were transparent, and coarser wavy structures were seen in the SEM. With increasing acid content, preferred orientation parallel to the bottom of the dish becomes more pronounced. At a ratio of 0.28, gelled monoliths were clear, slightly yellow and relatively strong. At an acid/butoxide ratio of 0.56, the gelled samples were thick and bright yellow with a mica-like cleaved appearance. After peptization, gelation takes place in about 1 day, even in a closed erlenmeyer flask. For a ratio of 1.12, gelation occurs immediately after adding the acid, and the final product has a whitish appearance. The diffraction patterns of all these gels correspond to boehmite, mixed with some bayerite for the 0.035 ratio gel. The only exception was the extreme acidic case where the crystalline product obtained has not been identified. The crystallite size (Scherrer formula) of all the boehmite gels is about 30 Å. For solid layers, a peak at about 1.52 Å is also present. It disappears immediately upon grinding, or after long time of (1 month) drying solid layers. One of the most striking features of the highly transparent gel $[0.28\ HNO_3/Al(OBu)_3]$ is a strongly preferred horizontal orientation of the boehmite layers. The (020) boehmite peak (Fig. 4) is strongly enhanced for solid layers, compared with a powder of the same sample, while the other peaks disappear.

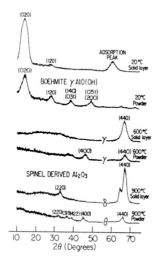

Fig. 4 X-ray diffraction patterns (CuKα) of the 0.28 mole HNO_3/AlO_3 gel.

The strong preferred orientation is maintained in the spinel related $\gamma-Al_2O_3$ at 600°C and $\delta-Al_2O_3$ at 900°C, the (440) spinel peak being considerably enhanced in the solid layers. The samples have a pronounced tendency to cleave extensively during heating. It is also observed that the splitting of the (440) peak characteristic of $\delta-Al_2O_3$ seems to disappear during grinding.

Structure of Low Temperature Gels. In basic and weakly acidic conditions, the products are clearly Bayerite, with well-defined (monoclinic) elongated crystals seen in the SEM (Fig. 5). The 0.14 HNO$_3$/Al(OBu)$_3$ gel is opalescent with a characteristic wavy structure, but maintains a well-defined (001) Bayerite peak. The 0.28 and 0.56 acidic mole gels are transparent, stable in air, brittle and hard to crush. They are also extremely X-ray amorphous, both in solid layers and in powder form (Fig. 6). SEM Observations up to a magnification of 50Kx do not show any distinct features.

Fig. 5 SEM of low T gels. Molar ratio HNO$_3$ or NH$_4$OH(*)/Al(OBu^s)$_3$: 0.035*(top), 0.14 (medium), 0.28 (lower)

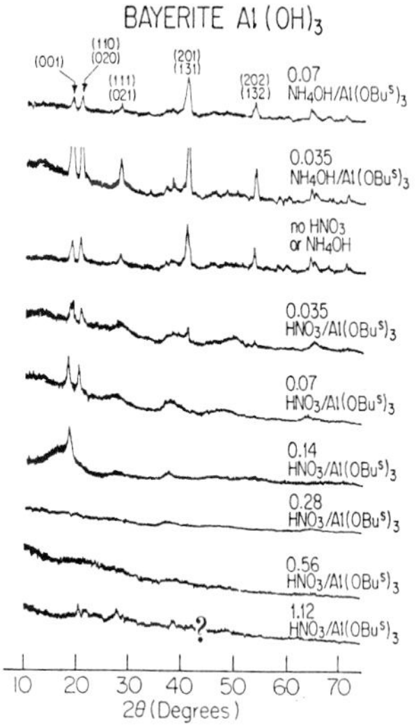

Fig. 6 X-ray diffraction patterns (CuKα) of low T gels

Structure of the pH8 Vacuum Product. These products are all Bayerite, even when dried as rapidly as possible by centrifugation followed by application of a vacuum. During centrifugation, a classification of the particles by size was observed. Upon drying, the deposit shrank to a monolith, the lower part of which (larger particles) was white and weak, while the upper part (smaller particles) was clear (like the 0.78 acidic 20°C gel), hard to crush, and extensively cracked. Attempts to X-ray a powder of the upper part gave only a Bayerite pattern, but it was not possible to eliminate completely the white lower part which was strongly bonded to the upper part.

Discussion

On heat treatment, the high T gels produce transition aluminas which are not super-amorphous, but feature a strong preferred orientation. In this regard, the only peak present on solid layers is the (440) spinel-derived peak which corresponds to plane passing through all the oxygen and all the octahedral Al positions, which is about at d=1.40 Å. During the transition to α alumina, this preferred orientation is lost. The monolithic macroscopic shape is maintained, but spherical pores are observed in the SEM, and all the peaks of α alumina are present in X-ray diffraction. The observation of a split (440) peak characteristic of δ alumina, strongly enhanced in solid layers, remains in detail unexplained. It is clear that the transition from boehmite to the spinel-derived transition aluminas (γ-alumina at 600°C, δ-alumina at 900°C) is such that the boehmite layers are transformed into the spinel (440) planes.

In all products, we have found a transition sequence boehmite →
γ-alumina → δ-alumina → α-alumina. A similar sequence is observed for the
very amorphous low temperature gels. This suggests that the "super
amorphous" X-ray aspect cannot be explained as a true "glassy
alumina" random network, of which the gels would be a precursor. Rather,
they may correspond to a "polymer type" gel of alumina, with a small
($\sim$ 30 Å) fiber texture. After drying, all the gels do not redissolve
completely in water, but expand by a factor about 20, while maintaining the
global initial shape before swelling.

In this regard, an extension of the theory developed by Tanaka [16]
for aqueous polymer gels should be possible for oxide gels, explaining the
transition between an expanded state and a contracted state. The coexist-
ence of these two states is possible, and is admitted for silica by Iler
[1].

ACKNOWLEDGEMENTS

Financial support for the present work was provided by IBM, DRET and
SNIAS. This support is gratefully acknowledged.

REFERENCES

1. R.K. Iler, The Chemistry of Silica (Wiley, New York 1979).
2. A. Cossa, Il Nuovo Cemento 3, 228-230 (1870).
3. W.H. Gitzen, Alumina as a Ceramic Material (American Ceramic Soc.,
   Columbus 1970).
4. J. Turkevich and J. Hillier, J. Anal. Chem. 21, 475-485 (1949).
5. J. Bugosh, R.L. Brown, F.R. McWhorter, G.W. Sears and R.J. Sippel,
   Ind. Eng. Chem. Prod. Res. Develop. 1[3], 157-161 (1962).
6. J. Bugosh, U.S. Patent 2,915,475, du Pont, December 1, 1959.
7. M. Le Peintre, C.R. Acad. Science, Paris 226, 1370-1371 (1948).
8. S. Teichner, C.R. Acad. Science, Paris 237, 810-812 (1953).
9. D. Papee, R. Tertian, R. Biais, Bull. Soc. Chim. Fr, 1301-1310 (1958).
10. W.O. Milligan, H.B. Weiser, J. Phys. Chem. Colloid 55, 490-496 (1951).
11. G. C. Bye, J.G. Robinson, Kolloid Zeit. 198, 53 (1964).
12. B. E. Yoldas, J. Mat. Sc., 10, 1856-1860 (1975).
13. D. E. Clark, J.J. Lanutti, Sol-gel Conference, Gainesville, Florida,
    Feb. 1983.
14. B. E. Yoldas, J. Am. Ceram. Soc. 65, 387-393 (1982).
15. S. M. El Mashri, A.J. Forty, Electr. Micros. Conf. 61, Cambridge
    (G.B.), Sept. 1971, p. 395-398.
16. T. Tanaka, G. Swislow, and J. Ohmine, Physical Rev. let. 42, 1556-1559
    (1979).

APPLICATIONS OF SOL-GEL PROCESSING

STRUCTURE AND PROPERTIES OF VANADIUM PENTOXIDE GELS

J. LIVAGE,
Spectrochimie du Solide - Université PARIS VI - 4, place Jussieu -
75230 Paris, France.

ABSTRACT

Transition metal oxide gels can be obtained, through a
polycondensation process, by acidification of aqueous solutions. Thin layers can be easily deposited onto a substrate.
Their electronic and ionic properties could lead to new
developments of the sol-gel process. The semiconducting
properties of V_2O_5 gels can be used for antistatic coatings
or electrical switching devices. These gels exhibit a lamellar structure and can be described as particle hydrates.
They are fast proton conductors and could behave as a host
lattice for intercalation.

INTRODUCTION

The sol-gel synthesis of glasses and ceramics has received significant
attention during the last decade (1-5). Dip-coating appears to be one of
the most promising application of the sol-gel process (6). Single oxide
coatings already appeared on the market in 1953 and multicomponent oxide
coatings in 1969 (7). Most of the layers deposited by dip-coating are based
on SiO_2 and TiO_2. They can be used for automotive rear-view mirrors, antireflection coatings or sunshielding windows (8). Silicate glass layers have
also been described (9). They exhibit laser damage threshold four times
greater than those of multilayer antireflection films (10). Coating of metals
with ceramics via colloidal intermediates provides greater corrosion resistance (11). Transparent electrically conductive layers (12) and superionic
conducting Nasicon thick films (13) have also been made by the sol-gel
process.

Transition metal oxide gels have been known for almost a century (14).
The first V_2O_5 gel for instance was reported by A. Ditte in 1885. They have
however almost never been used in the sol-gel technology. In this paper, we
would like to describe some properties of transition metal oxide coatings
derived from gels and show that they could lead to new developments of the
sol-gel technology (15).

Transition metal oxide are usually mixed valence compounds. Metal ions
may exhibit several oxidation states so that electron transfer, from low to
high valence states, can take place. This leads to specific electrical and
optical properties : semiconducting V_2O_5 layers (16), electrochromic WO_3
display devices (17). Magnetic interactions may also occur when the two
valence states have non zero magnetic moments as in ferrofluids (18).

Transition metal oxide gels can also be considered as hydrated oxides.
Ionic properties arise from the ionization of water molecules, trapped in
the gel, by the oxide surface. Some hydrous oxides show high proton conductivities at room temperature (19) and could behave as inorganic ion exchangers
(20). This is the case for vanadium pentoxide gels that exhibit a layered
structure and could be considered as a host lattice for intercalation of
ionic species.

SOL-GEL SYNTHESIS OF TRANSITION METAL OXIDES

Transition metal oxide gels can be obtained by hydrolysis and poly-condensation of metal alkoxides. This process has already been widely studied by Bradley (21) and applied to the synthesis of oxides such as TiO_2 or ZrO_2 (22)(23). In this paper, we shall rather focus our discussion on the synthesis of gels from aqueous solutions of inorganic salts.Such basic compounds are cheaper than the metal-organic ones and the resulting gels are free of carbon, leading to better glasses or ceramics.

As a starting point, let us consider a water molecule bonded to a metallic ion M^{Z+}. The formation of a $M-OH_2$ bond, by overlapping of the water σ orbital with the empty metallic d orbitals, draws electrons away from the O-H bonds, and weakens them. The coordinated water molecules then behave as stronger acids than the solvent water molecules (24). Depending on the σ transfer, we may have :

$$M - OH_2 \rightleftharpoons M - OH + H^+ \rightleftharpoons M - O + 2H^+$$

These equilibria depend on the metal ion, particularly its charge, size, electronegativity and ionization potentials. They also depend on the pH. For a given ion, we then have pH intervals in which water, hydroxide or oxide are common ligands to the central ion (25). Usually $M-OH_2$ bonds are observed for low valent states in acidic medium, $(Al^{3+}(H_2O)_6)$, while M-O bonds occur for high valence states in basic medium (VO_4^{3-}). In an intermediate pH range, M-OH bonds are formed.One of the main proper-ties of these hydroxo groups is that they could lead to polycondensation reactions such as :

$$\text{Olation} \quad 2M-OH \longrightarrow M \overset{OH}{\underset{OH}{\diamondsuit}} M$$

or oxolation 2 M-OH $\longrightarrow$ M-O-M + H_2O

These polymerization processes appear to be quite general. Polyca-tions can be obtained by increasing the pH of an aqueous solution of a low valent ion such as Al^{3+} (26).

$$2|Al(H_2O)_6|^{3+} \xrightarrow{20OH^-} |(H_2O)_4 Al \overset{OH}{\underset{OH}{\diamondsuit}} Al (H_2O)_4|^{4+} + 4 H_2O$$

After addition of 2.5 moles of hydroxide ion per mole of Al^{3+}, the $|Al_{13}O_4(OH)_{24}(H_2O)_{12}|^{7+}$ polycation with a Keggin structure (27) is observed (fig.1a).

(a) 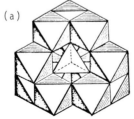   (b)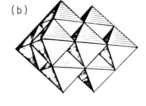

Fig.1 - Structure of condensed ions: a)- $(Al_{13}O_4(OH)_{24}H_2O_{12})^{7+}$
b)- $(V_{10}O_{28})^{6-}$

Polyanions can be obtained by decreasing the pH of an aqueous solution of a high valent ion such as VO_4^{3-}.

$$2\ VO_4^{3-} \xrightarrow{\ 2H^+\ } 2HVO_4^{2-} \longrightarrow V_2O_7^{4-} + H_2O$$

Decavanadate ions $(V_{10}O_{28})^{6-}$ are obtained around pH 6 (fig. 1b). Such isopolyanions have already been studied in great detail (24)(28). More or less condensed species are obtained depending on the concentration and pH, as shown in figure 2.

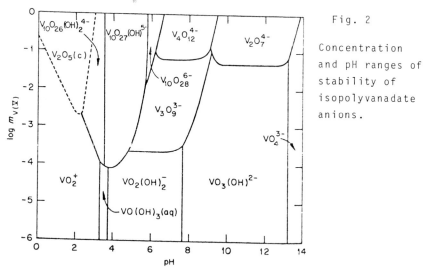

Fig. 2

Concentration and pH ranges of stability of isopolyvanadate anions.

Such a diagram shows that for a given vanadium concentration, for instance 0.1 M.l^{-1}, condensation increases as the pH decreases. The total negative charge per vanadium decreases upon protonation, leading finally to the positive VO_2^+ ion. Around pH = 2, we go through a point of zero charge. Around this point of zero charge, vanadium pentoxide would precipitate and V_2O_5 could be considered as an infinite polymer in which the total charge is zero. Stable colloidal species or gels could be obtained if the pH remains slightly higher than 2. These species are then negatively charged and electrostatic repulsions prevents further collisions and floculation.

Vanadium pentoxide gels are usually obtained by adding nitric acid to a vanadate salt. The gel then contains foreign ions such as NO_3^- and Na^+. Polyvanadic acid solutions can also be obtained by ion exchange in a sulfonic resin (Dowex 50 W-X2, 50-100 mesh) from sodium metavanadate solutions (29). The freshly prepared acid is yellow and decacondensed. Polymerization occurs spontaneously at room temperature, within a few hours, leading to a red viscous gel. Molecular weight determination, by light scattering and ultracentrifugation experiments, shows that the colloidal species are highly condensed ($M \sim 2.10^6$). They bear a negative charge of about -0.2 per vanadium. The viscosity of the gel depends on the concentration. A sol-gel transition occurs for a vanadium concentration of about 0.1 Mole.l^{-1} (30).

Vanadium pentoxide layers can easily be deposited onto a glass or polymeric substrate by dip-coating, spraying or screen-printing. After drying at low temperature, below 100°C, a rather hard coating is obtained that corresponds to the rough formula V_2O_5,nH_2O. Heating the layer above 300°C would lead to the crystallization of orthorhombic V_2O_5.

STRUCTURE OF V_2O_5 GELS

The amount of water nH_2O in the V_2O_5 gel depends on the experimental procedure. When dried at room temperature, a V_2O_5,1.8 H_2O xerogel is obtained. Thermal analysis shows that two main water departures occur upon heating (31)(32). The first one, corresponding to n=1.3, begins at room temperature up to 180°C. It is due to weakly bonded water molecules and appears to be quite reversible. Dehydration down to V_2O_5,0.5 H_2O can also be obtained under vacuum at room temperature or by blowing dry nitrogen gas above the sample. Rehydration occurs spontaneously as soon as the xerogel is left in air. The second water departure, corresponding to n=0.5, occurs above 180°C. It ends up with the crystallization of the xerogel into orthorhombic V_2O_5 above 300°C. It is due to strongly bonded H_2O or OH groups and can only be obtained upon heating. This water departure is not reversible and no rehydration is observed if a xerogel containing less than 0.5 H_2O is left in air.

An observation of the gel by electron microscopy shows that it is made of entangled fibres. These fibres actually look like flat ribbons about 10^3Å long and 10^2Å wide.

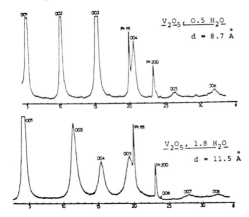

Fig. 3: X-ray diffraction patterns of $V_2O_5 \cdot nH_2O$ xerogels, showing the 001 peaks.

Electron and X-ray diffraction studies (33)(34) have shown that the internal structure of the ribbons corresponds to a 2.D cell : a = 27 Å, b = 3,6 Å, α = 90°. The vanadium and oxygen distribution inside this cell seems to be closely related to the lamellar structure of orthorhombic V_2O_5 except that the elementary cell contains five V_2O_5 units instead of two.

Some stacking of the ribbons occurs when the gel is deposited onto a substrate. X-ray diffraction patterns of the layers exhibit a series of 00ℓ peaks, typical of a 1.D order along a direction perpendicular to the surface of the ribbons (fig. 3). The basal spacing along this direction depends on the amount of water : d = 8.7 Å for a xerogel dried under vacuum (V_2O_5, 0.5 H_2O) and d = 11.5 Å for a xerogel dried at room temperature (V_2O_5, 1.8 H_2O). By comparison with similar layered clay systems, the 2.8 Å increase of the d-spacing could be attributed to the intercalation of one water layer.

Infra-red and Raman spectra have also been performed on V_2O_5 gels (35)(36). They exhibit several bands between 3700-3000 cm^{-1} and around 1620 cm^{-1}, some of them being strongly dichroïc. These bands are typical of more or less hydrogen bonded water molecules. According to the model suggested by M.T. Vandenborre (36), two kinds of water molecules are observed in the V_2O_5,1.8 H_2O xerogel (fig. 4). The first one is directly linked to a vanadium atom, along the c axis, opposite to the V=O short bond. The other one is linked through hydrogen bonds to the first water molecule on one side and to the apex oxygen of a VO_5 pyramid belonging to another ribbon on the other side.

Water molecules, in V_2O_5,0.5 H_2O, appear to be very weakly hydrogen bonded. They should correspond to H_2O trapped into the V_2O_5 lattice (fig. 5). Below the V_2O_5,0.5 H_2O composition, the water departure is no longer reversible. Raman spectra suggest that it could correspond to a further polycondensation process in wich V-O-V bridges between V_2O_5 ribbons are formed (37).

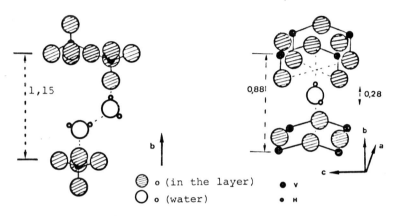

o (in the layer)

o (water)

● v

● H

Fig. 4 : Position of H_2O in the V_2O_5,1.8 H_2O xerogel (36).

Fig. 5 : Position of H_2O in the V_2O_5,0.5 H_2O xerogel (36).

ELECTRICAL PROPERTIES OF V_2O_5 GELS

Vanadium pentoxide gels, V_2O_5,nH_2O, are mixed conductors. Ionic conduction arises from proton diffusion between the fibres. It becomes predominant when water is intercalated in the layered structure of the gel (n>1.8). Electronic conduction arises from the hopping of unpaired electrons between vanadium ions in different valence states. It occurs within the oxide network and becomes predominant when the intercalated water is removed (n<0.5).

Fast proton conduction

D.c. conductivity measurements performed on a V_2O_5,1.8 H_2O xerogel show a strong decrease of the intensity when a d.c. voltage is applied. This transient regime appears to be quite long and a steady state can be obtained after a few days only. Such a behaviour is typical of ionic conduction. A.c. conductivity measurements, using blocking electrodes, were performed in the temperature range 320-200K.

The a.c. impedance diagrams are typical of a simple fast-ion conductor (38). They show Debye behaviour and Warburg contributions, i.e. semicircles at high frequencies and straight lines with slopes of 43-45° relative to the real axis at low frequencies. The a.c. conductivity was deduced from extrapolation of this linear low-frequency portion. The room temperature conductivity of the layers is quite high : $\sigma = 10^{-2}\Omega^{-1}.cm^{-1}$. Vanadium pentoxide gels appear to be as good proton conductors, as the well known H.U.P (HUO_2PO_4, $4H_2O$) (39). A plot of $\log(\sigma T)$ versus T^{-1} shows two linear regimes (fig. 6). A kink is observed around 260K. The activation energy above this temperature $E_1 = 0.35$ eV is smaller than the activation energy $E_2 = 0.44$ eV below 260K. This kink could correspond to the temperature at which the intercalated water layer is freezing.

A.c. conductivity of the V_2O_5,nH_2O layers strongly depends on the amount of water in the gel, and therefore on the partial water pressure above the layer (fig. 7). The conductivity decreases down to $10^{-6}\Omega^{-1}cm^{-1}$ when the sample is kept in a dry atmosphere with P_2O_5.

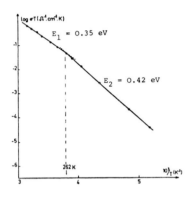

Fig. 6 : temperature dependance of the ac conductivity of a V_2O_5,1.8 H_2O xerogel

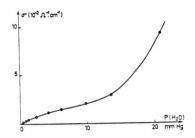

Fig. 7 : variation of the ac conductivity of a V_2O_5,n H_2O as a function of water pressure.

Fast proton conduction has already been reported for hydrous oxides (19). These materials can be separated into two main groups (20) :

. Framework hydrates with a strongly bonded three-dimensional network in wich mobile H_2O, OH^- or H_3O^+ species are incorporated.

. Particle hydrates in which charged particles are held together by water and either H_3O^+ or OH^- ions.

Vanadium pentoxide gels obviously belong to the particle hydrate group. A high proton conductivity necessitates a large concentration of mobile protons and a high proton mobility. The first factor is optimized in highly acidic or basic compounds. V_2O_5 gels appear to be good candidates. They exhibit a large oxide-water interface. They are negatively charged and produce acidic solutions (pH~2) when equilibrated with water.

It is more difficult to find a correlation between the proton mobility and the acid/base properties of the oxide. Two mechanisms are probably involved (40) :

. A vehicle mechanism corresponding to the drift of H_3O^+ ions between the fibres.

. A grotthus mechanism in wich a proton tunnelling occurs from a H_3O^+ ion to a H_2O molecule.

In both cases, the lamellar structure of the V_2O_5 gel provides a good conduction pathway for proton diffusion.

Small polaron hopping

The electrical behaviour of $V_2O_5,0.5\ H_2O$ layers appear to be quite ohmic. No transient regime is observed when a d.c. voltage is applied. The a.c. and d.c. conductivities are identical : $\sigma = 4.10^{-5}\Omega^{-1}cm^{-1}$ at 300K for a xerogel containing 1% of V(IV) ions. This conductivity actually increases quite fast when the amount of reduced ions increases : $\sigma = 3.10^{-3}\Omega^{-1}cm^{-1}$, at 300K, for a xerogel containing 10% of V(IV) ions. These experiments suggest that the conductivity is mainly electronic. The charge carrier mobility, deduced from $\sigma=n\mu e$, leads to a very low value indeed : $\mu \cong 4.10^{-6}cm^2.V.s$.

The temperature dependence of the d.c. conductivity, plotted as $\log(\sigma T)$ versus T^{-1} is shown in (fig. 8). The non-linear variation, together with the very low mobility, is typical of small polaron hopping in amorphous transition metal oxides (41)(42). The small polaron formation arises from a strong electron-phonon coupling and the conductivity can be expressed as (43) :

$$\sigma = \frac{\nu_0 e^2}{RkT}\ c(1-c)\ \exp(-2\alpha R)\ \exp(-W/kT)$$

where ν_0 is a phonon frequency related to the Debye temperature Θ by $h\nu_0 = {}^0k\Theta$. R is the average hopping distance. C is the ratio $V(IV)/V(IV)+V(V)$. The $\exp(-2\alpha R)$ term corresponds to a tunnelling transfer between two vanadium sites.

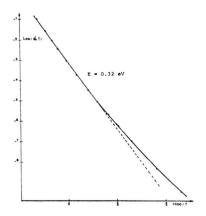

E = 0.32 eV

Fig. 8 : Temperature dependance of the dc conductivity of a $V_2O_5,0.5\ H_2O$ xerogel.

According to the small polaron model, hopping at high temperature is thermally activated by an optical multiphonon process. The activation energy W is given by $W = W_h + 1/2\ W_d$. W_h is a polaronic term and W_d corresponds to the potential energy distribution arising from the random disorder in the gel. A linear plot is actually observed above 220K, leading to $W = 0.32$ eV. As the temperature is lowered, the phonon spectrum freezes out and the polaronic term W_h continuously drops down leading to a decrease of the observed activation energy W below $\Theta/2$. An acoustical phonon assisted hopping takes place at lower temperatures ($T<\Theta/4$) and the activation energy should become $W = 1/2\ W_d$ (43).

Because of the high resistivity of the $V_2O_5,0.5\ H_2O$ gel a full analysis of the conductivity curve was not possible. However we noticed that the electronic conductivity increases quite fast when the amount of water becomes smaller than n = 0.5 ($\sigma \sim 10^{-1}\ ^{-1}cm^{-1}$ at 300K). As previously mentioned, the water departure is then no more reversible and V-O-V bridges between fibres

are formed. A detailed analysis, according to the models given by Holstein
(44) and Schnakenberg (45) leads to (46) : W_h = 0.15 eV and W_d = 0.06 eV.
More conductive gels can still be obtained by increasing the amount of
V(IV) ions, allowing conductivity measurements down to 30K, in a temperature
range where the disorder term W_d becomes predominant (47).

Semiconducting V_2O_5 layers

Several patents have already been taken, suggesting new applications
of the electrical properties of V_2O_5 layers deposited from gels. Their high
conductivity, at room temperature, could make them usefull as antistatic
coatings on the back of photographic films (48). Such coatings appear to
be much less sensitive to ambient moisture than the polymeric antistatic
coatings previously used in the photographic industry.

Switching devices, based on V_2O_5 layers made by the sol-gel process,
have recently been developped (49)(50). A coating, about 1μm thick is
deposited onto a glass substrate and gold electrodes are evaporated at the
surface of the xerogel. A typical Intensity-Voltage characteristic of such
a device is shown in (fig. 9).

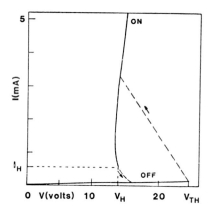

Fig. 9 : I = f(V) curve of
a switching device based on
a V_2O_5 layer deposited from
gels.

Switching toward an "ON" state occurs for a threshold voltage V_t of about
20 volts. No memory effect is observed and the devices switches back to the
"OFF" state when the intensity falls below a minimum value (holding current)
of about 2 mA. The ON/OFF ratio can reach two orders of magnitude and the
switching effect appears to be highly reversible. It can operate more than
10^8 times without failure. The switching characteristics depend on the
V(IV) content and the temperature. Switching disappears above 350K, or
when the V(IV)/V(V) ratio becomes larger than 0.06.

CATIONIC EXCHANGE PROPERTIES OF V_2O_5 GELS

Intercalation of guest species into host lattices has received an
increasing attention during the last few years. Intercalation usually occurs
with materials exhibiting strong 2D anisotropy such as sheet silicates,
graphite or transition metal dichalcogenides. It leads to an expansion of
the lattice perpendicular to the layer planes. Intercalation is usually
reversible and the host matrix retains its basic structure during the course
of forward and backward reactions (51)(52). Some energy has to be given in

order to compensate the lattice expansion. This is usually realized by ion or electron exchange between the host lattice and the guest species.

V_2O_5 gels appear to be good candidates for intercalation. They exhibit a layered structure with very weak interactions between adjacent layers (31). They are fast conductors and should therefore behave as good inorganic ion exchangers (20). They are mixed valence oxides and V(V) could easily be reduced into V(IV) (53). Ion exchange can be performed either chemically (54) or electrochemically (55).

Ion exchange with metallic cations can readily be obtained at room temperature by dipping a layer of $V_2O_5,1.8\ H_2O$ coated onto a glass substrate into an aqueous solution of a metallic chloride (0.1 Mole.l^{-1}). The pH of the solution decreases and about 0.3 M^+ or 0.15 M^{++} cations per V_2O_5 can be intercalated. The basal spacing d, between the layers, depends on the nature of the metallic ion, mainly its positive charge and ionic radius. Large monovalent cations (Na^+, K^+, Rb^+, Cs^+, NH_4^+,...) lead to a d spacing of about 11 Å, while divalent cations (Ca^{++}, Mn^{++}, Mg^{++}, Fe^{++}, Co^{++},...) or Li^+ lead to a d spacing around 13.7 Å.

The difference $\Delta d=2.7$ Å between these two basal spacings corresponds to the thickness of a water layer. This suggests that, because of the competition between the hydration energy of the cations and the energy required for lattice expansion, weakly polarizing cations can intercalate with one water layer only while more polarizing cations can be solvated with two water layers.

Larger ions, such as alkylammonium cations $C_nH_{2n+1}N^+(CH_3)_3$ can also be intercalated into the layered structure of V_2O_5 gels (56). Competition then arises between the electrostatic interactions of the oxide layers and the Van der Waals attractions of the alkyl chains. Depending on the number of carbon atoms, one interaction or the other prevails. Alkyl chain orientation in the interfoliar space can be deduced from the basal d-spacing variation with the chain length. The chains remain parallel to the layers up to C_7. They are perpendicular to them above C_{11}. In between they make an angle ranging from 42° to 53° with the layer plane.

REFERENCES

1. R. Roy, J. Am. Ceram. Soc. 52, 344 (1969).
2. S. Sakka and K. Kamiya, J. Non-Cryst. Solids, 42, 403 (1980).
3. S.P. Mukherjee, J. Non-Cryst. Solids, 42, 477 (1980).
4. D.P. Partlow and B.E. Yoldas, J. Non-Cryst. Solids, 46, 153 (1981).
5. H. Dislich, J. Non-Cryst. Solids, 57, 371, (1983).
6. H. Dislich and E. Hussmann, Thin Solid Films, 77, 129 (1981).
7. H. Dislich and P. Hinz, J. Non-Cryst. Solids, 48, 6 (1982).
8. H. Dislich, Angew. Chem. Int. Ed. Engl., 10, 363 (1971).
9. C.J. Brinker and S.P. Mukherjee, Thin Solid Films, 77, 141 (1981).
10. S.P. Mukherjee and W.H. Lowdermilk, J. Non-Cryst. Solids, 48, 177 (1982).
11. R.L. Nelson, J.D.F. Ramsay, J.L. Woodhead, J.A. Caiens and J.A.A. Crossley, Thin Solid Films, 81, 329 (1981).
12. H. Dislich, P. Hinz and G. Wolf, US-Pat. 4 229 491 (1978).
13. H. Perthuis, Ph. Colomban, J.P. Boilot and G. Velasco, Ceramic Powders, P. Vincenzi ed. (Elsevier-Amsterdam 1983) 575.
14. J. Livage and J. Lemerle, Ann. Rev. Mater. Sci., 12, 103 (1982).
15. J. Livage, Studies in Inorganic Chemistry, R. Metselaar, H.J.M. Heijligers and J. Schoonman eds. (Elsevier, Amsterdam 1983) 17.
16. C. Sanchez, F. Babonneau, R. Morineau and J. Livage, Phil. Mag. B, 47 279 (1983).
17. A. Chemseddine, R. Morineau and J. Livage, Solid State Ionics, 9-10, 357 (1983).

134

18. R. Massart, I.E.E.E. Trans. on Magnetics, 17, 1241 (1981).
19. D.J. Dzimitrowicz, J.B. Goodenough and P.J. Wiseman, Mat. Res. Bull., 17, 971 (1982).
20. W.A. England, M.G. Cross, A. Hamnett, P.J. Wiseman and J.B. Goodenough, Solid State Ionics, 1, 231 (1980).
21. D.C. Bradley, R.C. Mehrotra and D.P. Gaur, Metal Alkoxides, Academic Press, London, (1978).
22. K. Kiss, J. Magder, M.S. Vukasovich and R.J. Lockhart, J. Am. Ceram. Soc. 49, 291 (1966).
23. J.S. Smith, R.T. Dolloff and M.S. Mazdiyasni, J. Am. Ceram. Soc., 53, 91 (1970).
24. D.L. Kepert, The early transition metals, Academic Press, London (1972).
25. C.F. Baes and R.E. Mesmer, the hydrolysis of cations, John Wiley, New-York, (1976).
26. G. Johansson, Acta Chem. Scand., 16, 403 (1962).
27. G. Johansson, Acta Chem. Scand., 14, 771 (1960).
28. P. Souchay, Ions minéraux condensés, Masson Ed. Paris, (1969).
29. J. Lemerle, L. Nejem and J. Lefebvre, J. Chem. Res. 301 (1978).
30. N. Gharbi, C. Sanchez, J. Livage, J. Lemerle, L. Nejem and J. Lefebvre, Inorg. Chem. 21, 2758 (1982).
31. P. Aldebert, N. Baffier, N. Gharbi and J. Livage, Mat. Res. Bull., 16, 669 (1981).
32. L. Abello and C. Pommier, J. Chim. Phys., 80, 373 (1983).
33. J.J. Legendre and J. Livage, J. Colloids and Interface Science, 94, 75 (1983).
34. J.J. Legendre, P. Aldebert, N. Baffier and J. Livage, J. Colloids and Interface Science, 94, 84 (1983).
35. C. Sanchez, J. Livage and G. Lucazeau, J. Raman Spectra, 12, 68 (1982).
36. M.T. Vandenborre, R. Prost, E. Huard and J. Livage, Mat. Res. Bull., 18, 1133 (1983).
37. L. Abello, E. Husson, Y. Repelin and G. Lucazeau, J.Mol.Str.(submitted).
38. P. Barboux, N. Baffier, R. Morineau and J. Livage, Solid State Ionics, 9-10, 1973 (1983).
39. A.T. Howe and M.G. Shilton, J. Solid State Chem., 34, 149 (1980).
40. K.D. Krener, A. Rabenau and W. Weppner, Angew. Chem. Int. Engl. Ed., 21, 208-9 (1982).
41. L. Murawski, C.H. Chung and J.D. Mackenzie, J. Non-Cryst. Solids, 32, 91 (1979).
42. J. Livage, J. Phys., 42, C4, 981 (1981).
43. I.G. Austin and N. F. Mott, Adv. in Physics, 18, 41 (1969).
44. T. Holstein, Ann. Phys., 8, 343 (1959).
45. J. Schnakenberg, Phys. Stat. Sol.(b), 28, 623 (1968).
46. C. Sanchez, F. Babonneau, R. Morineau, J. Livage and J. Bullot, Phil. Mag. B, 47, 279 (1983).
47. J. Bullot, P. Cordier, O. Gallais, M. Gauthier and J. Livage, Phys. Stat. Sol.(a), 68, 357 (1981).
48. Kodak Pathé, French Patents BF 2318 442 (1977) and BF 2429 252 (1979).
49. J. Bullot and J. Livage, French Patent, 81, 13665 (1981).
50. J. Bullot, O. Gallais, M. Gauthier and J. Livage, Phys. Stat. Sol.(a), 71, K1-5 (1982).
51. R. Schollorn, Angew. Chem. Int. Engl. Ed., 19, 983 (1980).
52. M.S. Whittingham and M.B. Dines, Survey of progress in chemistry, 9, 55 (1980).
53. P. Aldebert and V. Paul-Boncour, Mat. Res. Bull., 18, 1263 (1983).
54. P. Aldebert, N. Baffier, J.J. Legendre and J. Livage, Rev. Chimie Minérale, 19, 485 (1982).
55. B. Araki, C. Mailhé, N. Baffier, J. Livage and J. Vedel, Solid State Ionics, 9-10, 439 (1983).
56. A. Bouhaouss and P. Aldebert, Mat. Res. Bull., 18, 1247 (1983).

THE ROLE OF GEL PROCESSING IN THE PREPARATION OF CATALYST SUPPORTS

J.A. CAIRNS, D.L. SEGAL AND J.L. WOODHEAD
Chemistry Division, AERE Harwell, Didcot, Oxfordshire, OX11 0RA, England.

ABSTRACT

Catalyst supports are required to meet a wide range of
objectives, including chemical purity and composition,
optimised surface area/porosity and thermal stability. Gel
processing offers means by which many of these requirements
can be achieved. In addition, the use of sols of controlled
rheology allows the support material to be prepared in a form
which facilitates its application to substrates (including
metals) as a coating, thereby greatly improving its
versatility for producing novel catalysts. The preparation
of typical support materials will be discussed, together with
examples of their use in catalysis.

INTRODUCTION

A catalyst is a substance which accelerates the rate of a chemical
reaction. It does so by adsorbing one or more of the reactants on to its
surface, thereby converting them to a more active form. Hence, the
specific activity of a catalyst (i.e. its activity/mass) can be increased by
converting it to a high surface area form. One of the most convenient ways
of doing this is to disperse one component of the catalyst (usually a metal
or metal oxide) over a high surface area material known as the support. The
support plays many roles, apart from acting as a dispersing medium. Thus it
minimizes the tendency of the dispersed catalyst phase to aggregate into
larger clusters, thereby losing activity; it can shield the dispersed phase
from poisons; and it can, by virtue of its structure (surface area and pore
size distribution), confer specificity on the catalytic reactions by imposing
a restriction on the dimensions of reactants or products.

Alumina and silica are two of the most widely used supports. The
purpose of this work is to describe the special attributes of sol-gel
technology in producing these materials (and others) in a form suitable for
use as versatile catalyst supports.

SYNTHESIS OF HETEROGENEOUS CATALYST SUPPORTS

The most widely used supports in heterogeneous catalysis are alumina
and silica. Therefore we begin with a brief description of how these may be
prepared conventionally. The simplest route to the former involves
precipitation of hydroxide or hydrous oxide from an aqueous salt solution [1].
The nature of the precipitate, i.e. whether amorphous or crystalline, is
determined by the precise precipitation conditions such as rate, concentration,
temperature and pH. Furthermore, the degree of crystallinity can be increased
by ageing in the aqueous environment, at either ambient or elevated
temperatures. For example, boehmite, a crystalline α-alumina monohydrate is
prepared by the addition of NH_4OH to an Al(III) salt solution under controlled
conditions of pH. Similarly silica may be prepared by mixing sodium silicate
solution with sulphuric acid at both alkaline and acid pHs to produce
precipitated silicas and silica gels respectively.

Mat. Res. Soc. Symp. Proc. Vol. 32 (1984) © Elsevier Science Publishing Co., Inc.

Oxide materials of high surface area can be prepared by the use of a process known as <u>flame hydrolysis</u>. In this case the oxide is produced by reacting a suitable volatile compound (typically an anhydrous chloride) with a mixture of hydrogen and oxygen in a flame, at about 1000°C. For example, $SiCl_4$ reacts under these conditions to produce an amorphous low density silica powder. The relative amounts of reactants determine the typical pore size and surface area of the product.

SOL-GEL ROUTES TO SUPPORT MATERIALS [2, 3]

These processes allow even greater control to be exercised over such features as primary particle size, density, pore structure and composition. There are two important <u>sol-gel</u> routes. The first uses metal-organic precursors, i.e. organic liquids [4], while the second involves the production of colloidal particles (typically in the size range 1 nm-1 μm) in an aqueous medium. An example of the first route is used in the production of boehmite ($\alpha Al_2O_3:H_2O$) by controlled hydrolysis of a long-chain aluminium alkoxide:

$$Al \begin{array}{l} OR_1 \\ OR_2 \\ OR_3 \end{array} + 3H_2O \rightarrow R_1OH + R_2OH + R_3OH$$

$$+$$

boehmite.

The alumina hydrate can be dispersed in dilute acid to form a concentrated aqueous dispersion.

The essentials of the second route can be summarised as follows:

The starting material can be a metal salt solution or a solid such as a hydrous oxide. For example, Al(III) nitrate solution can be converted to a dispersible hydrous oxide by addition of aqueous ammonia and washing the precipitate free from salts. This oxide may then be peptized to a sol on addition of a carefully controlled amount of mineral acid. Removal of water from the sol produces a gel. It should be noted that this sol-gel transition is often reversible.

Sols produced by this route can be classified under two broad headings, depending upon whether the constituent colloidal particles are aggregated or non-aggregated. This is illustrated schematically in Figure 1. The main point to note here is that sols containing aggregated particles dry down to porous gels which retain their porosity on calcination, thereby rendering them suitable for potential use as catalyst supports. On the other hand, non-aggregated sols dry down to dense gels and when calcined yield non-porous ceramics. For this reason they can be used to produce protective coatings on various substrates, including metals.

Incidentally, both dense [5] and porous [6] oxides can be obtained from the metal-organic route.

A common feature of the two sol-gel routes is their ability to make a wide range of mixed oxides, although in general the metal-organic route provides more effective interaction at the molecular level.

For completeness, we should also mention here <u>gel precipitation</u> which allows the production of spheres of controlled size. This is achieved by passing the feedstock (which may be an aqueous solution or a colloidal dispersion in an aqueous medium mixed with an organic gelling aid) through a vibrating jet and into a precipitation medium such as aqueous ammonia. This route is particularly attractive for the production of fast reactor fuel because it produces a well-defined spherical product (size range $\sim$ 80-800 μm) by a dust-free process [7].

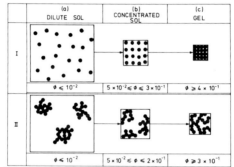

FIG. 1 DIAGRAM ILLUSTRATING THE FORMATION OF GELS OF LOW AND HIGH POROSITY FROM (I) UNAGGREGATED AND (II) AGGREGATED SOLS RESPECTIVELY.

APPLICATIONS

(a) The deposition of a catalyst on to a steel substrate

Catalysts must be fabricated into a suitable form before being loaded into a reactor where they interact with the feedstock. The most common physical forms are spheres, pellets and monoliths. In the latter case the catalyst is bonded to a ceramic body which usually incorporates a multitude of parallel channels to allow sufficient access of the reactants to the catalyst sites and to minimize pressure drop effects. The monolith substrate can be fabricated from ceramic or metal; the latter has the additional advantages of improved strength and versatility of shape. One type of metal which is particularly suitable for this purpose is an aluminium-containing steel such as Fecralloy^R steel (Cr: $\sim$ 15%, Al: $\sim$ 5%, Y: $\sim$ 0.3%). This

---

R Fecralloy is a Registered Trade Mark of the United Kingdom Atomic Energy Authority for a specific range of steels.

138

material on heating in air forms an alumina surface which endows it with
excellent oxidation resistance and facilitates its being coated with an
applied layer of catalyst [8]. Sol-gel technology constitutes an attractive
method of producing such a catalyst because of the ease with which sols can
be applied to substrates, including metals. Thus on subsequent drying and
firing, a porous, adherent catalyst support is bonded to the substrate. The
catalytic metal can then be applied to the support or incorporated at an
earlier stage into the sol. Catalysts produced in this way have been shown
to be active for a wide range of applications.

(b) The use of sols to prepare catalyst specimens for transmission
     electron microscopy

     The preparation of transmission electron microscopy specimens from
conventional catalyst pellets is a highly skilled operation. One of the
problems facing the microscopist is to ensure that the specimen is typical
of the catalyst as a whole. This implies that the composition of the
catalyst should be highly uniform. One way to approach this objective is to
prepare the catalyst from a sol of the support precursor containing a
dissolved salt of the catalytic metal. A sample transferred from the liquid
to an electron microscope grid can then be dried and fired to produce a
specimen which can be reproduced any number of times. This approach was used
to prepare specimens of Pt/Al$_2$O$_3$ to study the sintering of Pt as a function
of temperature [9].

(c) The use of sols to coat steels, and thereby prevent undesirable
     catalytic effects

     Metal surfaces in contact with gases or liquids can, in certain
circumstances, catalyse undesirable reactions. A particular case in point
arises when certain metals are exposed to hydrocarbon-containing gases at
temperature. Thus it has been shown that stainless steel, when exposed to
an RF plasma discharge of CO$_2$/CH$_4$ gas mixture at 650°C, becomes coated with
a filamentary carbonaceous deposit, resulting from the catalytic influence
of the steel. This phenomenon was prevented by coating the surface of the
steel by a catalytically inert layer of silica, using a silica sol [10].

CONCLUSIONS

     The special attributes of sol-gel technology enable it to produce a
wide variety of catalysts. It has the added attraction of being applicable
as coatings to both ceramic and metallic substrates.

REFERENCES

1. J.R. Anderson, Structure of Metallic Catalysts, Academic Press (1975).
2. J.L. Woodhead, Silicates Industriels, 37, 191 (1972).
3. J.L. Woodhead and D.L. Segal, Chemistry in Britain, to be published
    (1984).
4. H. Dislich and P. Hinz, J. Non-Crystalline Solids, 48, 11 (1982).
5. B.E. Yoldas, J. Materials Science, 12, 1203 (1977).
6. B.E. Yoldas, Amer. Ceram. Soc. Bulletin, 59, 479 (1980).
7. R.L. Nelson, N. Parkinson, W.C.L. Kent, Nuclear Technology, 53, 196
    (1981).
8. R.L. Nelson, J.D.F. Ramsay, J.L. Woodhead, J.A. Cairns and
    J.A.A. Crossley, Thin Solid Films, 81, 329 (1981).
9. P.J.F. Harris, E.D. Boyes and J.A. Cairns, J. Catal., 82, 127 (1983).
10. J.A. Cairns, J.P. Coad, E.W.T. Richards and I.A. Stenhouse, Nature,
    288, 686 (1980).

ELECTRICAL PROPERTIES OF Na_2O - SiO_2 DRIED GELS

D. RAVAINE,* J. TRAORE,* L. C. KLEIN** AND
I. SCHWARTZ**
*Laboratoire d'Energetique Electrochimique
ENSEEG BP 75 38402, Saint Martin d'Heres FRANCE
**Rutgers University, Department of Ceramics,
P.O. Box 909, Piscataway, NJ 08854, USA

ABSTRACT

Gels with compositions Na_2O-$(1-x)SiO_2$ were prepared
from tetraethylorthosilicate (TEOS) with sodium acetate
and from tetramethylorthosilicate (TMOS) with sodium
methoxide or TEOS with sodium ethoxide. After gelation
(10 min.) the samples were slowly dried for about one
month at room temperature. Thermogravimetric analysis
(up to 500°C) in air and O_2 atmosphere was performed.
The electrical properties were studied in air by impedance
spectroscopy during successive temperature cycles starting from room temperature dried gels of composition 0.10
$Na_2O \cdot 0.90$ SiO_2. Protonic conductivity appears to be
eliminated after reaching the 2nd stage of the drying
process (200°C). The activation energies for conduction
and the conductivities show small discrepancies from
known values for homogeneous glasses of equivalent
compositions. Considering the effect of water on the
conductivity of glasses prepared by the melting procedure, one can estimate the residual water content in
the so-called dried gels to be close to 0.5 wt %.

INTRODUCTION

Improving the ionic conductivity of glasses has been the subject of
numerous works these last ten years. Fast ion conducting glasses have now
entered the field as promising candidates for energy storage and other
applications [1-3]. The conductivity enhancement has been obtained through
progress in glass chemistry: sulfide glasses or HX (M: alkali or silver;
X: halogen) doped-glasses have lead to a considerable increase of conductivity compared to the previously investigated binary oxide glasses. Further
improvement is now limited by the choice of available compositions which
can be prepared by conventional melting. For instance, solid ionic salts
which decompose at low temperature, cannot be incorporated into traditional
glasses while they are known to enhance the conductivity when added into
ion-conducting solid polymers.

From this point of view, the sol-gel process appears as an attractive
method for preparing new glass compositions with improved ionic conductivities. Another interesting point comes immediately to mind: full densification may not be a necessary step to produce a suitable solid electrolyte. Typically in solid state ionic devices, the ionic conductor used as
an electrochemical membrane must fulfill the following requirements: 1)
to be free of any mobile water or solvent molecules which would be easily
electrolysed in high voltage cells; 2) to be single ion conducting in order
to prevent any spurious redox reactions at either electrode; 3) to exhibit
an extremely low electronic contribution to the conductivity.

Mat. Res. Soc. Symp. Proc. Vol. 32 (1984) Published by Elsevier Science Publishing Co., Inc.

This paper presents an investigation of the effect of thermal treatment on the conductivity of room temperature dried gels. Sodium containing silicate gels have been chosen for comparison with available conductivity data on the same conventional glass compositions. Impedance spectroscopy has been used for the electrical measurements since this method is able to provide information on the bulk properties independent of intergranular effects. The observed conductivity variations have been correlated with thermogravimetric analysis (TGA) in the same temperature range (ambient to the glass transition temperature, Tg).

Preparation and Characterization of Gels

The gels were synthesized by hydrolysis and polycondensation of organometallic compounds following different techniques:

Method A: TEOS, ethanol, water and acid (HNO_3) are mixed at $70^{\circ}C$ in a molar ratio 1 : 4 : 1 : 0.001. This initial water addition equals one fourth the stoichiometric amount required to fully hydrolyze the TEOS to monosilicic acid. After 1 h., a solution of sodium acetate in acetic acid (pH = 6) and excess water is added at room temperature to the previous solution. The proportions of sodium acetate are selected to yield x Na_2O - (1-x) SiO_2 (with x = 0.1, 0.2, 0.3 and 0.4) after loss of water and residual organics. The solution is vigorously stirred for one hour and then poured into transparent plastic Petri dishes and covered. The solutions are allowed to slowly dry at room temperature.
Method B: 25 c.c. of TMOS in methanol is mixed at room temperature with different quantities of saturated solution of sodium methoxide in methanol. After stirring, the solution is cooled down to $0^{\circ}C$ and maintained at this temperature during addition of distilled water at a rate of one drop per second. Mole ratios of n (H_2O) / n $Si(OCH_3)_4$ between 1.3 and 3.0 are used (4 c.c. < V(H_2O) < 9 c.c.).
Method C: This method is the ethanolic counterpart of method B. TEOS is used as a source of SiO_2 and sodium ethoxide for Na_2O.

For all samples, the gelation times are determined by observing their surfaces when tilting the plastic containers. These values are given in Table 1. Fig. 1 shows the strong dependance of the gelation times on the quantity of added water as already observed by PHALIPPOU et al. (4) on gels of 0.33 Na_2O - 0.66 SiO_2 composition.

TABLE 1. Gelation Times

Method	Composition Na_2O	SiO_2 m/o	Gelation Time	H_2O/TEOS	Transparent	Monolithic	Crystals
A	0.1	0.9	18hr	1:1	Yes	No	–
A	0.2	0.8	12hr	1:1	Yes	No	Yes
A	0.3	0.7	6hr	1:1	Yes	No	Yes
A	0.4	0.6	2hr	1:1	No	No	Yes
B	0.1	0.9	12hr	1.3:1	No	Yes	No
B	0.1	0.9	2hr	1.9:1	No	Yes	No
B	0.1	0.9	4min	3.0:1	No	Yes	No

The plastic covers were not airtight, so that the samples were allowed to slowly dry over a long period of time. When drying, the samples were characterized by radial shrinkage (Fig. 2). Table 1 also gives some information on gel shapes after complete drying at room temperature. It is observed that radial shrinkage is faster and more pronounced for method B.

Gelation is very fast and uncontrolled for samples prepared by method C. Moreover, sodium ethoxide is a very light powder and difficult to handle. No further attempts have been made to obtain gels using method C. Finally, only gels of 0.1 Na_2O - 0.9 SiO_2 compositions were available as monolithic pieces for the electrical investigations.

Electrical Study of Dried Gels

Flat samples about 2 mm in thickness and 1 cm^2 in surface area, were coated on both faces with platinum paint. The platinum paint was allowed to dry for a few minutes and used as dried in order to prevent organic solvent penetration. Good electrical contacts and damped stresses were expected from the introduction of small pieces of plasticized graphite between the sample and two slightly spring-loaded platinum electrodes. The cell was put under ambient atmosphere in a controlled heating rate furnace and the temperature was slowly raised (about 2°C per min.) after each measurement step. An HP 4192 A impendancemeter was used with available frequencies ranging between 5 Hz and 13 MHz. For temperatures lower than 50°C, a continuous drift is observed towards higher resistivity. Above 50°C, the measurements are stable over the whole frequency range and the results are then plotted in complex impedance diagrams. Analysis of the electrical impedance in the complex plane has been shown to provide the desired information including bulk properties, intergranular effects and electrode polarization phenomena. In this paper, the focus is on the high frequency domain for the analysis of the electrical relaxations occuring inside the bulk (intragranular effects).

A typical plot obtained for a dried gel in the high frequency range is shown on Fig. 3. The intercept of the circular arc shape curve with the real axis gives the bulk ohmic resistance. This assertion can be confirmed by measuring the capacitance of the sample from the top frequency of the circular arc. This quantity appears to be temperature independent with a value close to 5 pF. Correspondingly, the mean value for the relative dielectric permittivity is 10 which is in fairly good agreement with observed values of the permittivity in conventional glasses of the same composition [5]. These circular arcs, observed in the high frequency domain, are then due to the bulk electrical relaxations; consequently, the intragranular ohmic resistances are given by their intercepts with the real axis. The conductivities are deduced by taking into account the full size of the gel samples.

Arrhenius diagrams for the conductivities are shown in Fig. 4 and 5. For the dried gel prepared by method A, the investigated temperature range has been limited to 200°C due to the disruption of the electrical contacts on the sample. After removal of the cell, the gel sample was found to have many breaks and cracks: the heating rate was certainly too fast for this gel which still contains more than 50 w/o water and solvent after drying at room temperature. On the gel sample B, the conductivities exhibit a slope of constant value after reaching a temperature close to 250°C. After having kept the sample at 440°C for one week, the conductivities are measured again along a decrease in temperature. The same slope is observed over the full temperature range while the whole set of conductivity values are two times smaller than during the first heating.

The activation energy measured from the slope of the stabilized straight line is about 20.5 kcal/mole which is 2.5 kcal/mole larger than the activation energy for ionic conduction measured on traditional glasses of the same composition [5-9].

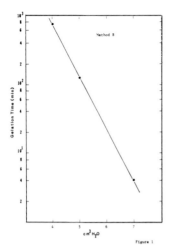

←Fig 1 - Gelation time vs cm^3 H$_2$O for 0.1Na$_2$O·0.9SiO$_2$ by Method B.

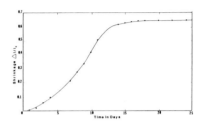

↑Fig 2 - Radial Shrinkage vs time for gel from Fig 1 with 5 cm^3 H$_2$O

→Fig 3 - Complex impedance plot at 278°C for gel from Fig 2. Frequencies in kHz.

↓Fig 4 - Arrhenius plot of the conductivity for increasing temperature with 0.1Na$_2$O-0.9SiO$_2$ by Method A. ($10^{-8} - 10^{-5}$ ohm·cm^{-1})

Fig 5 - Arrhenius plot of Method B sample. Numbered points follow thermal history. →

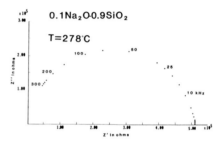

($10^{-8} - 10^{-5}$ ohm·cm^{-1}) ↓ Fig 5

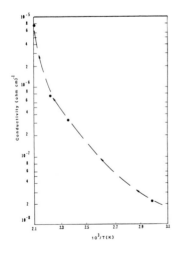

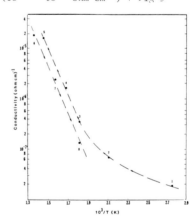

Thermogravimetric Analysis

Weight losses have been recorded on room temperature dried gels of the same surface area/thickness ratio and of 0.1 Na_2O-0.9 SiO_2 compositions. A SETARAM B70 thermobalance was used with heating rate of $2^{\circ}C$/min. Two samples prepared by Method A have been tested under air and oxygen, and one sample prepared by Method B has been tested under air only. Weight losses are shown vs. temperature in Figures 6 to 8. For each experiment, Table 2 gives the relative weight losses observed for each region along with their characteristic temperatures deduced from the intercept of the base line with the slope at the inflection point. The first region, up to about $50^{\circ}C$, is the continuation of the drying process observed at room temperature. Free water and solvent molecules are still removed at the very first stage of temperature rise. Gel samples prepared with methoxide are shown to contain much less residual water or solvent as expected from their stronger radial shrinkage. The weight loss in region II is smaller than that of region I and appears at about $200^{\circ}C$. This temperature slightly (but unambiguously) depends of the oxygen content. Thus, this region is likely due to the pyrolysis of the residual organics. Finally, region III appears at $400^{\circ}C$ and must be attributed to the more bonded hydroxide and alkoxide groups.

TABLE 2: Relative Weight Loss for Regions I, II, III.

	Region I		Region II		Region III		Total
	$T^{\circ}C$	m/m%	$T^{\circ}C$	m/m%	$T^{\circ}C$	m/m%	m/m%
Method A in air	30	38	210	7	400	9	54
Method A in O_2	50	35	170	7	400	10	52
Method B in air	65	15	250	1.5	-	0	16.5

Discussion

The temperature coincidence between the electrical study and the thermogravimetric analysis shows that the conductivity properties of alkali containing silicate gels are closely related to the different drying steps. The conductivity continuously drifts towards lower values when an appreciable amount of free water and free molecules are still present ($50^{\circ}C$). The gels exhibit most of the conduction properties of traditional glasses when residual organics are removed (above $200^{\circ}C$). Nevertheless, the activation energies for conduction and conductivities show small discrepancies from known values for homogeneous glasses of equivalent compositions. This can be assigned to the penetration of water molecules inside of the gel particles. TOMOZAWA et al. [10,11] have studied the effect of water content on ionic transport in traditional 0.25 Na_2O - 0.75 SiO_2 glass prepared under hydrothermal conditions. Both the conductivity and the activation energy for conduction appear to be dependent on the water content. From their results, one can estimate the residual water content in the bulk particles of the gel to be close to 0.5 w/o. For this low level of water content, only the sodium ion contributes to the conduction. Starting from room temperature dried gels, a $200^{\circ}C$ thermal treatment can then be considered as sufficient to reproduce the bulk electrical properties of traditional glasses.

144

ACKNOWLEDGEMENTS

This work was supported by the NATO Research Grants Program under Grant No. 293/82. The authors greatly appreciate the assistance of M. Pons for the TGA measurements.

REFERENCES

1. M. Ribes, B. Carette and M. Maurin, J. de Physique C9-12 (1982) 403.
2. J. P. Malugani and G. Robert; Mat. Res. Bull. 14 (1979) 1075.
3. J. P. Malugani, B. Fahys, R. Mercier, G. Robert, J. P. Duchange, S. Baudry, M. Broussely and J. P. Gabano, to be published in Proceedings of SSI 83, North Holland.
4. L. L. Hench, M. Prassas and J. Phalippou; J. of Non Cryst. Solids 53 (1982) 183.
5. D. Ravaine and J. L. Souquet; J. Chim. Phys. 5 (1974) 693.
6. K. Otto and M. E. Milberg; J. Am. Ceram. Soc. 51 (1968) 326.
7. R. J. Charles; J. Am. Ceram. Soc. 49 (1966) 55.
8. Y. Haven and B. Verkerk; Phys. Chem. Glasses 6 (1965) 38.
9. R. M. Hakim and D. R. Uhlmann; Phys. Chem. Glasses 12 (1971) 132.
10. M. Takata, J. Acocella, M. Tomozawa and E. B. Watson; J. Am. Ceram. Soc. 64 (1981) 719.
11. M. Takata, J. Acocella, M. Tomozawa and E. B. Watson; J. Am. Ceram. Soc. 65 (1982) 91.

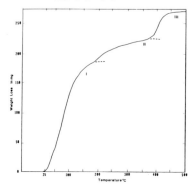

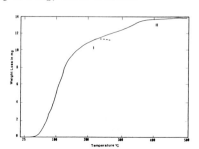

←Fig 6 - TGA curve in air for $0.1Na_2O \cdot 0.9SiO_2$ by Method A. $W_i=531.6mg$; $W_f=274.0mg$; $dT/dt=2°C/min$.

↓Fig 7 - TGA curve in O_2 for $0.1Na_2O \cdot 0.9SiO_2$ by Method A. W_i (Initial Weight) = 76.6mg; W_f (Final Weight) = 37.0mg; heating rate = 2°C/min.

↑Fig 8 - TGA curve in air for $0.1Na_2O \cdot 0.9SiO_2$ by Method B. $W_i=85.4mg$; $W_f=71.3mg$; $dT/dt=2°C/min$.

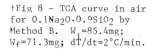

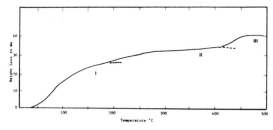

PREPARATION OF THIN COMPOSITE COATINGS BY SOL-GEL TECHNIQUES, *J. Martinsen, R. A. Figat and M. W. Shafer*, IBM T. J. Watson Research Center, New York, 10598

INTRODUCTION

The formation and characterization of thin ceramic coatings is currently an active area in materials science. Specifically, there has been increased interest in these materials as dielectric and optical layers for advanced microelectronic circuitry [1].This is in addition to their continued use and evaluation as protective coatings in a variety of applications [2]. One method of forming these coatings is via the sol-gel technique. However, only a few papers have been published in this area [3-7]. Dense silicon dioxide, aluminium oxide and various glass compositions have been deposited in thin film form on various substrates. Further, silicon dioxide films, formed by the sol-gel method, have been nitrided in ammonia at high temperatures and partially converted to an oxynitride [8-11]. These films were shown to be highly non-homogeneous, with a predominance of nitrogen at the surface. Despite the desirability of dense uniform well bonded coatings of the nitrided ceramics, i.e. Si_3N_4 and the oxynitrides, no method yet exists for the formation of nitrides by the sol-gel method.

This paper discusses some results of a completely different strategy for formation of silicon nitride films and the incorporation of nitrogen into ceramic coatings. The approach involves dispersing a solid particulate phase, i.e. Si_3N_4, in a sol and using a spinning technique to deposit the coatings. We report the preparation of a number of multi-component films on fused quartz and sapphire substrates, as well as a microstructural investigation of these films as a function of firing temperature.

Experimental Methods

The silica sols were prepared by combining (by volume) 39% tetraethoxysilane (TEOS), 55% ethanol, and 6% water, which corresponds to a 2:1 molar ratio of water to silicon. A small amount of nitric acid was added to catalyze the hydrolysis, with a resulting final pH of 2.1-2.8. The mixture was heated to reflux for 5 hours, cooled, and then allowed to age for one week. The alumina sols were prepared by mixing (by weight) 20% aluminum isopropoxide and 80% water, which corresponds to a 45:1 molar ratio of water to aluminum. The reaction was catalyzed with a small amount of nitric acid (pH ~ 2.4), and then heated to 95°C for 4 hours. No aging was necessary for this material. The silica borate sol is prepared by first reacting a mixture of 40% of TEOS, 57% ethanol, and 3% water (catalyzed with nitric acid) at 80°C for 16 hours. To this is added enough boron isopropoxide so that the molar ratio of silicon to boron is 9:1. Enough water is also added to attain a water to Si + B molar ratio of 2:1. The mixture is then again heated at 90° for 2 hours, cooled, and is then ready for use.

Mat. Res. Soc. Symp. Proc. Vol. 32 (1984) © Elsevier Science Publishing Co., Inc.

Silicon nitride slurries are prepared by combining the appropriate amounts of α-silicon nitride (UBE-SN-E-10; 0.5μ particle size) and the sol in a McCrone micronising mill containing agate pellets, milling the mixture for $1/2$ to 3 hours, and then immediately spin coating the mixture onto the corresponding substrate disk; fused quartz or sapphire. The disks are placed on a spin coater and cleaned with acidic ethanol just prior to slurry application. The slurry is then dropped onto the spinning substrate (2000-3000 rpm) until a uniform film is obtained. The samples are dried at 100°C overnight, and then fired at the designated temperature for 1 to 4 hours.

Results and Discussion

Thin, highly transparent films of the simple SiO_2 and Al_2O_3 sols were deposited on the corresponding substrates in order to study the microstructure prior to the incorporation of the particulates as a function of temperature. It was discovered that deposition of silica films on fused quartz using sols with a 4:1 or greater molar ratio of water to silicon invariably crack on drying or firing at relatively low temperature. This is not surprising since the structure of the higher water gel is more crosslinked than the lower water gel, thus favoring three dimensional rather than the quasi one-dimensional shrinkage necessary to maintain the integrity of the thin film [12]. Sols with ratios of 1:1 (H_2O:Si) or less often would not gel. Thus it was found that a ratio of 2:1 was optimal for the generation of thin films of silica whose integrity is maintained through drying and firing up to 1300°C.

The silica films are observed to be smooth and transparent at all temperatures up to 1300°C. Film thicknesses at a spinning rate of 2000 rpm are found to vary from 0.3μ to 0.6μ per application (1000°C), with the thicker films generally being obtained from older sols. This observation probably reflects an increase in the nature and extent of polymerization, and thus the viscosity, of the sol with time. There is no evidence based on our SEM studies of fractured surfaces and edges, of a distinct interface between the film and the substrate, presumably due to a facile reaction between the silanol groups on the surface of the substrates and the residual alkoxide groups in the sol. Thus, the film is indistinguishable from and indeed becomes an integral part of the substrate.

The choice of a 45:1 molar ratio of water to aluminum to form the alumina sol was an arbitrary compromise between generating a sol with a very low Al_2O_3 ratio and slow kinetics to produce the sol [13]. As was the case with silica, the alumina films on sapphire are smooth and transparent at low temperatures, with no detectable interface in the fracture surface between the film and the substrate. At some temperature between 1300°C and 1600°C the film turns opaque and develops a generally uniform porosity. (Fig. 1). The fracture surface shows (Fig. 2) that these pores are columnar in structure perpendicular to the film surface. X-ray diffraction patterns do not suggest any extensive crystalline ordering with respect to the sapphire substrate in this film.

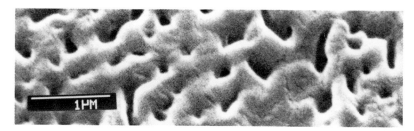

Fig. 1.   SEM micrograph showing the top of an Al_2O_3 fired at 1600°C for 2 hours.

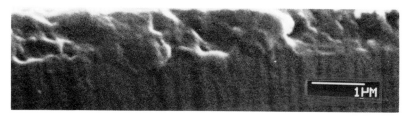

Fig. 2.   Fracture surface micrograph of the Al_2O_3 fired at 1600°C for 2 hours.

Films comprised of 44% by volume Si_3N_4 to SiO_2 in a silica sol were prepared and dried overnight at 500°C. Then two more thin coatings of the pure silica sol were deposited, and the samples were heated for one hour at temperatures of: 100°C, 500°C, 1000°C and 1300°C. Micrographs of these films (Fig. 3) reveal a generally uniform distribution of Si_3N_4 in a relatively

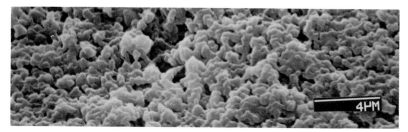

Fig. 3.   Top view of a film (44 volume percent Si_3N_4:SiO_2) fired at 1300°C for 1 hour.

dense structure, with SiO_2 filling in the interparticle spaces. The films are typically 7μ to 10μ thick, and show virtually no shrinkage up to a temperature of 500°C. At a temperature between 500°C and 1000°C there is an ~ 30% shrinkage in only the direction of the film, accompanied by an accumulation of the glassy phase near the substrate-film boundary. Between 1000°C and

1300°C this glassy phase flows into and becomes indistinguishable from the substrate. (Fig. 4). The surface roughness of these films is determined roughly by the Si_3N_4 particle size.

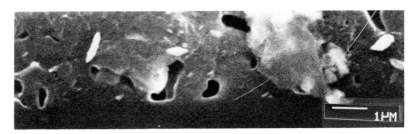

Fig. 4.   Cross section view of a film (44% Si_3N_4:SiO) fired at 1300°C for 1 hour.

Thin films containing 46 volume percent Si_3N_4 made from composite silica borate sols (initially 90% SiO_2: 10% B_2O_3) are comparable in thickness and low temperature surface features to those with only the silica glass. The glass phase however flows at a lower temperature, with the result that at 1000°C the surface is somewhat smoother in these films than in the silica films. (Fig. 5). Between 1000°C and 1300°C the apparent porosity increases, perhaps as a result of the aggregation of the micropores which are not visible at lower firing temperatures.

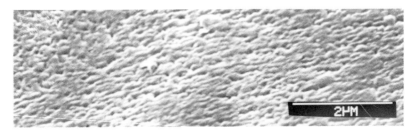

Fig. 5.   Top view of a composite film (46% Si_3N_4 to sol oxide; sol initially 90% SiO_2:10% B_2O_3) fired at 1000°C for 4 hours.

Silicon nitride suspended in alumina sol and coated on to sapphire substrate looks much the same as with the silica sols at low temperatures. Even at 1300°C there is very little evidence of any flow or softening in the glassy phase. At 1600°C, however, the surface has changed dramatically (Fig. 6) showing fibrous structure rather than the particulate structure at lower temperatures. Preliminary X-ray patterns suggest that a sillimanite type phase is formed as a result of a high temperature reaction between the alumina phase and the silicon nitride.

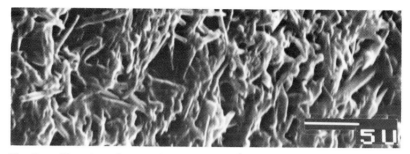

Fig. 6.   Top view of a film (49% $Si_3N_4:Al_2O_3$) fired at 1600°C for 2 hours.

Several preliminary attempts were made to determine the adhesion of the various composite films by epoxying steel pins to the surface and then measuring the force nesessary to pull them off. In the films fired at low temperature the films came off at the substrate with a minimal of force. The high temperature fired films, however, fractured not at the interface, but rather in the substrate below the pin. This behavior is likely due to stresses induced in the substrate arising from a thermal expansion mismatch between the film and the substrate.

Several experiments were done to attempt to increase the surface smoothness of the composite films. First, a silicon nitride containing sample in a silica sol was exposed to a xenon arc lamp for about five seconds. This sample, due to its relatively low absorbitivity, attained a surface temperature of only about 1200°C as detected by a pyrometer. This obviously was not high enough to have any observable effect and the sample appeared similar to one heated at 1300°C for one hour in air. The second experiment was to take a coating with a composition of $65Si_3N_4$ - $25Al_2O_3$ - $10Si$ (wt.%) which was heated at 500°C, and then surface heat it with a pulsed CO_2 laser. For an energy density of about 160-180mJ/mm² we were able to "reform" the surface to a considerable degree (Fig. 7). The spot heated by the laser (about

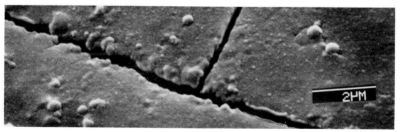

Fig. 7.   Top view of area of film exposed to the laser pulse.

2.5mm²) is seen to be much smoother but having cracked due to the thermal expansion mismatch and the apparent steep temperature gradient. A "densification" of about 50% is observed in the heated area, as seen by thickness measurements (Tally surf). However, microprobe analysis of this film showed a significant decrease in the silicon concentration and a somewhat smaller decrease in the nitrogen level in the heated area. From this it appears that some of the Si_3N_4 apparently vaporized or ablated and the diminished thickness is both a result of material loss and the densification. Further work is in progress in this area.

REFERENCES

1. A. Abraham, Thin-Film Technology for Microelectronics, FPH/OP, Israel, (1975).
2. H. Dislich and E. Hussman, Thin Solid Films, 77, 129-139 (1981).
3. H. Schroeder, Phys. Thin Films, 5, 87-141 (1969).
4. B. E. Yoldas and T. W. O'Keefe, Appl. Opt. 18, 3133 (1979).
5. B. E. Yoldas, Appl. Opt., 19, 1425-1429 (1980).
6. J. D. Machenzie, J. Non-Cryst. Solids, 48, 1-19 (1982).
7. C. J. Brinker and M. S. Harrington, Solar Energy Materials, 5, 159-172 (1981).
8. C. J. Brinker, J. Amer. Cer. Soc., 65, C4-5 (1982).
9. T. Ito, T. Nozaki and K. Kajiwara, J. Electrochem. Soc., 127, 2053-2057 (1980).
10. Y. Hayafuji and K. Kajiwara, J. Electrochem, Soc., 129, 2102-2108 (1982).
11. C. G. Pantano, P. M. Glaser and D. H. Armbrust, Proceedings of the International Conference on Ultrastructure Processing of Ceramics, Glasses, and Composites, Volume 1, chapter 13, (1983).
12. D. W. Shaefer, K. D. Keefer, C. J. Brinker, Presented at IUPAC MACRO'82, Amherst, MA. (1982).
13. B. E. Yoldas, Amer. Ceram. Soc. Bull., 54, 289-290 (1975).

DESIGN AND SYNTHESIS OF METAL-ORGANIC PRECURSORS TO ALUMINOSILICATES

ARLENE G. WILLIAMS AND LEONARD V. INTERRANTE
General Electric Corporate Research and Development, Schenectady, NY 12301

ABSTRACT

Diketonate aluminum alkoxides were prepared and allowed to react with acetoxyalkylsilanes. The conversion of the aluminosiloxanes thus obtained to aluminosilicates at 450°C was demonstrated. These aluminosiloxane ceramic precursors are glassy, oligomeric materials which readily dissolve in organic solvents to give viscous solutions ideal for casting films. Films of the metal-organic compounds prepared in this manner yield monolithic, crack-free aluminosilicate films directly on thermal curing, with a thickness limit of 3000Å for single crystal silicon wafer substrates. These aluminosilicate films have been found to be effective anti-corrosion barriers for various metal substrates. The chemistry of this organoaluminosilane system and the nature of the aluminosilicate films obtained has been investigated using a variety of chemical and physical methods. The results of this investigation will be described and the possible advantages of this direct conversion process over the sol-gel method as a means of obtaining aluminosilicate films will be discussed.

INTRODUCTION

The refractory character and chemical stability of aluminosilicates make them attractive candidates as corrosion resistant coatings for metals and other substrates. The preparation of aluminosilicates from metal-organic precursors has been previously demonstrated [1,2]. This process involves the formation of aluminosiloxanes, and the subsequent hydrolysis of these materials, followed by heat treatment to yield pure aluminosilicates. Although monolithic materials can be obtained by this technique, it does not provide a convenient method for homogeneous thin film (1000 – 5000Å) deposition on substrate surfaces. We report now the development of an organoaluminosiloxane ceramic precursor which may be cast as a film on a substrate surface by dip-coating from a viscous organic solution. The resultant glassy precursor film may be directly converted, by controlled pyrolysis, to a monolithic, homogeneous aluminosilicate coating.

DESIGN AND SYNTHESIS OF THE PRECURSOR

A ceramic precursor for solution-based thin film deposition must dissolve readily in organic solvents to produce a solution from which a coherent film may be cast onto substrate surfaces by dip, spin, or spray coating techniques. It is desirable that this initial film be a homogeneous, glassy coating that maintains its coherence during thermal conversion to a ceramic material. For an aluminosilicate precursor of a specific composition it is essential that the conversion to ceramic take place cleanly, with complete loss of organic content, avoiding fragmentation and subsequent removal of volatile organoaluminum or siloxane components.

The above considerations suggest that an appropriate precursor would be a polymeric compound, consisting of Al-O-Si linkages. Polymers are more likely to exhibit a glassy consistency and dissolve to give viscous

Mat. Res. Soc. Symp. Proc. Vol. 32 (1984) © Elsevier Science Publishing Co., Inc.

solutions than their monomeric counterparts. The construction of an Al-O-Si backbone would limit the possibilities for the formation of volatile silicon or aluminum compounds.

Aluminum alkoxides are known to exhibit extensive oligomerization through bridging alkoxide ligands to achieve the preferred coordination numbers of four and six for each trivalent aluminum center. As a consequence it is difficult to accomplish a displacement and polymerization sequence to produce a soluble, aluminum rich aluminosiloxane polymer of a specific composition. Substitution of one alkoxide ligand for a chelating diketonate has been introduced [3] as a means of inhibiting the bridging interaction and producing linear aluminum oxide based polymers upon hydrolysis. Mono-diketonatealuminum alkoxides were therefore selected as our starting point in the preparation of organoaluminosiloxane ceramic precursors as illustrated in Scheme I.

SCHEME I

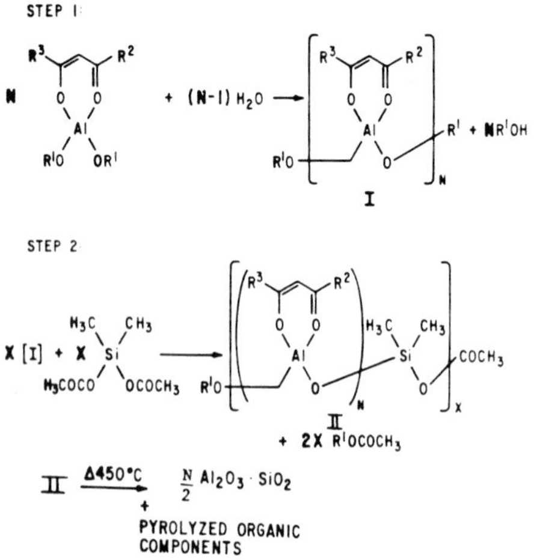

STEP 1

$$ N \quad \underset{\underset{R^1O \quad OR^1}{\overset{\displaystyle Al}{|}}}{\overset{R^3 \diagup\diagdown R^2}{\underset{O \quad O}{}}} + (N{-}1)\,H_2O \longrightarrow \left[\underset{R^1O}{\overset{R^3 \diagup\diagdown R^2}{\underset{\overset{\displaystyle Al}{|}}{O \quad O}}} \right]_N R^1 + NR^1OH $$

I

STEP 2

$$ X\,[I] + X \underset{\underset{H_3COCO \quad OCOCH_3}{}}{\overset{H_3C \diagup\diagdown CH_3}{\underset{\overset{\displaystyle Si}{}}{}}} \longrightarrow \left[\left(\underset{R^1O}{\overset{R^3 \diagup\diagdown R^2}{\underset{\overset{\displaystyle Al}{|}}{O \quad O}}} \right)_N \underset{}{\overset{H_3C \diagup\diagdown CH_3}{\underset{\overset{\displaystyle Si}{}}{}}} \right]_X COCH_3 $$

II
+ 2X R^{1}OCOCH$_3$

$$ II \xrightarrow{\Delta 450^\circ C} \frac{N}{2} Al_2O_3 \cdot SiO_2 $$
+
PYROLYZED ORGANIC COMPONENTS

Mono-diketonate aluminum alkoxides were subjected to a controlled partial hydrolysis, then allowed to react with diacetoxydimethylsilane. Products obtained in this manner were brittle glassy solids, which were infinitely soluble in organic solvents such as toluene and chloroform. Coherent aluminosilicate coatings of up to 3000Å in thickness were obtained on silicon wafers, a variety of metal alloy flats, and quartz plates by dip coating them in solutions of these aluminosiloxanes followed by thermolysis above 300°C.

Although the quality of the aluminosilicate coatings obtained from the procedure in Scheme I is quite good, the Al/Si composition of the starting material is not maintained in the ceramic product. Homocondensation of the diacetoxysilane compound competes effectively with the desired heterocondensation, resulting in the formation of volatile siloxanes that are removed with the solvent in the final stage of the aluminosiloxane preparation. The production of siloxanes was verified by gas chromatography (GC) of the solvent. Further evidence that the condensation does not proceed entirely according to the equations in Scheme I was obtained by the quantitative identification using GC, of the alkylacetate formed as a side product of the aluminosiloxane formation. Yields of 45-50% of the theoretical amount expected for a complete reaction according to Scheme I were observed.

Elemental analysis results listed in Table I reveal that not only is there a significant difference between the Al/Si ratio for the reactants and that found in the product, but that this difference is not constant. This implies that neither of the competing condensation reactions are favored, and that the ratio of the two products should be sensitive to small changes in the reaction conditions. Indeed, slight variations in conditions for runs 5-7 in Table I show significant variation in the Al/Si ratios obtained. Stoichiometric control of the ceramic material during precursor preparation is thus quite limited in this system.

TABLE I. Variation of Al:Si ratio from starting materials to product, indicating an overall loss of silane during the reaction.

Run	Al:Si Ratio Starting Materials	Product
1	5.00	11.46
2	4.00	4.61
3	3.00	3.76
4	2.00	3.62
5[a]	1.00	1.23
6[a]	1.00	2.15
7[a]	1.00	3.07

[a]Differences in operating conditions (i.e. concentration, additional heating period after solvent removal) are attributed to the inconsistency of the results obtained for runs 5-7.

Structural analysis [4] of the mono-diketonatealuminum alkoxides has shown that intermolecular exchange of the ligands takes place to form the dimeric and trimeric compounds illustrated below, for acetylacetonatealuminum diisopropoxide:

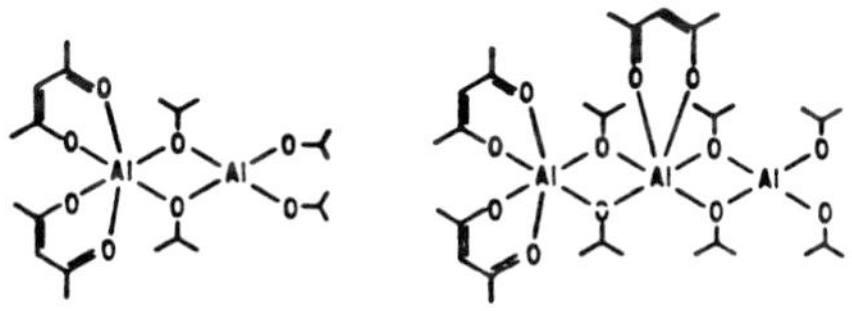

A   B

Both compounds A and B above were identified by X-ray crystallography, and the dimer to trimer conversion as written was characterized by NMR. The identification of these compounds suggests that the chemistry of the diketonate aluminum alkoxides is more complex than is implied by Scheme I.

Compound A above was allowed to react with four equivalents of trimethylacetoxysilane to produce a quantitative heterocondensation, yielding product C, which was fully characterized by ^{1}H NMR and X-ray

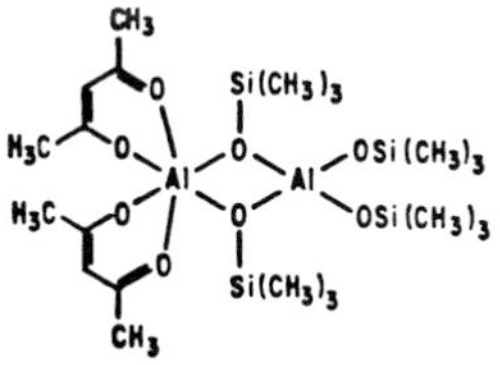

C

crystallography. This suggests that the homocondensation of siloxanes observed previously may be eliminated by the employment of monofunctional trimethylsilanes. A modification of the original system based on this information may be described by the equation below (Scheme II).

Scheme II

$$[Al(chel)(OR)_2]_x + nSi(CH_3)_3OC(O)CH_3 \rightarrow$$

$$[Al(chel)(OR)_{2-n}]_x[OSi(CH_3)_3]_n + nROC(O)CH_3$$

$$\xrightarrow{\Delta 450^\circ C} \quad \frac{x}{2} Al_2O_3 \cdot SiO_2$$

$$+ \text{ Pyrolyzed Organic Components}$$

Prehydrolysis of the aluminum compound is no longer necessary. The stoichiometry of the aluminosiloxane product may be established precisely by the number of equivalents of silane introduced initially.

Scheme II was originally carried out with a 3:1 ratio of aluminum complex to silane, yielding indeed an aluminosiloxane with a 3:1 Al/Si ratio as required for a mullite precursor. This new precursor, which was fully characterized by chemical analysis, retains the favorable physical characteristics of the initial Scheme I product. It is a glassy material which readily dissolves in organic solvents to give viscous solutions. The amount of silane was varied in subsequent reactions, to demonstrate the stoichiometric control allowed by this quantitative reaction. The Al/Si ratios of these products are listed in Table II, as determined by elemental analysis.

TABLE II. Comparison of Al:Si ratios in starting materials to that in product for Scheme II.

Run	Al:Si Starting Materials	Product
8	5.00	5.10
9	4.00	4.07
10	3.00	3.12
11	2.00	2.02
12	1.00	1.04

CONVERSION OF PRECURSOR TO CERAMIC COATINGS

Thermal conversion of this precursor has been effected at 450°C, with the total weight loss observed within 2% of the calculated theoretical value. No further weight loss, or structural changes are detected on additional heating up to 1000°C by differential thermal analysis. At 1300°C, samples with a 3:1 Al/Si ratio convert from the initially amorphous aluminosilicate to crystalline mullite. This has been demonstrated for precursors obtained from both Schemes I and II. The Al/Si ratio for each aluminosiloxane was maintained during the conversion to ceramic. This was established by X-ray fluorescence measurements on the final aluminosilicate products.

The thickness and coherence of the aluminosilicate coatings obtained by this process is highly dependent on the solution concentration, substrate choice, condition of the substrate surface, and rate of heating in the thermal conversion process. Solution concentrations below 10% deposit a coating too thin to effectively cover the surface during the subsequent heat treatment. In the case of metal alloy substrates, surface oxidation results, breaking up the aluminosilicate coating. Concentrations above 40%

produce a thick coating of mudflats, leaving large uncoated areas between each flat. Concentrations in the range of 10-40% were found to produce coherent coatings from 1000-3000Å in thickness, with good adherence to a variety of substrates. Preliminary experiments have suggested that surface treatments to enhance the interaction between the substrate and the precursor solution may be carried out to improve the adhesion of the coating, when necessary in cases where large expansion coefficient differences cannot be avoided. We have observed improved adhesion of aluminosilicate coatings on nickel-based alloys when their surface was chemically roughened followed by oxidation in air at 800°C prior to treatment with the precursor solution. It is crucial that the substrate surface be clean, dry, and free of particulates, prior to immersion in the precursor solution, to insure homogeneous deposition. Finally, the heating rate for optimal coating conditions has been established experimentally to be 100°C/h. Faster heating rates resulted in extensive flaking and peeling of the coating, and incomplete pyrolysis due to trapped carbonaceous residues. Heating rates slower than 100°C/h did not offer an improved coherence of the coating.

Coated surfaces were examined by SEM to determine their thickness and overall homogeneity. A cross sectional view of a silicon wafer with a 3000Å coating of aluminosilicate (Al/Si=3:1) deposited by this method is displayed in Figure 1. In Figure 2 a higher magnification scanning electron micrograph is shown which illustrates the lack of porosity for a monolithic aluminosilicate coating on a quartz plate. Coherent 3000Å coatings obtained by this technique were shown to be effective in inhibiting high temperature surface oxidation of metal alloys.

FIG. 1. Cross sectional SEM of aluminosilicate on silicon wafer.

FIG. 2. 3000Å layer of monolithic aluminosilicate on quartz.

Samples of a nickel-based alloy were dipped into precursor solutions to coat only half their surface, followed by thermal treatment at 450°C to produce ceramic coatings. These partially coated samples were heated to 1000°C in air for 30 minutes, and examined by SEM. This procedure allows a convenient side by side comparison of coated and noncoated surfaces, as shown in Figure 3. A definite inhibition of the surface oxidation in the coated region is apparent. Preliminary tests on cobalt-based alloys in a SO$_x$ environment indicate significant inhibition of high temperature sulfidation for similarly coated samples.

156

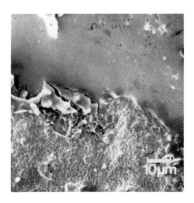

FIG. 3. Aluminosilicate on nickel alloy, testing for protection from high temperature oxidation.

CONCLUSIONS

Aluminosiloxane oligomers were prepared by the quantitative reaction of trimethylacetoxysilane with mono-diketonatealuminum alkoxides. The products are glassy materials with excellent physical characteristics for metal-organic precursors to ceramic coatings. Conversion of these metal-organic compounds to aluminosilicates takes place cleanly at $450^{\circ}C$ in air, with an overall retention of the initial Al/Si ratio. The preparation of organic solutions of the aluminosiloxanes enables dip coating of desired substrates to obtain films which may be directly converted to ceramic coatings with heat treatment. Monolithic, crack-free ceramic films of up to 3000Å may be produced in this manner. These coatings provide significant protection of metal substrates from high temperature surface oxidation and chemical corrosion.

REFERENCES

1. B.E. Yoldas, J. Mater. Sci. 12, 1203 (1977).
2. B.E. Yoldas, Ceram. Bull. 59, 479 (1980).
3. T.R. Patterson, F.J. Pavlik, A.A. Baldoni, R.L. Frank, J. Amer. Chem. Soc. 81, 4213 (1959).
4. J.H. Wengrovius, R.G. Going, M.F. Garbauskas, J.S. Kasper, submitted for publication.

Preparation of Barium Titanate Films Using Sol-Gel Techniques

R. G. Dosch, Sandia National Laboratories, P. O. Box 5800, Albuquerque, New Mexico 87185

ABSTRACT

Sol-gel techniques provide a convenient method of applying barium titanate films on solid substrates at ambient temperature. A significant advantage in using this technique is the capability of maintaining compositional homogeneity while coating large areas of surfaces with complex geometries. Solutions prepared by reacting a titanium alkoxide with barium hydroxide in methanol were used for preparing thin films (1000-3000Å) of barium titanate on silicon and metal substrates. Electrical properties of the films depended both on solution parameters and subsequent heat treatment of the films. Highest field strengths ($8-9 \times 10^6$V/cm) were observed in amorphous films prepared from partially hydrolyzed solutions.

Introduction

A variety of techniques have been used to prepare barium titanate films[1-4], but most involve either relatively sophisticated apparatus or area and geometrical constraints. Hydrolysis of mixtures of barium napthenate and titanium alkoxides has been successfully used[5] for this purpose, but requires relatively high temperatures to remove organics. In this work, solutions of titanium isopropoxide and barium hydroxide were used to form barium titanate films on Si, Ni, and Ti substrates. This chemical system appears to provide a simple means for coating large areas while maintaining stoichiometric homogeneity. This system also minimizes the organic content of films dried at ambient temperature and provides the option of subsequent alteration of film composition via ion exchange reactions.

In this paper, the preparation and associated chemistry of precursor solutions and films are presented. Solution parameters and subsequent thermal history which affect the electrical properties of the films are also discussed.

Experimental

Titanium isopropoxide (TPT) was purified by vacuum distillation. Sodium methoxide was added to prevent Cl^- carryover into the distillate. $Ba(OH)_2$ was prepared by adding stoichiometric amounts of H_2O to a BaO slurry in MEOH. The BaO was prepared by precipitating recrystallized $Ba(NO_3)_2$ as barium oxalate which was then heated to form $BaCO_3$. The $BaCO_3$ was converted to BaO at high temperature in glassy carbon crucibles.

*This work performed at Sandia National Laboratories supported by the U.S. Department of Energy under contract number DE-AC04-76DP00789.

158

Alcohol solutions containing Ba:Ti mole ratios <1 were then prepared by
adding TIPT to dilute solutions of Ba(OH)$_2$ in methanol (MEOH).

Films were prepared by applying solution through a 0.2μM filter onto
a substrate which was then spun to remove excess solution. All films
discussed herein were formed on 2 inch diameter, 10-12 mil thick discs of
single crystal, p-type silicon (100 orientation). Ni and Ti substrates
were prepared by RF and DC sputtering, respectively, onto the Si discs.
All heat treatments subsequent to film deposition were done in air.

Film thickness was generally determined by gravimetric techniques
which were found to be in good agreement with Rutherford backscattering
and transmission electron microscope results on selected samples.
Electrical contacts to the films were made using a 50 mil diameter Hg
spot under slight pressure. Contact to the uncoated side of the Si disc
was made by contact of a metal electrode to Ag paint applied to a freshly
abraded area.

Results and Discussion

When TIPT is added to a MEOH solution of Ba(OH)$_2$, an exothermic
alcohol exchange reaction occurs forming insoluble titanium methoxide
which reacts with OH$^-$ forming a soluble species. The OH$^-$ is quantitatively
consumed at OH$^-$:Ti mole ratios of <0.5. At mole ratios >0.5, dissolution
of titanium methoxide is complete and a clear solution is obtained. The
soluble Ti species is anionic in nature and continues to consume OH$^-$ as
the OH:Ti ratio increases (Fig. 1). Thus, this chemical system is not
applicable to forming films with Ba:Ti mole ratios less than 0.25. When
the OH:Ti ratio exceeds 0.5, free OH$^-$ is present which enhances the good
wetting properties of the solutions.

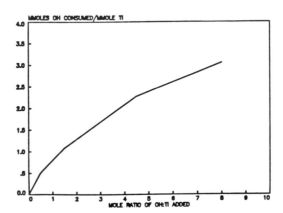

Fig. 1. Reaction stoichiometry of titanium isopropoxide and barium
hydroxide in MEOH solution.

Addition of the MEOH Ba(OH)$_2$-TIPT solutions to excess H$_2$O results in
the formation of solid hydrolysis products. For solutions with OH:Ti
mole ratios <1, the stoichiometry of the solution is maintained in the
hydrolysis product (Fig. 2). Thus, barium titanate films prepared from
solutions with mole ratios of OH:Ti between 0.5 and 1 can be
compositionally altered by aqueous ion exchange as in Eq. 1.

$$Ba(Ti_2O_5H)_2 + xSr^{++} \rightleftharpoons (Ba_{1-x})Srx(Ti_2O_5H)_2 + xBa^{++} \qquad (Eq.\ 1)$$

where an empirical representation, Ti_2O_5H, is used for hydrated titania.
Greater than 90% exchange of Ba^{++} for other alkaline earth ions has
been achieved by soaking films in aqueous solutions containing soluble
alkaline earth salts.

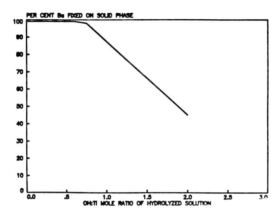

Fig. 2. Hydrolysis of hydrous barium titanate in aqueous solution.

Unless stated otherwise, the electrical properties discussed below
were measured for films with $BaTiO_3$ stoichiometry prepared from MEOH
solutions with OH:Ti mole ratios of 2 and were approximately 1000Å
thick. Current-voltage curves are shown in Fig. 3 as a function of subsequent processing temperature. The formation temperature of crystalline
$BaTiO_3$ is believed to be in the range of 450-500°C based on x-ray diffraction (XRD) results with powders formed by hydrolyzing $Ba(OH)_2$-TIPT
solutions.

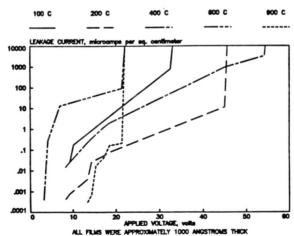

Fig. 3 Current voltage curves of $BaTiO_3$ films on Si as a function of
processing temperature.

The field strengths of the films increase after heat treatments in the 100-400°C range and show a definite decrease after 600°C or above. X-ray diffraction identified the films heated above 600°C as crystalline $BaTiO_3$. A film heated to 700°C was examined by TEM which showed the total thickness to vary between 1200-1400Å. A 120-150Å thick amorphous interaction layer was found between the $BaTiO_3$ and Si substrate. The $BaTiO_3$ phase, identified from selected area pseudo-ring diffraction patterns, had preferentially oriented, layered, bimodal grains. The 500-600Å nearest the substrate was relatively large grained, 500Å in the largest dimension. The typical grain size in the outer 600-800Å of the film was 100-200Å.

Substrate materials were also found to be important in determining electrical properties. Films formed on Ti showed higher leakage current and lower breakdown voltage than those formed on either Ni (Fig. 4) or Si (Fig. 3). The leakage currents of films on Ni were lower than those of films formed on Si.

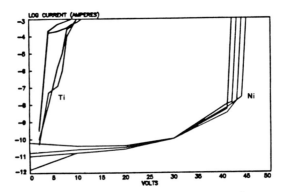

Fig. 4   Current-voltage curves for $BaTiO_3$ films (1000Å thick) on Ti and Ni substrates after processing at 400°C.

Breakdown voltages of the films were increased by prehydrolysis of $Ba(OH)_2$-TIPT solutions. This was done by adding a dilute solution of H_2O in MEOH to the solutions prior to use. Current voltage curves of $BaTiO_3$ films prepared with and without prehydrolysis from the same MEOH solution of $Ba(OH)_2$-TIPT are shown in Fig. 5. Four moles of H_2O were added per mole of Ti in the prehydrolyzed solution. Films prepared by this technique showed the best insulating properties with respect to electrical breakdown; films processed at 400°C withstood fields on the order of 9×10^6 V/cm prior to breakdown. Ferroelectric properties of the films have not been determined as yet.

CONCLUSIONS

Sol-gel techniques provide a convenient means of applying thin $BaTiO_3$ films to solid substrates. The $Ba(OH)_2$-TIPT precursor solution has good wetting characteristics and can produce films whose compositions can be altered by ion exchange reactions. The electrical breakdown strength of amorphous films is typically greater than that of crystalline films. Field strengths on the order of 9×10^6 V/cm were obtained using precursor solutions which had been prehydrolyzed.

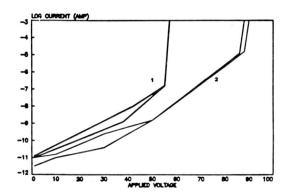

Fig. 5   Current-voltage curves of BaTiO₃ films (1000Å thick) on Si
substrates prepared from: 1) MEOH solution precursor, and
2) pre-hydrolyzed MEOH precursor solution using a 4:1 mole ratio
of H₂O to Ti.

REFERENCES

1. C. Feldman, J. Appl. Phys. 27, 870-73(1956).
2. V. Oharmadhikari and W. Grannemann, J. Appl. Phys. 53(12), 8988-92
    (1982).
3. S. Yamanaka, J. Electrochem. Soc. Jpn. 78, 908(1958).
4. J. Panitz and C. Hu, Ferroelectrics 27, 161-4(1980).
5. J. Fukushima, K. Kodaira, and T. Matsushita, Amer. Ceram. Soc., Bull.
    55(12), 1064-5(1976).

RHEOLOGICAL PROPERTIES OF AN ALKOXIDE DERIVED HLW SLURRY FEED†

L. H. CADOFF, D. SMITH–MAGOWAN, D. E. HARRISON
Westinghouse R&D Center, Pittsburgh, PA 15235
and J. M. POPE
West Valley Nuclear Services Co., Inc., West Valley, NY 14171-0191

ABSTRACT

Coagulation of HLW feed slurry can result from the interaction between acid metal cations and soluble silicates/silica. Feed slurry viscosity depend on a number of factors: (a) concentration and composition of the waste slurry electrolytes, (b) type and concentration of silica and silicate constituents in the glass former, (c) pH of the slurry feed, and (d) degree and order of mixing of the glass former with the waste sludge. Coagulation can be avoided by mixing to achieve a good dispersal of glass former globules in the waste slurry even when pH and electrolyte content favor it.

INTRODUCTION

The reference process for immobilizing high level nuclear waste (HLW) in borosilicate glass at West Valley is to melt a slurry of glass frit and waste sludge in a liquid fed ceramic melter. An alternative to glass frit is to utilize a dispersion of glass forming additives for blending with the waste sludge[1],[2]. A method for doing this based on the hydrolysis of metal alkoxides demonstrated that it had several potential advantages over the glass frit method. These include: (a) intimate mixing of the glass forming components with waste particles to achieve more rapid melting into a more homogeneous, defect-free waste glass, and (b) a feed slurry which is non-abrasive, resistant to segregation, and readily transportable by air lifting.

Success of the reactive mixing technique depends on producing a suspension of glass formers that will readily blend with neutralized waste sludge to form a weak coagulum. Bench scale experience was that weak promoters in dilute solutions would hydrolyze a boron/silicon alkoxide sufficiently to yield a suspension of borosiloxane that would produce a weak coagulum when blended with neutralized waste sludge. Thus, Na(NO₃) or NaCl gave a better dispersion of glass formers than NaOH. However, compositions based on either nitrates or hydroxides made better glasses than did chlorides. For these reasons a sodium and lithium nitrate glass composition was selected for the first continuous liquid fed melting trials at PNL. When this glass additive composition was mixed with neutralized Purex waste sludge, the result was a stable, readily air liftable feed slurry which could be concentrated up to ∼200 g/l while retaining adequate flow over the surface of the melt. However, it proved to be an unsatisfactory liquid melter feed because of the formation of an insulating bubble layer at the melt/cold cap interface, apparently due to the decomposition of the nitrates. To overcome this problem, the glass additive composition was altered by replacing the sodium nitrate with sodium silicate and the lithium nitrate with lithium hydroxide. This new composition formed a stable, readily air liftable feed slurry which melted without foaming. However, at oxide loadings above 150 g/l, it would become as stiff as toothpaste. The only other change from the

---

†This work was supported by West Valley Nuclear Services Company under DoE Contract #DEAC-0781NE44139.

first feed slurry was that the sludge was a blend of neutralized Purex and acid Thorex waste.

The purpose of studying the rheology of this system was to establish the cause of strong coagulation, and to discover a means for eliminating it at oxide loading levels that are comparable to glass frit systems.

EXPERIMENTAL PROGRAM

Materials

Borosilicate glass compositions were prepared from either a borosiloxane copolymer obtained from Stauffer Chemical Company or a mixture of tetraethylorthosilicate (TEOS) and triethylborate (TEB). Both materials give indistinguishable suspensions of glass former when prepared as follows: (1) Add 36.5g LiOH to 242g $Na_2SiO_3 \cdot 5H_2O$, followed by 8g MgO, 4g TiO_2, 2g ZrO_2, and 2g La_2O_3 to 6000 ml H_2O; (2) Slowly blend in 400g borosiloxane copolymer* and 451g TEOS; and (3) Heat and stir mixture at $\approx 90°C$ to reduce volume to 2200 to 3100 ml. The oxide equivalent of this FEB glass former composition is given in Table I. Glass former suspensions have viscosities of between 1 and 2 cp, pH's of 11.1 to 11.6, and a fraction solids, after calcining at 700°C, of between 0.11 and 0.16. Upon standing, the glass former separates into a translucent supernatant and a fine precipitate.

TABLE I
FEB Glass Former Composition

OXIDE	w/o	OXIDE	w/o
SiO_2	63.9	MgO	1.7
B_2O_3	12.6	TiO_2	0.9
Na_2O	15.2	ZrO_2	0.4
Li_2O	4.9	La_2O_3	0.4

A typical simulated, washed Purex/Thorex waste sludge composition is given in Table II. This slurry separates on standing into a brown solid fraction and a green colored supernatant. A typical simulated, washed Purex waste sludge is given in Table III. Purex sludge separates in a brown precipitate and a clear supernatant. In general, Purex/Thorex wastes are more acidic (pH 3.1-3.6) than Purex waste (4.2-4.5). The fraction solids in both wastes is between 0.11 and 0.16.

Experimental Procedures

Rheological measurements of glass former and waste slurries and their mixtures were determined with digital Brookfield viscometers, models LVTD and RVT. Viscosity measurements of coagula utilized bar type spindles in conjunction with the Brookfield "Helipath" stand. The viscosity of glass former slurries as a function of electrolyte addition was monitored with a Brookfield UL adapter. Test solutions were prepared by rapidly stirring a 15 ml charge of glass former in a beaker while a small aliquot (≤ 0.5 ml) of electrolyte was slowly pipetted in over a period of several minutes. This procedure was necessary since slow stirring and/or rapid additions of electrolyte invariably led to the formation of large globules or buttons of reaction product and erratic non-reproducible viscosity data. The well stirred test solution was then poured into the UL adapter and steady-state readings

*Stauffer Chemical Company, 455-9-81E, 14.65 w/o B_2O_3, 24.69 w/o SiO_2.

were taken at 50 rpm; a speed that insured against phase separation. Next, the solution was poured back into a beaker, a new aliquot of electrolyte was stirred in and the entire procedure was repeated. Measurements of pH were made with a Broadley-James pH electrode and a Horizon Ecology Company model 5996 pH meter. The reproducibility of pH values for difficult to measure heavy slurries and coagula were no better than ±0.5.

TABLE II
Composition of Simulated
Purex/Thorex Waste Sludge

TABLE III
Composition of Simulated
Purex Waste Sludge

COMPOUND	CONC., g/l	COMPOUND	CONC., g/l
$Fe(OH)_3$	73.7	$Fe(OH)_3$	53.7
$FePO_4 \cdot 2H_2O$	48.7	$FePO_4 \cdot 2H_2O$	38.5
$Re(NO_3)_3^{(•)}$	47.1	$Cr(OH)_3$	4.30
$NaNO_3$	27.5	$NiO \cdot H_2O$	2.20
$Cr(OH)_3$	7.10	$Al(OH)_3$	1.00
$MnCO_3$	4.16	R.E. N Mix	4.10
$Sr(NO_3)_2$	2.06	MnO_2	2.40
$CaNO_3$	1.44	Na_2SO_4	2.30
$Zr(NO_3)_4$	3.70	$Nd(NO_3)_3 \cdot 6H_2O$	10.0
$Ni(OH)_2$	2.93	$NaNO_3$	18.1
$Al(OH)_3$	3.00	$NaOH$	0.30
$Na_2MoO_4 \cdot 2H_2O$	2.52	$NaCl$	0.01
KNO_3	0.50	$CsNO_3$	0.62
Na_2SO_4	3.00	$Sr(NO_3)_2$	0.34
		$Na_2MoO_4 \cdot 2H_2O$	1.70
		NaI	0.20

(•)Rare Earth Nitrate Mix

RESULTS AND DISCUSSION

Coagulum Formation

Simulated Purex/Thorex waste and a sodium silicate containing glass former will set up within a few minutes after mixing into a rigid mass that is strong enough to support its own weight. On the other hand, simulated Purex waste results in only a weak coagulum. Repeating these experiments using only the supernatants from gravity settled glass former and waste sludge gives similar results showing that the electrolyte from the waste and the soluble fraction from the glass formers are responsible for the coagulation reaction. The principal difference between these simulated Purex/Thorex and Purex waste sludges is the large contribution of soluble polyvalent cations introduced by the Thorex waste (compare Tables II and III).

The effect of polyvalent cations on the coagulation of soluble silicates was determined by monitoring the viscosity and pH of a sodium silicate solution. Several sodium silicate solutions in the concentration range of the Feb glass were examined. These included 7.4 w/o, and 9.1 w/o $Na_2SiO_3 \cdot 5H_2O$, and 7.4 w/o $Na_2SiO_3 \cdot 5H_2O$ + 1.1 w/o LiOH aqueous solutions. The pH of these solutions was >12.6. Figure 1 shows the effect of 1.2M $Fe(NO_3)_3 \cdot 9H_2O$ additions on the viscosity of these solutions. The first few aliquots of electrolyte causes globules of ferric silicate and ferric hydroxide to precipitate [3]. The gelatinous nature of these solids gradually increases the viscosity of the solution. Above a critical electrolyte concentration, however, the viscosity increases abruptly (Curve B). This abrupt increase in viscosity is believed to be due to the precipitation and gelling of silicic acid that occurs when the pH decreases to 10.8, the solubility limit. A lowering of the pH is due mainly to the hydrolysis of water according to the reaction:

$$Fe^{3+} + nH_2O \Leftrightarrow Fe(OH)_3 + nH^+$$

The hydrogen ion effect is further demonstrated by curve D. This solution contains the same amount of sodium silicate as before plus 1.1 w/o LiOH. Because of the increased hydroxide content, more ferric nitrate is required to induce precipitation of silicic acid. However, just as before the onset of gelling occurs at a pH of 10.8. A higher sodium silicate concentration further increases the viscosity of the solution and shifts the peak to the right because silicate content is higher and the solution is more basic (compare curves B and C). A similar electrolyte/viscosity dependence is observed when electrolyte is added to a Feb glass former composition (curve A) except that a much stiffer coagulum develops. This result suggests that an additional component such as colloidal silica is contributing along with soluble silicates to the coagulation phenomenon.

Values of electrolyte concentration needed to induce coagulation for various electrolytes are given in Table IV. With the exception of Na^+ which produces a rather weak coagulum at pH 11.1, all of the other cations produced stiff coagula at a pH of 10.8. The concentration of rare earth ions alone in the Purex/Thorex waste approaches their threshold limits in agreement with the strong propensity of this waste to form a strong coagulum. On the other hand, Purex waste, which has only one fourth of the rare earth ion concentration, gives only a weak coagulum.

TABLE IV

Electrolyte Concentrations to Coagulate A Typical Glass Former Slurry[+]

Electrolyte	Moles/liter
$ZrO(NO_3)_2 \cdot 6H_2O$	0.049
$Cr(NO_3)_3 \cdot 9H_2O$	0.021
$Al(NO_3)_3 \cdot 9H_2O$	0.052
$Fe(NO_3)_3 \cdot 9H_2O$	0.059
$Nd(NO_3)_3 \cdot 6H_2O$	0.082
$Ce(NO_3)_3 \cdot 6H_2O$	0.082
$Ni(NO_3)_2 \cdot 6H_2O$	0.123
$NaNO_3$	0.368
HNO_3	0.140

[+] Initial pH of glass former = 11.1.

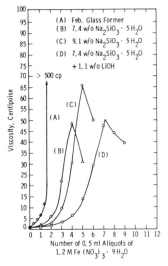

Fig. 1 — Dependence of viscosity on $Fe(NO_3)_3 \cdot 9H_2O$ additions of FEB glass former and sodium silicate solutions (15 ml charge).

Coagulum Breakage

When sufficient caustic is added to raise the pH above 12.5, a coagulum will exhibit a dramatic decrease in viscosity and within one hour it will break down into a tractable, soft, redispersible solid and light brown supernatant.[†] Below pH 12.5, there is a gradual decrease in coagulum strength as caustic content increases, but very little tendency for coagulum breakage. Essentially similar results are obtained when a 25 w/o solution of $(CH_3)_4NOH$ is substituted for NaOH. Strong base in large amounts can break the coagulum by redissolving or depolymerizing the precipitated silicic acid. Sufficient base may also dissolve some of the insoluble metal silicates as suggested by the high coloration of the supernatant following these additions.

Small additions of strong acids proved to be an ineffective means of decreasing coagulum strength. However, when large quantities of either hydrofluoric or formic acid are used to decrease the pH to the 3–3.5 range, a marked decrease in coagulum viscosity followed by coagulum breakage is observed. Unfortunately, within 5 to 24 hours after coagulum breakage, the supernatant will gel firmly. In general, formic acid tends to inhibit regelation longer than hydrofluoric acid.

Coagulum Prevention

Experimentation with various combinations of components suggested that mixing itself plays an important role in coagulation. Mixing experiments were run in which glass former was slowly added in either small increments or in a slow continuous manner to waste sludge while stirring. This procedure completely eliminated coagulation and actually produced slurries that were more fluid than the waste sludge itself. This is illustrated in Figure 2 by the viscosity of an optimally mixed feed slurry compared to that of waste sludge.

Both the degree and order of mixing are important: When either the glass former is rapidly poured into the waste, allowed to set for several minutes, then mixed, or when waste is added slowly to the glass former, a strong coagulum results. The viscosity dependence typical of these coagula is given in Figure 3. These viscosities are between 50 and 100 times greater than those of an optimally mixed slurry even though the two compositions are identical.

Tests with a variety of whole and supernatant surrogate waste fractions demonstrated that coagulation could be avoided by using a mixing sequence that prevents the formation of a continuous glass former phase. However, waste supernatants produced feed slurries with consistently higher viscosities than did their corresponding "whole" wastes even when the volume of waste to volume of glass former was held constant. This result suggests that solids may have a role in weakening coagula possibly by further helping to prevent recombination of the dispersed glass former globules.

Oxide Loading

With constant stirring, a concentrated glass former (with $f_s = 0.3$) was slowly added to waste ($f_s = 0.16$) to produce a coagulum-free feed slurry with a total oxide loading of 300 g/l. This loading is twice that previously achieved with alkoxide glass formers and approaches that of a glass frit system. It further demonstrates the general applicability of the process of intimately dispersing the glass forming additive in the waste on inhibiting coagulation. This concentrated feed slurry was melted at 1050°C into a homogeneous glass with no evidence of extraneous phases.

[†]An undesirable aspect of using caustic is that the final sodium content of the glass is raised above 18%.

168

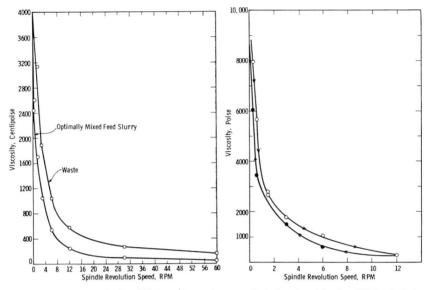

Fig. 2 — Dependence of viscosity on RPM of purex/thorex waste and optimally mixed feed slurry

Fig. 3 — Dependence of viscosity on RPM of poorly mixed coagulated feed slurry

ACKNOWLEDGMENTS

Thanks are due to Stauffer Chemical Company for supplying the borosiloxane copolymer and G. Mellinger of BPNL for supplying simulated waste sludge.

REFERENCES

[1]   J. M. Pope and D. E. Harrison, **Advanced Method for Making Vitreous Waste Forms** in Proc. Symp. on 'Science Underlying Radioactive Waste Management,' Materials Research Society Annual Meeting, Boston, MA, November 17–20, 1980.

[2]   J. M. Pope and D. E. Harrison, **Alkoxide Derived Vitreous Forms in Proc. Symp. on High Temperature Chemistry,** Electrochem. Soc., October 11–16, 1981.

[3]   R. K. Iler, **The Chemistry of Silica,** John Wiley and Sons, New York, 1979.

FERROELECTRIC CERAMICS--THE SOL-GEL METHOD VERSUS CONVENTIONAL PROCESSING

EDWARD WU, K.C. CHEN and J.D. MACKENZIE
Department of Materials Science and Engineering, University of California,
Los Angeles, CA 90024.

ABSTRACT

The sol-gel method for the preparation of ceramics and glasses has frequently been mentioned as more advantageous as compared to conventional methods. However, there are few known examples of a direct comparison for the same material. In the present work both the processing and resultant properties for ferroelectrics such as $BaTiO_3$, $KTaO_3$, $KNbO_3$ and $K(Ta,Nb)O_3$ made by both methods are directly compared. The uniformity is evaluated by high-angle x-ray diffraction, electron microscopy and EDAX and the dielectric properties are compared. The advantages and disadvantages of the sol-gel method are discussed.

INTRODUCTION

In the past few years, the so-called "sol-gel" methods have been actively studied as possibly superior routes for the preparation of glass and ceramics. However, there is no known publication in which a direct comparison has been made for ferroelectric ceramics made by sol-gel methods and the same material made by conventional ceramic powder methods. This paper is concerned with the direct comparison of some ferroelectric ceramics by these two different types of techniques. In the sol-gel methods, the starting raw materials are solutions which can be intimately mixed. Ultrafine solids can be caused to precipitate from such liquid solutions [1-5]. The fine powder is subsequently sintered or hot-pressed. In contrast, the sol-gel method involves first the formation of a bulk amorphous gel which is then crystallized [6-10]. A comparison of these gel methods and the conventional approach is shown in Fig. 1. The present study is confined to the sol-gel method III of Fig. 1.

EXPERIMENTAL

$BaTiO_3$, $KNbO_3$, $KTaO_3$ and $K(Ta,Nb)O_3$ ceramics were prepared according to the sol-gel method III of Fig. 3. The starting raw materials were $K(OEt)$, $Ta_2(OEt)_{10}$, $Nb_2(OEt)_{10}$, $Ba(OPr)_2$ and $Ti(OPr)_4$. With the exception of $Ba(OPr)_2$ which was prepared in our laboratory, all raw materials were obtained from Alfa Chemicals, Inc. Gelation was carried out at room temperature. Crystallization of the bulk gels was done at $550^\circ-700^\circ C$. Sintering was carried out at $1100^\circ-1350^\circ C$.

DISCUSSIONS

1. The Sol-Gel Process and Its Advantages

The advantages of the sol-gel process, compared to conventional methods for polycrystalline ceramics are: (a) better homogeneity, (b) better distribution of dopant, (c) better control of stoichiometry because of lower reaction temperatures, (d) less contamination, (e) formation of ultrafine powder for subsequent sintering, and (f) ease of preparation of thin films. In Fig. 2 it is seen that for the formation of $BaTiO_3$, the conventional method uses $BaCO_3$ and TiO_2 powders of micron sizes. These react at elevated temperatures and the formation of crystalline $BaTiO_3$ necessitates the interdiffusion of the relevant ions. The bottom part of Fig. 2 shows the unit cell of $BaTiO_3$ in which the separation distances between Ba and Ti ions are of the order of a few angstroms. The difficulty of attaining homogeneity is easily appreciated. If one were to mix

I. CONVENTIONAL METHOD:

II. SOL-PRECIPITATE METHOD:

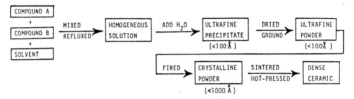

III.SOL-GEL METHOD:

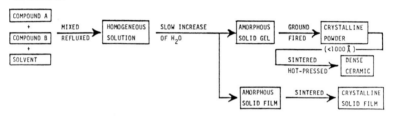

FIG. 1 Comparison of conventional method, sol-precipitate method and sol-gel method.

$Ba(OPr)_2$ and $Ti(OPr)_4$ liquids in propanol, however, the average separation distance between in Ba and Ti is much smaller than the 10,000 Å for $BaCO_3$ and TiO_2. The formation of $BaTiO_3$ is thus much easier. Table 1 shows that gels derived from $Ba(OPr)_2$ alone would crystallize in air at 300°C to give $BaCO_3$. (all heat-treatment time is 2 hr.)

Table 1 Crystallization of $BaCO_3$ TiO_2 and $BaTiO_3$ from sol-gel solutions.

Heat-treatment temperature	Crystalline phases		
	$Ba(OPr)_2$	$Ti(OPr)_4$	$Ba(OPr)_2 \cdot Ti(OPr)_4$
200°C	—	—	—
300°C	$BaCO_3$	TiO_2 (Anatase)	—
400°C	$BaCO_3$	TiO_2	trace $BaCO_3$
500°C	$BaCO_3$	TiO_2	trace $BaCO_3$
550°C	$BaCO_3$	TiO_2	$BaTiO_3$
600°C	$BaCO_3$	TiO_2	$BaTiO_3$

For Ti(OPr)$_4$ alone, TiO$_2$ would crystallize also at 300°C. However, for gels made from equal molar ratio of the two alkoxides, no crystallization was observed at 300°C and only traces of BaCO$_3$ were formed at 400°C. BaTiO$_3$ was formed readily at 550°C. The superior mixing from the use of liquid solutions is therefore evident. The infrared absorption spectra of liquid mixtures of K(OEt) + Ta$_2$(OEt)$_{10}$, K(OEt) + Nb$_2$(OEt)$_{10}$ and K(OEt) + Ta$_2$(OEt)$_{10}$ + Nb$_2$(OEt)$_{10}$ are significantly different from those of the individual components and suggests the formation of complexes such as K[Ta(OEt)$_6$] in solution. Again, intimate mixing can be achieved [11].

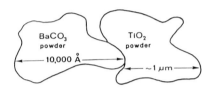

REACTION OF FINE POWDERS TO FORM CERAMIC BARIUM TITANATE

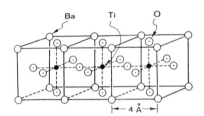

CRYSTAL UNIT CELLS OF BARIUM TITANATE

FIG. 2   Simple depiction of the difficulty of forming BaTiO$_3$ crystals from micron size BaCO$_3$ and TiO$_2$.

2.   Properties of Gel-Derived Ceramics

K(Ta,Nb)O$_3$ can be prepared by the sol-precipitate method or by the sol-gel method. We have found the sol-precipitate method to be difficult to control in that the chemical composition of the precipitate always deviated significantly from the intended composition. The conventional ceramic powder process was also difficult. Figure 3. shows that KNbO$_3$ and KTaO$_3$ form a continuous series of solid solutions. If mixing is not sufficiently homogeneous, then the resultant ceramic made can be highly inhomogeneous. Different grains can now have different Nb:Ta ratios. No longer can there be a single Curie temperature. The usefulness of KTN as a pyroelectric material is thus decreased. Figure 4 shows the high angle x-ray diffraction patterns of three KTN solid solutions with different Nb:Ta ratios. The gel-derived powder was fired at 700°C for three hours. The homogeneity of each sample is clearly seen. The same three intended solid solutions made by the ordinary ceramic powder process all gave only one broad band in this high angle region. The relative inhomogeneity of these ceramics was confirmed by EDAX studies.

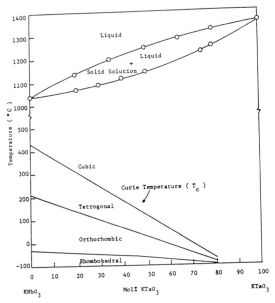

FIG. 3 The $KNbO_3$-$KTaO_3$ system and the Curie temperatures of the solid solution [12,13].

Figure 5 shows a comparison of the dielectric constants of various samples of $BaTiO_3$ as a function of temperature. The single crystal [14] and our sol-gel-derived $BaTiO_3$ ceramics fired at $1350^{\circ}C$ for two hours are seen to give very sharp peaks at the Curie temperature. The two samples made by conventional ceramic methods gave broad peaks at slightly lower Curie temperatures [15].

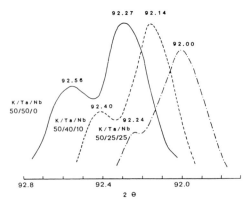

FIG. 4 Cu-Kα x-ray diffraction patterns of cubic KTN at (321) plane.

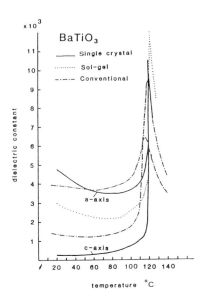

FIG. 5 Comparison of dielectric constants of various samples of BaTiO$_3$.

We are grateful for the partial support of this research by the Directorate of Chemical and Atmospheric Sciences, Air Force Office of Scientific Research under Contract No. P49620-83-K-0003 and by the "MICRO" Program of the University of California with joint sponsorship of Hughes Aircraft Company.

REFERENCES

1. K.S. Mazdiyasni, R.T. Dolloff and J.S. Smith, J. Am. Ceram. Soc. <u>52</u>, 523-526 (1969).

2. H.C. Graham, N.M. Tallan and K.S. Mazdiyasni, J. Am. Ceram. Soc. <u>54</u>, 548-553 (1971).

3. K.S. Mazdiyasni, L.M. Brown, J. Am. Ceram. Soc. <u>54</u>, 539-543 (1971).

4. K. Kiss, J. Magder, M.S. Vukasovich and R.J. Lockhart, J. Am. Ceram. Soc. <u>49</u>, 291-295 (1966).

5. L.M. Brown and K.S. Mazdiyasni, J. Am. Ceram. Soc. <u>55</u>, 541-544 (1972).

6. B.E. Yoldas, Bull. Am. Ceram. Soc., <u>54</u>, 286 (1975).

7. P.F. Becher, J.H. Sommers, B.A. Bender and B.A. MacFarlane in Materials Sci. Research, <u>11</u>, Processing of Crystalline Ceramics, H. Palmour III et al., ed. (Plenum Press 1978) pp. 79-86.

8. H. Schroder, in Physics of Thin Films <u>5</u>, G. Hass and R.E. Thun Eds. (Academic Press 1978) pp. 87-141.

9.   J. Fukushima, K. Kodaira and T. Matsushita, Bull. Am. Ceram. Soc. 55, 1064 (1976).

10.   H. Dislich, J. Non-Cryst. Solids, 48, 11 (1982).

11.   E. Tao-Hung Wu, Ph.D. thesis, University of California, Los Angeles (1983).

12.   P.D. Garn and S.S. Flaschen, Anal. Chem., 29, 275 (1975).

13.   S. Triebwasser, Phys. Rev., 114 63-70 (1959).

14.   W.J. Merz, Phys. Rev. 76 1221-1225 (1949).

15.   G.H. Jonker and W. Noorlander in Science of Ceramics 1, 255-265, G.H. Stewart Ed. (Academic Press, 1962) pp. 255-264.

CHEMICAL SYNTHESIS OF CERAMIC POWDERS

CHEMICAL SYNTHESIS OF SINGLE AND MIXED PHASE OXIDE CERAMICS

K. S. MAZDIYASNI
Air Force Wright Aeronautical Laboratories
Wright-Patterson Air Force Base, Ohio 45433 USA

ABSTRACT

Thermal and hydrolytic decomposition of metal alkoxides, $M(OR)_n$, is employed to obtain ultra-high-purity submicron-size < 50 nm refractory, ferroelectric, and electro-optic oxide powders. The high surface activity associated with these powders makes possible relatively low temperature processing of the powder compacts to near theoretical density with uniform fine-grain-size bodies. Transmission electron microscopy, high-temperature X-ray, and infrared and Raman spectroscopy are used to observe nucleation, growth, crystallite morphology of the synthesized powder, and microstructural features.

INTRODUCTION

High-purity submicron-size particulate refractory boride, carbide, nitride, and oxide powders have received much attention during the past two decades. In part, this intensive effort has been prompted by new applications for these compounds in the synthesis of high-density refractory, nuclear, electronic, laser, and optical bodies, thin-film electronic components, fibers, fillers for ablative and other composite structures, catalysts and catalyst carriers, phosphors, propellant additives, and adsorbent materials. Also, advances in technology, instrumentation, and data interpretation have contributed much to the understanding of fine-particulate technology as a whole.

In the classical method of preparing fine-particulate oxides, salts such as the oxalates, acetates, and carbonates are thermally decomposed to the oxides [1]. Hydroxides have also been commonly employed [2]. The hydrolysis of compounds such as metal chlorides is another convenient method which has received recent study in the vapor-phase reaction of metal halides with steam [3]. Relatively novel methods such as freeze-drying, the plasma arc with induction coupling [4], and the Vitro electric arc process [5] have been used because of the current interest in high-purity particulates. While each approach has definite merits, none of the aforementioned methods has achieved consistent results in producing refractory oxides in both high purity and submicron powder size. The newer techniques do not achieve higher purities than classical thermal decomposition and precipitation methods, although particle size may consistently be 100 nm or less.

Specifically selected "tailor-made" metal organic compounds can be made which will thermally or hydrolytically decompose to form familiar inorganic nonmetallic materials such as the refractory, ferroelectric, piezoelectric, and electro-optic oxides [6]. This approach to high-purity fine particulates, 3-150 nm, offers almost unlimited flexibility because of the wide variety of organic compounds that have become available in recent years. Unique processes described here invariably result in ceramic powders and bodies with improved thermophysical and thermomechanical properties over conventional ceramics.

Mat. Res. Soc. Symp. Proc. Vol. 32 (1984) Published by Elsevier Science Publishing Co., Inc.

EXPERIMENTAL PROCEDURE

Ammonia Method

The most economic and simple method for the large-scale production of alkoxides involves the addition of commercial-grade anhydrous metal halide to a mixture of 10% anhydrous alcohol in a diluent (benzene or toluene) in the presence of anhydrous ammonia [7-9].

$$MCl_4 + 4ROH + 4NH_3 \xrightarrow[5°C]{C_6H_6} M(OR)_4 + 4NH_4Cl$$

$$M = Ti, Zr, Hf \text{ and } R = i-C_3H_7$$

(1)

The group IV B metal tetrakis isopropoxide is readily purified by fractional distillation or recrystallization. Because the removal of NH_4Cl by filtration is always cumbersome and time-consuming, the ammonia method may be carried out in the presence of an amide or nitrile. In this particular method the metal alkoxide separates out as the upper layer, while the ammonium chloride remains in solution in the amide or nitrile in the lower layer; thus, the filtration step is eliminated [9-10].

An improved method for the preparation of titanium alkoxides, based on a recent report for the preparation of Zr and Hf analogs, has been suggested by Anand, et al. [11]. This procedure consists of treating anhydrous metal halide saturated in benzene with HCl and various esters such as ethyl formate, ethyl acetate, diethyl oxalate, n-propyl acetate, isopropyl acetate, n-butyl acetate, and isobutyl acetate in the presence of anhydrous ammonia. The alcohol formed in the reaction is itself consumed in the formation of metal alkoxides.

$$MCl_4 + 4HCl + 4CH_3COOR + 4NH_3 \longrightarrow M(OR)_4 + 4CH_3COCl + 4NH_4Cl \tag{2}$$

Ester Exchange Reaction

A valuable method for converting one alkoxide to another is an ester exchange reaction [12]. The method is particularly suited for the preparation of the tertiary butoxide from the isopropoxide and t-butyl acetate. The reaction is as follows:

$$M\left[O-CH(CH_3)_2\right]_4 + 4CH_3COOC(CH_3)_3 \longrightarrow M\left[O-C(CH_3)_3\right]_4 + 4CH_3COOCH(CH_3)_2 \tag{3}$$

Since there is a large difference between the boiling points of the esters, the fractionation is simple and quite rapid. Another advantage appears to be the lower rate of oxidation of the esters as compared to that of alcohol.

Alcoholysis Reaction

Substitution of other branched R groups with the lower straight chain alcohols has been carried out in an alcohol interchange reaction as shown in Eq. (4).

$$M(OR)_4 + 4R'OH \longrightarrow M(OR')_4 + 4ROH \tag{4}$$

The distillation of azeotrope drives the reaction to completion. The interchange becomes slow in the final stages with highly branched alcohols, probably owing to steric hindrance. The rise in temperature to the higher boiling alcohol, however, reliably signals the end of the reactions. In this method of preparation, major roles are played by speed, reversibility, and absence of side reactions.

YTTRIUM AND LANTHANIDE ALKOXIDES

The electronegativity of yttrium and lanthanides places these elements between metals such as aluminum which forms covalent alkoxides and sodium which forms ionic alkoxides. The degree of ionic character of the M-O bond, which is dependent on the size and electronegativity of the metal atom, is important in determining the character of the alkoxide.

The study of alkoxides of yttrium and the lanthanides has been limited due to expensive starting materials and preparation and handling difficulties. Successful characterization of most of the alkoxides is complicated by their extreme sensitivity to moisture, heat, light, and atmospheric conditions.

SYNTHESIS

The yttrium and lanthanide tris-isopropoxides are prepared by the reaction of metal turnings with excess isopropyl alcohol and a small amount of $HgCl_2$ (10^{-4} mol per mol of metal) as a catalyst [13].

$$Ln + 3C_3H_7OH \xrightarrow{HgCl_2} Ln(OC_3H_7)_3 + 3/2\ H_2 + Hg + 2HCl \tag{5}$$

After filtration, the crude product is purified by recrystallization from hot isopropyl alcohol or vacuum sublimation. Yields of 75% or better are realized with this method. For some of the larger metal ions (lanthanum through neodymium), the reaction rate and percentage yield are increased by using a mixture of $HgCl_2$ and $Hg(C_2H_3O_2)$ or HgI_2 [14] for the catalyst.

Substitution of other R groups for the isopropoxy groups is accomplished by the alcohol interchange technique, similar to a procedure described previously.

ALUMINUM ALKOXIDE

Similarly, aluminum tris-isopropoxide is synthesized by the method of Adkins and Cox [15].

$$Al + 3C_3H_7OH \xrightarrow[\text{exothermic}]{HgCl_2} Al(OC_3H_7)_3 + Hg + 2HCl + 1/2\ H_2 \tag{6}$$

The alkali or alkali earth metal alkoxides are prepared by the reaction of metal with alcohol. The reaction is highly exothermic with evolution of excess hydrogen.

$$M + ROH \xrightarrow[\text{exothermic}]{} M(OR)_n + H_2 \tag{7}$$

where M is the metal, R is the isopropyl radical, and n = 1 or 2.

PYROLYSIS MECHANISM

When zirconium tertiary butoxide is pyrolyzed, the mechanism of reaction is one in which Olefin and alcohol are split in successive steps

$$Zr(OR)_4 \longrightarrow ZrO_2 + 2ROH + Olefin \tag{8}$$

and the suggested mechanism is as follows:

$$Zr(OC_4H_9)_4 \longrightarrow Zr(OC_4H_9)_3OH + CH_3-\underset{\underset{CH_3}{|}}{C} == CH_2 \qquad (9)$$

$$Zr(OC_4H_9)_3OH \longrightarrow ZrO_2 + 2C_4H_9OH + CH_3-\underset{\underset{CH_3}{|}}{C} == CH_2 \qquad (10)$$

After the initial decomposition steps, the reaction is very fast; therefore, no intermediates found are isolated in the decomposition. The products found are Olefin, alcohol, and oxide. The rate of reaction is dependent on the concentration of the starting material, and the rate equation is given as a function of the concentration.

$$V = kK_1 [Zr(OR)_4] \qquad (11)$$

A representative reaction is the decomposition of zirconium tetratertiary butoxide to zirconium oxide. Vaporization of the alkoxide is carried out at 190-210°C at 760 mm Hg and decomposition at 325-500°C:

$$Zr(OC_4H_9)_4 \longrightarrow ZrO_2 + 2C_4H_9OH + 2C_4H_8 \qquad (12)$$

The particle size of this powder was determined by Mazdiyasni, et al. [16], from an alcohol dispersion of the powder sprayed onto a carbon substrate on a copper mesh screen and viewed in an electron microscope. The results are shown in Fig. 1.

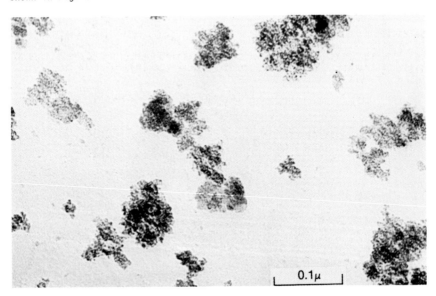

FIG. 1. Electron micrograph of as-prepared dispersed zirconia powder.

The average particle size was found to be 3-5 nm, with evidence that the larger particles were agglomerates of individual particles of 2 nm or less. The particle size range was generally $\sim$ 2 nm, with only an occasional large particle of 0.1 μ to 0.5 μ (100 to 500 nm). It is likely that these particles are also agglomerates. The powder as formed was amorphous to X-rays, but electron diffraction patterns showed strong lines for cubic zirconia. The high purity (> 99.95%) of the powder is indicated in Table I. The table lists purities from the starting tetrachloride and its subsequent conversion to the isopropoxide $Zr(OC_3H_7)_4$, to the butoxide $Zr(OC_4H_9)_4$, and finally to the oxide. Three batches are given in separate analyses for the oxide, demonstrating the reliability of high-purity levels on a continuing basis. Even the hafnium content is sharply decreased as an impurity in this process. However, if isopropoxide or, for this matter, t-butoxide is used in a closed system, one can have the following mechanism. In this instance the rate-controlling step may be the dehydration of alcohol. A molecule of H_2O produced from one molecule of alcohol would then regenerate two molecules of alcohol by hydrolysis of zirconium alkoxide, and hence the chain reaction would be set up in accordance with the following equations. In this case an intermediate product of hydrolysis--zirconium oxide alkoxide--will proceed under rapid disproportionation.

$$Zr(OR)_4 \xrightarrow{K_1} ZrO_2 + 2ROH + R = R \qquad (13)$$

$$ROH \xrightarrow{K_2} R = R + H_2O \qquad (14)$$

$$2 \rightharpoondown Zr - OR + HOH \longrightarrow \rightharpoondown Zr - O - Zr \leftharpoondown + 2ROH \qquad (15)$$

$$2ZrO(OR)_2 \longrightarrow ZrO_2 + Zr(OR)_4 \qquad (16)$$

This mechanism is used to produce very large quantities of oxide by the hydrolytic decomposition technique.

The basic requirement for the alkoxides is that the hydrolysis reaction be rapid and quantitative. Where the alkoxy (-OR) group may be quite stable to hydrolysis, a small amount of a mineral or organic acid may be added to catalyze the reaction. The oxides are, at this point, in a finely divided state of submicron (maximum agglomerate size 0.1 μm) particle size and of extremely high purity.

TABLE I. Emission spectrographic analyses of $ZrCl_4$ to $Zr(OR)_4$ to $Zr(OR')_4$ to ZrO_2.

Elements	$ZrCl_4$ (ppm)	$Zr(OR)_4$ (ppm)	$Zr(OR')_4$ (ppm)	ZrO_2 (ppm)	ZrO_2 (ppm)	ZrO_2 (ppm)
Al	500	180	5	ND < 5	ND < 5	ND < 5
Ca	80	70	ND < 5	ND < 5	ND < 5	ND < 5
Fe	500	160	20	ND < 10	ND < 10	ND < 10
Hf	< 10,000	5,000	1000	500	< 200	< 200
Mg	25	200	ND < 10	ND < 10	ND < 10	10
Mo	50	ND < 10	ND < 10	ND < 5	ND < 5	ND < 5
Na	500	500	50	< 100	< 100	< 100
Si	1900	700	50	300	300	300
Sn	100	ND < 10	ND < 10	ND < 10	ND < 10	ND < 10
Ti	100	20	5	5	5	5
W	< 50	< 50	< 50	< 50	< 50	< 50
Cu	200	20	< 10	< 10	< 10	< 10

ND < = Not detected, less than.

PURITY

The values shown in Table I are taken from results of emission spectroscopic analysis of the starting material to the final oxide product. The increased purity obtained through the successive steps of preparation of the isopropoxide, the ester exchange reaction to form tertiary butoxide, and the distillation and decomposition of this product to the oxide is remarkable.

Very high purity metal chloride as a starting material should result in an even purer final product. Economic considerations, however, would make it desirable to use a technical-grade metal chloride rather than a higher purity starting material in any large scale process. As long as the alcohols and solvents are recovered at each step of the reaction, the costs of such a purification method should not be excessive. The results produce a zirconia which is far purer than any commercially available high-purity product. The most interesting aspect of oxide preparation in this manner, however, is not the very high purity but the unusually small particle size.

HIGH-TEMPERATURE X-RAY STUDIES

The X-ray diffraction pattern of Zr alkoxide hydrolyticaly decomposed to submicron particulate ZrO_2 was observed by Mazdiyasni [17] over the temperature range 250 to 1200°C upon heating using CuKα radiation. The following sequence of phases was observed in the alkoxy-based powder as the temperature increased to 1200°C: amorphous, to metastable cubic, to metastable tetragonal, to tetragonal + monoclinic, to monoclinic + tetragonal, and finally to tetragonal. Calcination studies were undertaken to observe the growth of primary crystallites into larger particles and the influence upon growth of tumbling in different atmospheres such as air, oxygen, hydrogen, and vacuum. No measurable change in size and morphology of the crystallites was observed as a result of varying the atmospheric conditions during calcination. Figures 2 and 3 are typical electron micrographs showing, respectively, the effect of calcination at 500°C in air for 24 hr. without tumbling and calcination at 750°C in air for 24 hr. with tumbling. Tumbling is seen to be an effective method of controlling crystallite growth, while atmospheric conditions have a negligible effect during the pre-sintering stage of processing these powders.

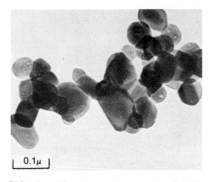

FIG. 2. Electron micrograph of ZrO_2 powder calcined at 500°C in air for 24 hr. without tumbling.

FIG. 3. Electron micrograph of ZrO_2 powder calcined at 750°C in air for 24 hr. with tumbling.

INFRARED AND RAMAN SPECTRA OF ZIRCONIA POLYMORPHS

Infrared and Raman spectroscopy may be used for the characterization of the ZrO_2 phases and as a method of non-destructive analysis. McDevitt and Baun [18] clearly distinguished the ir spectra of the monoclinic and stabilized cubic phases of ZrO_2; at the same time, they showed that monoclinic samples with the same X-ray diffraction pattern could have considerably different ir spectra. These workers suggested that ir spectra might be more sensitive than X-ray diffraction patterns to small changes in the crystal lattice of the phases of ZrO_2. This suggestion implies that ir and Raman spectroscopy may yield practical new information on destabilization or transformation toughening phenomena associated with ZrO_2, because both methods can be used non-destructively.

Phillipi and Mazdiyasni [19] used high-purity (> 99.99%) submicron alkoxy-derived metastable ZrO_2 powder as a starting material to follow the phase transformation sequence in unstabilized ZrO_2 at the lower temperatures. Commercially available monoclinic CaO and MgO-stabilized ZrO_2 were also investigated, as was a 6.5 mol% alkoxy-derived Y_2O_3-stabilized ZrO_2 "Zyttrite. "

Powder transmission was used for ir spectra. Phillipi and Mazdiyasni's [19] investigation focused on the fundamental vibration frequencies in the 900 to 200 cm^{-1} region, rather than on the weak bands found in some samples at higher frequencies.

The ir spectra of cubic ZrO_2 stabilized with Y_2O_3, CaO, and MgO were examined. No significant differences attributable to the type of concentration of the stabilizing agent were detected.

To determine the effects of thermal history, portions of alkoxy-derived pure ZrO_2 were heated for 24 hr. each in air at approximately 100°C intervals between 384 and 1078°C in a resistance furnace. Each sample in its Pt boat was withdrawn, cooled quickly to room temperature, and blended with CsI without grinding. Infrared spectra of these samples are presented in Fig. 4. An orderly progression of spectral changes with increasing temperature was observed which corresponds to the low-temperature metastable ZrO_2 polymorphic transitions. Changes in the basic character of the spectra occurred near 600 and 1000°C, in agreement with previous X-ray diffraction work described by Mazdiyasni [17]. The X-ray diffraction pattern of the sample calcined at 478°C indicated tetragonal ZrO_2 with a small amount of monoclinic, whereas the sample calcined at 766°C was pure monoclinic. Both patterns had diffuse lines in the back-reflection region, which indicated lattice strain. The spectrum of the sample calcined at 1078°C closely resembled that of monoclinic ZrO_2.

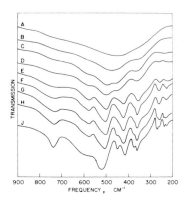

FIG. 4. Powder transmission spectra of metastable cubic ZrO_2 heated for 24 hr. at temperatures indicated: (a) as received, (b) 384°C, (c) 478°C, (d) 573°C, (e) 668°C, (f) 766°C, (g) 870°C, (h) 974°C, and (j) 1078°C. Ordinate scales displaced for clarity.

RAMAN SPECTRA

In contrast to ir spectra, in which the overall band contours provide as much information as the exact band frequencies, Raman spectra ordinarily consist of discrete lines. Accordingly, the spectral data are summarized in Table II. The Raman spectrum of a pure alkoxy-derived submicron metastable cubic ZrO_2 showed only a weak broad line at 490 ± 20 cm^{-1}. A well-annealed Y_2O_3-stabilized cubic ZrO_2 had a distinct asymmetrical line at 625 cm^{-1} and several other weak lines, whereas for a highly strained cubic sample, several of the weaker lines were better developed. Tetragonal and monoclinic zirconias had many more lines. The poorly resolved tetragonal pair around 640 and 615 cm^{-1} were well resolved in the monoclinic phase. The tetragonal and monoclinic phases have many lines in common, but two strong lines at 263 and 148 cm^{-1} are peculiar to the tetragonal phase, whereas the strong line at 348 cm^{-1} and two weaker ones are peculiar to the monoclinic phase.

The number of ir bands in the powder spectra increases from the cubic phase to the tetragonal and monoclinic phases. This observation is consistent with the predictions of group theory applied to single-crystal spectra. The ir spectrum of a material depends on its crystal structure; if it is inhomogeneous, a different spectrum is associated with each local structure. There is evidence that additional ir-active modes may be activated in inorganic spectra by mechanical damage to the surfaces of single crystals and particles. In particular, the work of Barker [20] showed that forbidden modes could be made to appear in the reflection spectrum of single-crystal corundum as a result of relaxation of certain of the selection rules by strain.

TABLE II. Raman frequencies of zirconia polymorphs (cm^{-1}).

Metastable cubic	Stabilized cubic	Tetragonal	Monoclinic
490 w,b	625 m,b	640 s	630 m
	480 w,b	615 ?	617 m
	360 w,b	561 w	559 w
	250 w,b	536 w	538 w
	150 w,b	473 s	502 w
		380 m	476 s
		332 m	382 m
		263 s	348 m
		223 w	337 m
		189 s	307 w
		179 s	223 w
		148 s	192 s
			180 s
			104 m

NOTE: w = weak, m = medium, s = strong, and b = broad.

DYNAMIC CALCINATION STUDIES

Table III shows that heat treatment of the as-formed powder (for BET studies) in a tumbling furnace resulted in marked differences in specific surface areas of the powder. The great reduction in surface area with increased temperature and time is, to a large extent, a result of the crystallite growth, the particles having grown from 1-5 nm to $\sim$ 100-150 nm in size after heat treatment at 320°C for 150 hr.

The solid-state reactions of ceramic powders depend strongly on the characteristics of the initial powder. Nucleation and crystallite growth of 1-5 nm particulates of alkoxy-based zirconium oxide are particularly illustrative of this strong dependence, as shown previously by Mazdiyasni, et al. [21]. Literature references indicate that there are at least three major possible mechanisms by which nucleation and crystal growth take place--those of Burke and Turnbull [22], Coble [23], and Anderson and Morgan [24]. Since a definitive choice of mechanism in the alkoxy-based system is not yet feasible on the basis of available data, only the experimental observations are presented.

TABLE III. Specific surface area of submicron ZrO_2.

No.	A Single Point BET		B Nelson and Eggersten BET	
	T-outgassing 1/2 hr. °C	Specific surface m^2/g	T-outgassing 24 hr. °C	Specific surface m^2/g
1	60	222	66	217
2	90	230	100	200
3	120	210	-	-
4	150	198	150	195
5	180	186	-	-
6	210	157	200	165
7	-	-	300	135.7
8	320°C/152 hr.	50	480	48.1
9	-	-	940	0.96

SYNTHESIS OF BULK CERAMIC POWDERS

Hydrolytic Decomposition: The alkoxides react rapidly with traces of water to form alkyl alkoxides

$$M(OR)_n + H_2O \longrightarrow MO(OR)_{n-2} + 2ROH \qquad (17)$$

where M is Ti, Zr, Hf, Th; R is an alkyl group, and n is the valence of the metal. The alkyl alkoxides then decompose to the oxides.

$$2MO(OR)_{n-2} \longrightarrow MO_2 + M(OR)_4 \qquad (18)$$

This hydrolysis of alkoxides has been used to prepare a wide variety of high-purity mixed oxides such as mullite, $3Al_2O_3 \cdot 2SiO_2$, and "Zyttrite," 6 mol% $Y_2O_3 \cdot ZrO_2$, and 7 mol% $Y_2O_3 \cdot HfO_2$, $BaTiO_3$, and PLZT, etc.

Fundamental understanding of the relationship between microstructural effects and the behavior of refractory, ferroelectric, or piezoelectric ceramic materials is still quite limited when compared, for example, with that for metals and metal alloys. Several problems experienced with pure oxides and mixed oxides may be inherent in the material, but other problems, just as serious and limiting, can be cited. Of particular concern is the lack of control over the many variables during processing from powder synthesis to final compact firing. Of major importance are impurities present in the starting powders or inadvertently introduced during processing and their influence on the properties of refractory or electronic materials. Although the presence of such impurities may not be a priori adverse, it is important to understand their influence on the final properties or to eliminate them completely, if necessary.

Mazdiyasni [25] has demonstrated the relative ease with which the major impurities in various oxides and mixed oxides are eliminated. Spectrographic analysis for impurities in the mixed oxides (Table IV) of the powders calcined at 500-800°C for 1/2 hr. showed no measurable impurities present in two or more representative batches of mixed alkoxides decomposed to the mixed oxides. Separate analyses for the oxides synthesized thereafter demonstrated that high purity levels were maintained on a continuing basis. None of the unlisted impurities were higher in concentration that those shown in Table IV.

PRESSURE-GREEN DENSITY RELATIONSHIPS:

Rhodes and Haag [26], Mazdiyasni, et al. [8], and Hoch and Nair [27] have observed that in practice, no advantage is gained in the final density of a green compact made of ultrafine ceramic powders by exceeding 45000 psi.

TABLE IV.   Emission spectrographic analysis of fine particulate mixed oxides.

Elements	(Mullite)	(ppm) HfO_2 . Y_2O_3	$BaTiO_3$	PLZT
Na	< 300	ND < 3	ND < 10	500
Si	Major	< 10	100	20
Ba	ND < 1	ND < 1	Major	ND < 5
Mg	1	ND < 3	ND < 30	ND < 30
Mn	ND < 1	ND < 10	ND < 30	30
B	1	ND < 10	ND < 10	ND < 10
Sn	ND < 3	ND < 10	30	ND < 30
Pb	ND < 1	ND < 10	ND < 10	Major
Al	Major	ND < 3	ND < 10	ND < 30
Fe	ND < 5	ND < 10	ND < 100	ND < 100
Ni	ND < 5	ND < 10	ND < 5	ND < 30
Co	ND < 30	ND < 3	ND < 5	ND < 10
Cr	ND < 10	ND < 10	30	ND < 30
W	ND < 5	ND < 10	10	ND < 10
Ca	ND < 10	ND < 10	10	30
Ti	ND < 10	ND < 5	Major	Major
Zr	ND < 5	100	ND < 5	Major

NOTE: ND < = not detected, less than.

Sintered parts pressed at 10-45 ksi all reach $\sim$ 95% of theoretical density in a matter of a few minutes at several hundred degrees lower than necessary for the conventional ceramics.

LOW-TEMPERATURE SINTERING OF ALKOXY-BASED POWDERS

Yttria-Stabilized Zirconia and Hafnia

Figures 5 and 6 show the microstructures of a 6 M% Y_2O_3-stabilized ZrO_2 and a 7 M% Y_2O_3-stabilized HfO_2 body, respectively, prepared by this technique. The electron micrographs show negligible internal porosity and insignificant porosity at the grain boundaries of the materials. Both materials are fine grained, having a microstructure in which the grains are 2-5 μm across in their longest dimensions. Such a grain structure ordinarily cannot be obtained for oxides as refractory as ZrO_2 or HfO_2 without heat treatment at temperatures ranging from 1800 to 2200°C and firing times of 24 to 48 hr. or longer. This makes achievement of low porosity concomitant with maintaining a fine, uniform grain size extremely difficult. It is apparent that the mixed oxide approaches in behavior, in the solid state, reaction conditions indicative of high mobility which result in extremely rapid reactions. Aggregates which are equivalent to several molecules of the particular oxide are possibly the particles entering into reaction.

CONCLUSION AND SUMMARY

The alkoxide decomposition process for the preparation of single-phase and homogeneous mixed oxides of high purity and fine grain size has been demonstrated. The extremely high purity in excess of 99.95% can be obtained for a wide variety of oxides. A particle size range below 50 nm is easily achieved while maintaining this exceptionally high level of purity for refractory and electronic oxide ceramics. The oxide powders have extremely high activity during subsequent solid state reactions, making feasible the low-temperature firing processes in the preparation of high-density oxide ceramics. The alkoxy technique is applicable to all mixed oxides where the metal alkoxides are readily prepared or commercially available. The advantage of the process lies in the intimate mixing of the highly active powders as prepared, facilitating further processing steps while maintaining stoichiometry and purity throughout the forming process.

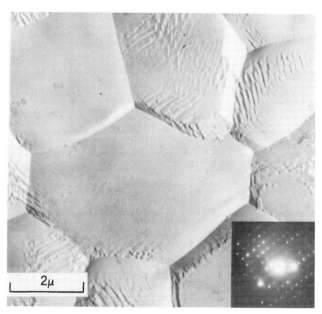

FIG. 5. Electron micrograph of cold-pressed and sintered 6 M% Y_2O_3 . ZrO_2.

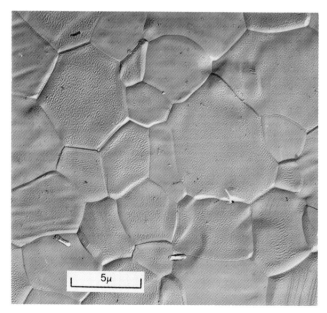

FIG. 6. Electron micrograph of cold-pressed and sintered 7 M% Y_2O_3 . HfO_2.

Thermal and hydrolytic decomposition of metal alkoxides provides a particularly unique approach to the production of thin films on various substrates, while hydrolytic decomposition provides an inexpensive, economical route toward the production of high-purity fine particulate powders on a large scale.

ACKNOWLEDGMENTS

The author would like to thank all of his co-workers for their contributions; special thanks go to Miss Sylvia R. Hatch (AFWAL/MLLM) and Mrs. H. L. Henrich (Systems Research Laboratories, Inc.) for their work in preparation of this manuscript and to Mrs. M. Whitaker (Systems Research Laboratories, Inc.) for editorial assistance.

REFERENCES

1. M. L. Nielsen, P. M. Hamilton, and R. J. Walsh in: Ultrafine Particles, W. E. Kuhn, H. Lamprey, and C. Sheer, eds. (Wiley, New York 1963) pp. 181-195.
2. R. C. Rau, J. Amer. Ceram. Soc. 47, 179 (1964).
3. L. J. White and G. J. Duffy, Inc. Eng. Chem. 51, 232-238 (1959).
4. H. J. Hedger and A. R. Hall, Powder Met. 8, 65-72 (1961).
5. C. Sheer and S. Korman in: Arcs in Inert Atmospheres and Vacuum, W. E. Kuhn, ed. (Wiley, New York 1956) p. 169-183.
6. K. S. Mazdiyasni, Ceram. Int. 8, 42-56 (1982).
7. D. C. Bradley, R. C. Mehratra, and D. P. Gaur, Metal Alkaxides (Academic Press, New York 1978) p. 411.
8. K. S. Mazdiyasni, C. T. Lynch, and J. S. Smith, J. Amer. Ceram. Soc. 50 532 (1967).
9. D. C. Bradley, R. C. Mehratra, and J. Wardlaw, Chem. Soc. (London) 1963 (1953).
10. D. F. Herman, U.S. Patents 2,654,770 (1953), 2,655,523 (1953).
11. S. K. Anand, J. J. Singh, R. K. Multani, and B. D. Jain, Israel J. Chem. 7, 171 (1969).
12. R. C. Mehratra, J. Amer. Chem. Soc. 76, 2266 (1954).
13. K. S. Mazdiyasni, C. T. Lynch, and J. S. Smith, Inorg. Chem. 5, 342 (1966).
14. L. M. Brown and K. S. Mazdiyasni, Inorg Chem. 9, 2783 (1970).
15. H. Adkins and J. Cox, J. Amer. Chem. Soc. 60, 1151 (1938).
16. K. S. Mazdiyasni, C. T. Lynch, and J. S. Smith, J. Amer. Ceram. Soc. 48 372 (1965).
17. K. S. Mazdiyasni in: Proceeding of the VI International Symposium on Reactivity of Solids, R. W. Mitchell, R. C. Devries, R. W. Roberts, and P. J. Cannon, eds. (Wiley, New York 1968) pp. 115-125.
18. N. T. McDevitt and W. K. Baun, J. Amer. Ceram. Soc. 54, 254 (1971).
19. C. M. Phillipi and K. S. Mazdiyasni, J. Amer. Ceram. 54, 254 (1971).
20. A. S. Barker, Phys. Rev. 132, 1471 (1963).
21. K. S. Mazdiyasni, C. T. Lynch, and J. S. Smith, J. Amer. Ceram. Soc. 49, 286 (1966).
22. J. E. Burke and D. Turnbull, Prog. Met. Phys. 3, 220 (1952).
23. R. Coble, J. Appl. Phys. 32, 787 (1961).
24. P. J. Anderson and P. L. Morgan, Trans Farad. Soc. 60, 930 (1964).
25. K. S. Mazdiyasni in: Proceeding of the Second International Conference on Fine Particles, W. E. Kuhn, ed. (Electrochemical Society, Princeton, NJ 1974) pp. 3-27.
26. W. H. Rhodes and R. M. Haag, AFML-TR-70-209 (Air Force Materials Laboratory, Wright-Patterson Air Force Base, OH September 1970).
27. M. Hoch and K. M. Nair, Ceramurgia Int. 2, 88 (1976).

SYNTHESIS, CHARACTERIZATION, AND PROCESSING OF MONOSIZED CERAMIC POWDERS

Bruce Fegley, Jr.[*] and Eric A. Barringer[*,**]
[*] Ceramics Processing Research Laboratory and Materials Processing Center,
[**] Dept. of Materials Science and Engineering, MIT, Cambridge, MA 02139

ABSTRACT

Controlled alkoxide hydrolysis reactions for the synthesis of monodispersed oxide powders are described. The chemical and physical properties of representative monodispersed powders of TiO_2, doped TiO_2, ZrO_2, doped ZrO_2, SiO_2, doped SiO_2, and $ZrO_2-Al_2O_3$ are described. The importance of surface chemistry for control of powder dispersion and packing is discussed and related to the control of sintered microstructures.

CONTROLLED CHEMICAL SYNTHESIS OF OXIDE POWDERS

Thermal decomposition and hydrolysis of metal alkoxides have been used to prepare a variety of high-purity oxide powders [1]. However, the major objective of this work was not to control the size distribution and shape of the resulting powders, which were generally very fine (< 100Å) and highly agglomerated. Thus, the subsequent processing of the powders to uniform fine-grained, low porosity ceramic bodies was generally not achieved and the full advantages of the alkoxide synthesis techniques for microstructure control were not realized.

Controlled hydrolysis reactions of metal alkoxides have been utilized to prepare monodispersed oxide powders of controlled size, shape, and composition, e.g., TiO_2, doped TiO_2, ZrO_2, doped ZrO_2, SiO_2, doped SiO_2, and $ZrO_2-Al_2O_3$. The synthetic methods used to prepare these materials involve fairly simple solution chemistry, but give a high degree of control and reproducibility. Basically, a dilute solution (~ 0.2 to 0.4 M) of the respective metal alkoxide in 200 proof anhydrous ethanol is hydrolyzed by adding an equal volume of a solution of deionized water in anhydrous ethanol. The second solution is poured into the first solution with stirring. The hydrolysis reactions are conducted in a glove box under a dry N_2 atmosphere and generally at room temperature (~ 25°C); precipitation of a white powder occurs in several seconds to several minutes. The precipitation time increases as the concentration of either the alkoxide solution or the water solution is decreased. The hydrolysis reaction may be schematically represented by the general equation

$$M(OR)_x + \frac{x}{2} H_2O = MO_{x/2} + xROH \qquad (1)$$

where x is a function of the valence of the metal cation.

After precipitation, the powder is washed by centrifuging and redispersing in distilled water (or other solvent); this cycle is repeated two to three times. The powder is then further processed by adding cation dopants or is sedimented into compacts for sintering studies. The synthesis and processing of the various oxide powders are briefly reviewed in the following sections.

Pure TiO_2

Barringer and Bowen [2] described the synthesis of monodispersed, spheroidal TiO_2 by the controlled hydrolysis of dilute solutions of titanium tetraethoxide, $Ti(OC_2H_5)_4$. In a typical experiment, a 0.30 M solution of

Mat. Res. Soc. Symp. Proc. Vol. 32 (1984) © Elsevier Science Publishing Co., Inc.

188

$Ti(OC_2H_5)_4$ in 200 proof anhydrous ethanol was hydrolyzed by adding a 1.2 M solution of deionized water in ethanol. Precipitation of TiO_2 powder in ~83% yield occurred in ~5 seconds (at 25.5°C). The mean size of the TiO_2 particles, which was visually estimated from SEM micrographs, was 0.38μm. Figure 1a is a TEM micrograph of TiO_2 made by the hydrolysis of $Ti(OC_2H_5)_4$.

Titania was also prepared by the hydrolysis of several other titanium alkoxides: Titanium tetraisopropoxide, $Ti(^iOC_3H_7)_4$, titanium tetrabutoxide, $Ti(OC_4H_9)_4$, and titanium tetra-2-ethylhexoxide, $Ti(OC_8H_{17})_4$. The latter two reactions were conducted in ethanol, while the $Ti(^iOC_3H_7)_4$ was hydrolyzed in isopropanol and in ethanol. The hydrolysis reactions gave TiO_2 powders which were equiaxed, but were multinuclear and agglomerated. Typically, 0.30 M alkoxide solutions were hydrolyzed by adding 1.8 M water solutions with stirring. Precipitation occurred in ~20 seconds at 25°C. Equiaxed particles ranging in size from ~0.3μm to 0.7μm were produced and doublets, triplets, and more agglomerated particles were also formed. The agglomeration may be due to the failure of these alkoxide hydrolysis reactions to simultaneously satisfy the conditions required for formation of monodispersed, spheroidal particles [2]. A TEM micrograph of TiO_2 made by the hydrolysis of $Ti(^iOC_3H_7)_4$ is shown in Barringer and Bowen [2].

Pure ZrO_2

We have previously reported the preparation of monodispersed, spheroidal ZrO_2 by the controlled hydrolysis of zirconium normal- and isopropoxides [3]. Typically, $Zr(^nOC_3H_7)_4$ in ethanol (~ 0.10 M) was hydrolyzed with an excess of water (~ 0.5 M in ethanol). The hydrolysis was done at 50°C and precipitation of a white powder occurred in approximately two minutes. $Zr(^iOC_3H_7)_4$ in ethanol (~ 0.11 M) was hydrolyzed by an excess of water (~ 0.92 M in ethanol); at 25°C precipitation occurred in ~11 seconds and gave ZrO_2 in ~95% yield. Figure 1b is a TEM micrograph of ZrO_2 made by the hydrolysis of $Zr(^nOC_3H_7)_4$. ZrO_2 made from hydrolysis of Zr isopropoxide is approximately the same size (visually estimated from SEM micrographs), but is equiaxed rather than spheroidal and is more agglomerated [12].

Pure SiO_2

Monodispersed SiO_2 was prepared by a refinement of the procedure of Stober et al [4]. Tetraethylorthosilicate (TEOS) was hydrolyzed by a mixture of NH_4OH and H_2O (in ethanol); typically final solution concentrations were 0.3 M TEOS, 2.0 M NH_3, and 8.0 M H_2O. For these concentrations, turbidity was observed after approximately four minutes and powder with an average diameter of 0.55μm (standard deviation = 0.04μm) was obtained [5].

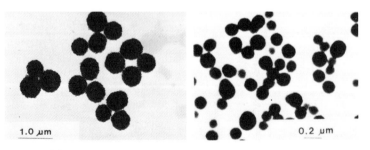

FIG. 1. (a) TEM micrograph of TiO_2 from $Ti(OC_2H_5)_4$, 0.30M; $[H_2O]=0.90M$, 25°C. (b) TEM micrograph of ZrO_2 from $Zr(^nOC_3H_7)_4$, 0.09 M; $[H_2O]=0.5M$, 50°C.

Doped TiO$_2$

Singly- and doubly-doped TiO$_2$ powders were prepared by the alkoxide hydrolysis process and by inorganic salt precipitation [6,7]. TiO$_2$ powders singly doped with either Ta$_2$O$_5$ or Nb$_2$O$_5$ were made by the controlled cohydrolysis of dilute solutions of Ti(OC$_2$H$_5$)$_4$ (~ 0.2 M) and either Ta(OC$_2$H$_5$)$_5$ or Nb(OC$_2$H$_5$)$_5$ (~ 5x10^{-4} M). Again, ethanol was used as a solvent and a water solution in ethanol was poured, with stirring, into the alkoxide solution. TiO$_2$ powders doubly doped with BaO, CuO, or SrO and either Ta$_2$O$_5$ or Nb$_2$O$_5$ were made by precipitating the respective metal carbonate onto the singly-doped particles, which were dispersed in a dilute aqueous solution of the respective metal chloride. The precipitation was conducted by adding an excess of (NH$_4$)$_2$CO$_3$ to the dispersion with stirring. Figure 2a is a SEM micrograph of Nb$_2$O$_5$, BaO-doped TiO$_2$.

The alkoxide cohydrolysis reactions involved in the synthesis of singly-doped TiO$_2$ are schematically represented by the general equation

$$Ti(OC_2H_5)_4 + M(OC_2H_5)_5 + H_2O = TiO_2 \cdot M_2O_5 \cdot XH_2O \qquad (2)$$

where M is Ta or Nb and X ~ 1/2. The carbonate precipitation reactions involved in the synthesis of doubly-doped TiO$_2$ powders are schematically represented by the general equation

$$M'Cl_2(aq) + (NH_4)_2CO_3(aq) = M'CO_3 \text{ (on particles)} \qquad (3)$$

where M' is Ba, Cu, or Sr. Doped TiO$_2$ powders containing SrO were also made by the cohydrolysis of Ti(OC$_2$H$_5$)$_4$ and Sr(^{i}OC$_3$H$_7$)$_2$.

Doped ZrO$_2$

Y$_2$O$_3$-doped ZrO$_2$ was made by the cohydrolysis of Zr(^{n}OC$_3$H$_7$)$_4$ and Y(^{i}OC$_3$H$_7$)$_3$. The solid yttrium isopropoxide was dissolved in the Zr(^{n}OC$_3$H$_7$)$_4$ and this solution was then dissolved in ethanol. The resulting solution was ~ 0.08 M Zr(^{n}OC$_3$H$_7$)$_4$ and ~ 0.01 M Y(^{i}OC$_3$H$_7$)$_3$, appropriate for making 6.3 mole % Y$_2$O$_3$-doped ZrO$_2$. Precipitation of a white powder was instantaneous at 50°C. Figure 2b is a SEM micrograph of Y$_2$O$_3$-doped ZrO$_2$ [13].

Doped SiO$_2$

Monodispersed B-doped SiO$_2$ particles [7,8] were made by partially hydrolyzing TEOS, adding tri-n-butyl borate (TBB), B(OC$_4$H$_9$)$_3$, and then completely hydrolyzing this mixture. A range of TEOS/H$_2$O concentrations and

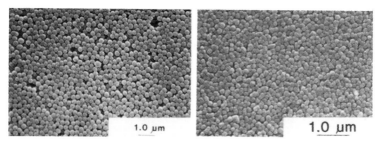

FIG. 2. (a) SEM micrograph of the top surface of a gravity sedimented compact of Nb$_2$O$_5$ (0.5 wt. %), BaO(0.2 wt. %) -doped TiO$_2$. (b) SEM micrograph of Y$_2$O$_3$ (6.5 mol %) -doped ZrO$_2$ powder.

B/Si ratios were investigated [8]. Typically, the final concentrations were 0.01-0.20 M TEOS, ~5 M H_2O, and ~2 M NH_3; the TBB/TEOS molar ratios were 0.5-2.0. Figure 3a is a TEM micrograph of B-doped SiO_2 made from 0.01 M TEOS and having a nominal B/Si ratio of 1:1 [8].

ZrO_2-Al_2O_3

Two-phase particles containing ZrO_2 and Al_2O_3 were made by the hydrolysis of $Zr(^nOC_3H_7)_4$ in an ethanolic dispersion of Al_2O_3 particles (Alcoa XA139 superground) [9]. Typically, several grams of Al_2O_3 powder, which had previously been classified to give a 0.2-0.3µm size fraction [10], was dispersed in 500 ml ethanol and mixed with a $Zr(^nOC_3H_7)_4$ solution in ethanol giving an approximately 0.03 M alkoxide solution. This was hydrolyzed with an excess of water (~ 0.3 M in ethanol) at 50°C. No observable change in the dispersion, which was already white, occurred after the hydrolysis. Figure 3b shows a TEM micrograph of ZrO_2-Al_2O_3 powder [9].

CHEMICAL AND PHYSICAL CHARACTERIZATION OF OXIDE POWDERS

Characterization of the chemical and physical properties of the oxide powders is critical for control of powder dispersion, packing and sintering. Some of the important physical properties are particle size distribution, shape, surface area, density, and crystal structure. Among the important chemical properties are bulk composition, dopant levels, dopant homogeneity, impurity content, and surface/liquid interfacial chemistry. A variety of techniques are available to measure these properties; some yield acceptable results, while others do not. The techniques utilized in our research and results for the different oxide powders, which are summarized in Table I, are briefly reviewed in the following sections. More detailed descriptions, which cannot be presented here because of space limitations, are given elsewhere in the literature [2, 6-13].

Particle Size, Size Distribution and Shape

Assessments of particle shape, state of agglomeration, and qualitative size distributions were obtained from SEM and TEM micrographs. Quantitative size distributions were obtained for >500 particles from TEM micrographs using a histogram method [11]; image analysis and direct counting were used to generate the histograms. Although this technique is precise, especially when >1000 particles are counted, it is time consuming and tedious. In addition, differentiation between hard agglomerates (aggregates) and flocs formed during sample preparation and drying was often difficult.

500 Å   0.3 µm

FIG. 3. (a) TEM micrograph of B-doped SiO_2 with a nominal B/Si ratio of 1:1. (b) TEM micrograph of ZrO_2-Al_2O_3 powder with a nominal ZrO_2 content of 20 volume percent.

More rapid techniques based on particle sedimentation and laser light scattering were also utilized. The Sedigraph unit (Micromeritics), based on X-ray absorption for sedimenting particles, worked well for several powders having a mean diameter $>0.3\mu m$. Dynamic laser light scattering (photon correlation spectroscopy, PCS) using the Coulter Model N4D gave rapid and accurate results for the uniform-size powders. This instrument utilized the cummulant analysis method to determine the Brownian diffusion coefficient and the hydrodynamic size.

Figures 1-3a illustrate that to a good first approximation the oxide particles produced by the controlled alkoxide hydrolysis reactions are monodispersed, spheroidal, and predominantly singlets. Table I lists representative values of the average particle size and size range (for ranges of reagent concentrations) for eight oxide powders. These data were obtained by a variety of techniques; correlation between techniques for a powder was generally excellent.

Physical Measurements (Surface Area, Density, Crystallinity)

The data on other physical properties such as surface area, density, and crystallinity are also summarized in Table I. All particles produced by the alkoxide hydrolysis technique were amorphous, as determined by X-ray and electron diffraction. The densities of the amorphous powders, measured by a stereopycnometer (Quantachrome Corp.) using He gas, were generally less than (by $>$ 20%) the densities of the stable low-temperature modifications of the crystalline oxides. However, the silica particles were only 5% less dense than fused silica.

Specific surface areas were measured by multipoint BET (Quantachrome Corp.), using N_2 gas as the adsorbate and assuming a cross-sectional area of 16.2Å^2. The B.E.T. surface areas were found to depend upon the initial washing of the powders. Powders washed in ethanol had lower surface areas (larger equivalent spherical diameters) than powders washed in water. Barringer [11] discussed this observation for pure TiO_2 and attributed it to a surface coating of fine spherical precipitates resulting from the rapid hydrolysis of residual (unreacted) $Ti(OC_2H_5)_4$ on the particle surface upon contact with water in the initial washing step (see Fig. 1a). TEM micrographs of ethanol-washed and water-washed ZrO_2 suggest that an analogous effect causes the higher surface areas of water-washed ZrO_2. The equivalent spherical diameters of the ethanol-washed ZrO_2 are in the range $0.12-0.22\mu m$ and span the mean particle sizes determined by TEM and PCS.

Chemical Analyses

Several methods have been used to determine dopant and impurity levels and volatile contents of the oxide powders. The results of the various analytical techniques are briefly reviewed below. More complete descriptions are given elsewhere in the literature [6-13].

Volatile contents of the powders were generally estimated by simultaneous DTA/TGA analysis. Water washed TiO_2 contained 10-16% (weight) water; carbon analysis of the as-synthesized powder showed 0.18% (weight) carbon [2]. The doped TiO_2 powders showed variable weight losses which correlated with their processing prior to DTA/TGA analysis. Powders that were not washed or not dried at 80°C lost the most weight, 21%, while powders which had been washed and then calcined at 450°C for 30 minutes lost the least weight, 2-5%. Most of the observed weight loss was due to the loss of water of hydration, which was removed below 200°C. DTA/TGA analysis of the pure ZrO_2 showed 14-20% weight loss, depending on the temperature at which the hydrolysis reaction was conducted and the subsequent processing of the powder. Carbon analyses of the water-washed ZrO_2 made by hydrolysis of Zr n-propoxide showed 1.02-1.07% carbon, presumably due to residual alkoxide.

Dopant contents and cation impurity levels in the oxide powders were mainly determined by inductively coupled plasma emission spectroscopy (ICP), and also by direct coupled plasma emission spectroscopy (DCP), instrumental neutron activation analysis (INAA), and semiquantitative emission spectroscopy. Sintered ceramics were analyzed by ICP, electron probe microanalysis (EPMA), and proton induced X-ray emission (PIXE).

The cation impurity levels in the pure TiO_2 made by hydrolyzing $Ti(OC_2H_5)_4$ were determined by semiquantitative emission spectroscopy and ICP. The former method showed <10 ppm Al, Ca, Cu, and Mg and <100 ppm Si, while the latter gave 40 ppm Ca and 80 ppm Si. EPMA wavelength dispersive analysis gave 90 ppm Si in a sintered TiO_2 ceramic. Cation impurities in pure ZrO_2 made by hydrolysis of Zr n-propoxide and isopropoxide were determined by ICP analyses of powders and EPMA analyses of sintered ceramics. ZrO_2 made from the n-propoxide contains ~ 1.3% (weight) Hf, while ZrO_2 made from the isopropoxide contains <100 ppm Hf. (PIXE analysis of a sintered ZrO_2 ceramic made from the isopropoxide gave 57 ppm Hf, ICP analysis of the powder gave <93 ppm Hf). Otherwise, the major cation impurities in the ZrO_2 are Al (<160 ppm), Fe (100-150 ppm), and Si (60-400 ppm). (Ti is present at ~ 350 ppm, but may be due to Ti contamination from the ultrasonic probe tip used to disperse the powder during processing.) The rather high and variable Si contents may be contamination from bottles used to store the alkoxides or from glassware used for the hydrolysis reactions.

The dopant levels in the doped TiO_2 powders and ceramics were determined by ICP, DCP, EPMA and INAA analyses. The Y_2O_3 level in the Y_2O_3-doped ZrO_2 ceramics was determined by EPMA and PIXE analyses. The boron level in the doped SiO_2 was determined by ICP. The results of these extensive analyses demonstrate that the controlled alkoxide hydrolysis reactions give exceptional control over the desired chemistry, with a low level of contaminants. This feature has also been emphasized by Mazdiyasni [1]. However, more importantly, this control over the chemistry is coupled with control of the particle size distribution and morphology. Results presented by Fegley et al [4] show that the doped TiO_2 powders are homogeneously doped with the Nb^{5+} or Ta^{5+} cations and that the Ba^{2+}, Cu^{2+}, Sr^{2+} cations are present on the surface of the TiO_2 particles. EPMA analyses of Y_2O_3-doped ZrO_2 ceramics show that the grain to grain homogeneity of the Y_2O_3 dopant is actually better than the homogeneity of the HfO_2 impurity.

SURFACE CHEMISTRY, POWDER DISPERSION AND PACKING

The state of aggregation of a dispersed powder and the subsequent packing into green bodies, both of which significantly affect the sinterability and final microstructure, depend on the stability of the dispersion against coagulation. Coagulation processes are controlled by the inter-particle forces which are dependent on the physical characteristics and the powder/solvent interfacial chemistry. An understanding of the interfacial chemistry (particle-solvent-solute interactions) is critical to the development and control of reliable fabrication processes. For instance, the casting of an aqueous, surfactant-stabilized Al_2O_3 slip may yield dense, uniformly-packed bodies ($\rho \approx 70\% \rho_{th}$) at pH $\approx$ 7-8, and yet yield porous bodies at pH $\approx$ 3. However, for aqueous dispersions without the surfactant this behavior is reversed.

Dispersion and Stability

The stability of a dispersion against coagulation depends on the sign and magnitude of the particle interaction energies. The frequency of

Brownian encounters determines the maximum rate of coagulation in the absence of forces; however, coagulation is retarded by the presence of repulsive interactions. The general equation describing two-body interactions consists of attractive and repulsive terms:

$$V_T = V_A(\text{van der Waals}) + V_R(\text{electrostatic}) + V_R(\text{steric}) + V_R(\text{others}). \quad (4)$$

The van der Waals attractive forces between particles, due to electronic fluctuations of the atoms within particles, depend on the dielectric properties of the particle and solvent through the Hamaker constant [14]. This force can be modified by the presence of an adsorbed solute layer, e.g., polymer surfactant [15].

The electrostatic repulsion is caused by the interaction of electrical double layers surrounding the dispersed particles, which arises from the particle surface/solvent acid-base reactions and electrolyte redistribution around the particles. The magnitude of the repulsion depends on the solvent dielectric constant and pH (surface charging) and the indifferent electrolyte concentration (electrostatic shielding); excellent reviews exist in the literature [16,17].

The nature of steric forces, due to the interaction of macromolecules adsorbed onto particle surfaces, is not as well understood as electrostatic forces; however, it is presently the topic of extensive investigation. These forces depend on the macromolecule structure and size (MW), mode of adsorption and adsorption density, and conformation. The interplay of solvent-polymer (solvation), solvent-surface (wetting), and surface-polymer (adsorption) interactions control steric stabilization; excellent reviews are given by Tadros [18] and Sato and Ruch [19]. The chemistries in the systems used in manufacturing must be better understood for reproducible fabrication of ceramic materials; the Al_2O_3 example previously cited is representative of such systems.

Although no precise rules can be stated regarding the formation of stable dispersions, some guidelines can be stated. In polar, protic solvent electrostatic forces may be the dominant repulsive component; in such systems stability requires a low electrolyte concentration (< 0.01 M) and a solution pH several units above or below the isoelectric point (IEP). In nonpolar and aprotic solvents steric stabilization is required. The most effective surfactants (dispersants) have strongly adsorbed functional groups and strongly solvated segments in the macromolecule. In mixed systems (polar solvents and surfactants) the interactions are complex; simple guidelines are not possible. In all cases, though, the stability of the dispersions can be assessed qualitatively by sedimentation techniques [20] or quantitatively by measuring the change in average particle size with time [21].

POWDER PACKING AND SINTERING

The goal of ceramic processes is to control particle packing in green microstructures, and thus control sintered microstructures and properties. The critical relationship between green and sintered microstructures in obtaining fine-grain, dense ceramics has been established [2,22,23]. In all cases studied (TiO_2, ZrO_2, SiO_2, B-SiO_2), stable dispersions were formed in deionized water with a pH several units above or below the IEP: TiO_2 at pH = 4, 8-10 (IEP = 5-6); ZrO_2 at pH = 4, 9-11 (IEP = 6-7.5); SiO_2 and B-SiO_2 at pH $\geqslant$ 5 (IEP = 2-3). Figures 4 (ZrO_2, pH = 10) and 5 (SiO_2, pH = 7) show representative examples of top and fracture surfaces of compacts formed by sedimentation from stable dispersions. The packing is extremely uniform and reproducible (from sample to sample); the voids present are about one particle diameter in size (0.2-0.4 μm).

The sintering process not only depends on the particle size (scaling laws), but also on the particle packing density and uniformity. This dependence was demonstrated for pure TiO_2 by Barringer and Bowen [2,11,23,24]; identical behavior was observed for all powder systems (the details of these studies are discussed in later publications). Figure 6 shows the microstructure of a ZrO_2 compact sintered for 1.5 hours at 1160°C; a dense ($\rho > 98\%$ ρ_{th}), uniform, fine-grain ($\bar{d} \simeq 0.2$ μm), structure is observed. Essentially no grain growth occured and all pores are intergranular and small relative to the grain size. The sintering temperature of this compact, which had an initial density of ~65% of theoretical, was 400-500°C below that required for conventional ZrO_2 powders. Figure 7a demonstrates that B-SiO_2 can be densified at temperatures approaching 700°C [8].

Porous compacts, when sintered under conditions similar to those used for dense compacts, yielded uniformly porous bodies. Such structures have potential as filter membranes, chromatography substrates, catalytic substrates and sensor elements. As an example, Figure 7b shows a porous SrO-doped TiO_2 body (0.52 mole percent SrO) sintered for 2.5 hours at 1100°C. The grains ($\bar{d} \simeq 0.4$-0.6 μm) and pores are uniform in size and spatial distribution. The discrete particles on the larger TiO_2 grains may be $SrTiO_3$ precipitates [25].

FIG. 4. (a) SEM micrograph of the top surface of a gravity sedimented compact (pH = 10) of ZrO_2. (b) Fracture surface of the same compact.

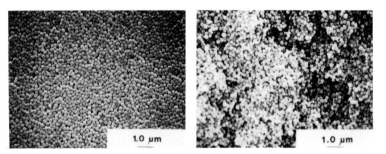

FIG. 5. (a) SEM micrograph of the top surface of a gravity sedimented (pH = 7) SiO_2 compact. (b) Fracture surface of the same compact.

SUMMARY

Several important points are demonstrated by this research:
1. The <u>controlled</u> hydrolysis of alkoxides is a very powerful technique for controlling the bulk composition and dopant distribution in oxide powders. The technique is versatile and can be used to produce homogeneous or heterogeneous dopant distributions by paying careful attention to the synthetic procedures and the phase solubilities for the specific system. More importantly, controlled hydrolysis allows control over particle size and shape. Thus, "ideal" powders in the 0.1-1.0μm size range can be made as monodispersed spheroids of controlled chemical composition.

2. Careful attention to surface/liquid interfacial chemistry is a prerequisite for controlling the dispersion and packing of "ideal" powders into uniform, dense green microstructures. Green pieces with >68% of theoretical density are routinely obtained. SEM micrographs show that packing defects and/or voids are on the order of a particle diameter. Conversely, the improper dispersion and packing of the "ideal" powders results in green microstructures having a nonuniform and highly variable density; agglomerated regions and voids with a typical size of many (tens) particle diameters are typically observed.

3. Sintering experiments on uniformly packed compacts show that the mono-dispersed powders can be sintered at temperatures hundreds of degrees (400-500°C) lower than commercial powders, that the resulting microstructures have 98+% theoretical density, that the grains are small and are of the order of the initial particle size, and that no abnormal grain growth occurs.

FIG. 6. (a) SEM micrograph of the top surface of a sintered ZrO_2 compact. Sintered for 1.5 hours at 1160°C. (b) SEM micrograph of the fracture surface of the same compact.

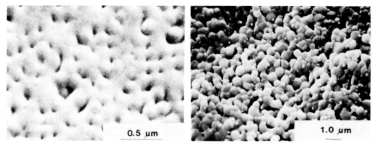

FIG. 7. (a) SEM micrograph of the top surface of a sintered (5 hours, 740°C) B-doped SiO_2 compact. (b) SEM micrograph of a sintered (2.5 hours, 1100°C) porous SrO-doped TiO_2 compact.

4. The use of monodispersed, spheroidal powders presents several practical advantages:
   a) Particle size and size distribution are easily determined by techniques such as light scattering.
   b) Monodispersed particles can be formed into ordered dispersions which can be studied by light diffraction techniques.
   c) Ability to test the sintering models developed over the past 3 decades by quantitatively studying sintering kinetics.
   d) Ability to control the sintered microstructure allows the careful study of microstructure-property relationships for electrical, magnetic, and mechanical properties.

ACKNOWLEDGMENTS

We acknowledge helpful discussions with H. K. Bowen and R. L. Pober. Also, we want to thank L. V. Janavicius, T. Kramer, D. Lange, P. Normile, L. Rigione, D. Todd, P. White, S. Woodhull, and W. Zamechek for analyses, photos, powder preparation, typing, and discussions.

REFERENCES

1. K. S. Mazdiyasni, Ceramics Intl. 8, 42 (1982).
2. E. A. Barringer and H. K. Bowen, Comm. Am. Ceram. Soc. 65, C199 (1982).
3. B. Fegley, H. K. Bowen, L. J. Rigione, and D. Todd, Bull. Am. Ceram. Soc. 62, 374 (1983).
4. W. Stober, A. Fink, and E. Bohn, J. Coll. Interf. Sci. 26, 62 (1968).
5. T. C. Huynh, A. Bleier, and H. K. Bowen, Bull. Am. Ceram. Soc. 61, 336 (1982).
6. B. Fegley, E. A. Barringer, and H. K. Bowen, Comm. Am. Ceram. Soc., in press (1984).
7. E. A. Barringer, N. Jubb, B. Fegley, R. L. Pober, and H. K. Bowen in: Int. Conf. on Ultra-Structure Processing of Ceramics, Classes and Composites, L. L. Hench and D. R. Ulrich, eds. (Wiley and Sons, NY 1984) pp. 000-000.
8. L. V. Janavicius, S.M. thesis, MIT (1984).
9. E. A. Barringer, B. Fegley, H. Okamura, P. Debeley, in preparation (1984).
10. R. L. Pober, R. Hay, M. L. Harris, in preparation (1984).
11. E. A. Barringer, Ph.D. thesis, MIT (1983).
12. B. Fegley, E. A. Barringer, D. Todd, in preparation (1984).
13. B. Fegley, P. White, H. Okamura, in preparation (1984).
14. H. C. Hamaker, Physica, 4, 1058 (1937).
15. B. Vincent, J. Coll. Interf. Sci., 42, 274 (1973).
16. J. Th. G. Overbeek, J. Coll. Interf. Sci., 58, 408 (1977).
17. R. H. Ottewill, J. Coll. Interf. Sci., 58, 357 (1977).
18. Th. F. Tadros in: The Effect of Polymers on Dispersion Properties, Th. F. Tadros, ed. (Academic Press, NY, 1982) pp. 1-38.
19. T. Sato and R. Ruch, "Stabilization of Colloidal Dispersions by Polymer Adsorption", (Marcel Dekker, NY, 1980), pp. 65-119.
20. M. V. Parish, R. R. Garcia, H. K. Bowen, in preparation (1984).
21. E. A. Barringer, B. E. Novich, and T. A. Ring, Accepted for publication in J. Coll. Interf. Sci.
22. W. H. Rhodes, J. Am. Ceram. Soc., 64, 19 (1981).
23. E. A. Barringer and H. K. Bowen in Int. Inst. for the Science of Sintering, 16, Belgrade, Yugoslavia, 1983.
24. E. A. Barringer, R. Brook, and H. K. Bowen in: Sixth Int. Conf. on Sintering and Related Phenomena Including Heterogeneous Catalysts, Notre Dame, 1983.
25. H. C. Ling and M. F. Yan, J. Mat. Sci., 18, 2688 (1983).

TABLE I. Chemical and Physical Properties of Monodispersed Oxide Powders

Material Property	TiO_2 [2,11]	Singly-Doped TiO_2 [6]	Doubly-Doped TiO_2 [6]	ZrO_2 [12]	Doped ZrO_2 [13]	ZrO_2-Al_2O_3	SiO_2 [5]	Doped SiO_2 [8]
Chemical Precursor(s)	$Ti(OC_2H_5)_4$	$Ti(OC_2H_5)_4$ $M(OC_2H_5)_5$ [a]	$Ti(OC_2H_5)_4$ $M(OC_2H_5)_5$ [a] $M'Cl_2$	$Zr(OC_3H_7)_4$ [c]	$Zr(OC_3H_7)_4$ [c] $Y(OC_3H_7)_3$ [i]	$Zr(OC_3H_7)_4$ [c] sized Al_2O_3	$Si(OC_2H_5)_4$	$Si(OC_2H_5)_4$ $B(OC_4H_9)_3$
Dopant Level		0.1-1.0%(wt.)	0.1-1.0%(wt.)		6.5 mole %	≤20 vol. %		≤25%
Typical % Yield	70-80%	~70%	~70%	95% [d]	~45%	-	>95%	-
Average Size Range and Standard Deviation	d=0.3-0.7μm σ_z =1.09	d=0.3-0.7μm [e] σ_z ~1.2 [e]	d=0.3-0.7μm [e] σ_z ~1.2 [e]	d=0.2-0.3μm σ_z =1.2	d=0.2-0.3μm σ_z =1.2	d=0.25μm [f] σ_z =1.22 [f]	d=0.1-0.7μm σ_z =1.03	d=0.09-0.17μm [g] σ_z =1.04
Surface Area (m²/g)	8-9 [h] 170-200 [i]	200-250 [i]	100-300 [i]	9-17 [h] 28-57 [i]	66-72 [i]	34-65 [i]	30-40 [i]	21
Shape	Spheroidal	Spheroidal	Spheroidal	Spheroidal	Spheroidal	Equiaxed	Spheroidal	Spheroidal
Structure	Unagglom-erated	Unagglom-erated	Unagglom-erated	Unagglom-erated	Unagglom-erated	Unagglom-rated	Unagglom-erated	Unagglom-erated
Crystal Form	Amorphous	Amorphous	Amorphous	Amorphous	Amorphous	Amorphous (ZrO_2)	Amorphous	Amorphous
Density (g/cm³)	3.0-3.2	3.0-3.2	3.0-3.2	3.1	-	-	2.1	1.94
% Volatiles (H_2O+Carbon)	~12%	~20%	~20%	~20%	-	-	~8%	10-15%
Cation Impurities	Ca,Si	Ca,Si,Zr	Ca,Si,Zr	Al,Fe,Si,Hf	Al,Fe,Si, Hf	Ca,Fe,Si	-	Ca,Fe,Na

Notes and Abbreviations for Table I.

a) M = Nb, Ta

b) M' = Ba, Cu, Sr

c) n-propoxide and isopropoxide

d) for Zr-isopropoxide

e) Visually estimated from SEM micrographs

f) For sized Al_2O_3 powder before coating, see text.

g) B/Si = 1.0, 0.01 M TEOS concentration

h) Alcohol washed powder

i) Water washed powder

Lead Zirconate-Lead Titanate (PZT) Ceramics From Organic-Derived
Precursors

R. G. Dosch, Sandia National Laboratories, P. O. Box 5800, Albuquerque,
New Mexico 87185

ABSTRACT

Organic-derived precursors can facilitate the
preparation of PZT ceramics, particularly in the areas
of processing and batch reproducibility. The use of
partially alkoxide-derived precursors and completely
organic-derived precursors has been studied. Two
methods of alkoxide hydrolysis were evaluated for
preparing zirconia which included direct hydrolysis in
water and hydrolysis of alkoxide-base reaction products
in acetone-H_2O mixtures. The chemistry of these methods
is discussed along with pertinent aqueous surface
chemistry of the hydrated zirconia products. This work
suggests that alkoxide-derived precursor properties are
related to the solution chemistry used in their preparation.

INTRODUCTION

The use of chemical preparation techniques for making ceramic
precursors is generaly stimulated by specific requirements which cannot
be readily met by conventional ceramic processing methods. Examples
encountered in our laboratory have included needs for improved batch
reproducibility, homogeneous distribution of trace constituents, reduction
of processing complexity, and for amorphous ceramic compositions. The
former is of particular interest for lead zirconate-lead titanate (PZT)
materials as batch variation in physical and electrical properties has
been a recurring problem over the years.

A number of options are apparent in applying organic-derived precur-
sors to PZT fabrication. The composition $Pb_{.99} Nb_{.02} (Zr_{.96} Ti_{.04})_{.98}O_3$
(PNZT) is of particular interest and precursor routes investigated have
included the use of metal alkoxide-derived ZrO_2 and $ZrO_2 - TiO_2 - Nb_2O_5$
mixtures in combination with the appropriate commercially available oxides
and also a completely organic-derived precursor prepared via a non-aqueous
reaction of alkoxide mixtures with lead lactate. The properties and
processing characteristics of PNZT materials prepared using these precursor
options have been reported elsewhere[1].

Work in our laboratory has shown that one of the most important
variables relative to PZNT properties is the characteristics of the ZrO_2
raw maerial. Similar observations have been reported for other PZT
compositions[2]. In this paper an attempt is made to relate differences
in properties of the ZrO_2 powders and the PZNT prepared from them to the
solution chemistry used in the ZrO_2 preparation.

*This work performed at Sandia National Laboratories supported by the
U.S. Department of Energy under contract number DE-AC04-76DP00789.

Experimental

Two hydrous ZrO_2 precursors were prepared from tetra-n-propyl zirconate, TPZ, (28.79% ZrO_2 equivalent) as follows: For ZrO_2-I, TPZ (355g) was added to 2 liters of boiling water. After stirring 15 minutes, the solids were separated by filtration, washed with acetone, and dried at 22°C under vacuum. On the other hand, for ZrO_2-II, TPZ (349.2g) was added to 200.9g of a 8.12% by weight solution of NaOH in methanol (MEOH). After stirring to obtain a clear solution, this mixture was hydrolyzed in 2 liters of acetone containing 88.1g H_2O. Sufficient HNO_3 was added to give a pH of 1.9 - 2.0 in the stirred slurry for a two hour period following the last acid addition. Solids were separated and dried as described above. Residual Na content was typically <0.01%. The alkoxide-derived ZrO_2 materials were calcined at 700-750°C and used to prepare PNZT by standard mixed oxide techniques. Final sintering of PNZT was done at 1300°C for 2 hours.

Aqueous pH measurements were made using an Altex Model 71 pH meter (Beckman Instruments, Fullerton, CA) and a Ross combination pH electrode (Orion Research, Cambridge, MA). Reference electrode/analyte contact for non-aqueous measurements was made through a quartz fiber junction bridge filled with a solution of 0.1M tetramethylammonium chloride in MEOH. A standard curve of electrode potential versus NaOH concentration in appropriate methanol-n-propanol mixtures was used to estimate OH^- concentrations in NaOH-TPZ reaction mixtures.

RESULTS AND DISCUSSION

Calcination Temperature

Calcination reaction temperatures for PNZT were significantly reduced when organic-derived precursors were used. DTA/XRD indicated conversion to crystalline PNZT at 550°C and 700°C for materials prepared from a complete organic-derived precursor and from alkoxide-derived ZrO_2 plus mixed oxides, respectively. These effects are most likely related to the increased surface area of organic-derived materials[3] which typically results in higher reactivity.

Microstructure

Differences in microstructure of PZNT prepared in an identical manner except for the source of ZrO_2 were of more interest. In one application, PNZT of relatively low density (7.3g/cm^3) with fine, uniformly distributed closed porosity provides the highest resistance to dielectric and mechanical breakdown. Samples prepared from ZrO_2-I had an average grain size of 11.2 μM, a density of 7.31 g/cm^3, and an open pore structure as compared to those prepared from ZrO_2-II which had an average grain size of 7.2 μM, similar density, and a closed pore structure. Based on these results, it it reasonable to assume that ZrO_2-I and ZrO_2-II differ in some properties related to reactions with the other oxide constituents, and further, that these differences were probably introduced via solution chemistry used in their preparation.

Examination of ZrO_2-I and ZrO_x-II with TEM showed both materials to consist of agglomerates of <0.1 M particles. The ZrO_2-II material appeared to contain a smaller fraction of sub-micron agglomerates. BET surface areas of as-prepared powders were 170m^2/g and 265m^2/gm for Zro_2-I and ZrO_2-II, respectively. A nearly linear decrease in surface area with increasing calcine temperature was observed up to 600°C where values of 29m^2/g and

$50m^2/g$ were determined for ZrO_2-I and ZrO_2-II materials, respectively. Pore volume distributions also differed as shown in Fig. 1. As discussed below, the solution chemistry used in the zirconia preparations may have contributed to the observed differences in physical properties.

Base Reactions

The reactions used for ZrO_2-I and ZrO_2-II are empirically represented as follows:

ZrO_2-I

$$Zr(OC_3H_7)_4 + 4H_2O \longrightarrow Zr(OH)_4 + 4C_3H_7OH \qquad (1)$$

ZrO_2-II

$$2Zr(OC_3H_7)_4 + NaOH + 4H_2O \longrightarrow NaZr_2O5H + 8C_3H_7OH \qquad (2a)$$

$$NaZr_2O5H + HNO_3 + 3H_2O \longrightarrow 2Zr(OH)_4 + NaNO_3 \qquad (2b)$$

Both preps involve hydrolysis of TPZ, however, ZrO_2-II prep involves prior addition of NaOH and subsequent neutralization with HNO_3 and one would expect differences in ZrO_2 produced by these methods to be initiated by these reactions. Reaction 2a involves two steps where TPZ is added to a MEOH solution of NaOH and this mixture is then hydrolyzed in an acetone-water mixture.

Conductivity data in isopropanol (IPOH) shown in Fig. 2 suggest a nearly quantitative reaction of NaOH with TPZ at mole ratios up to 0.5. Due to limited solubility of NaOH in IPOH, larger quantities are prepared using MEOH solutions which significantly changes reaction stoichiometry (Fig. 2). This may be due to alcohol exchange reactions where substitution of methoxide groups results in species which do not react in the same manner with NaOH. In the pure MEOH solutions used for ZrO_2-II, measurement of OH^- concentration in the reaction mixtures shows that only 0.03 moles of OH^- are consumed per mole of Zr (Fig. 3).

The NaOH–TPZ reaction in MEOH has been shown to influence the chemistry of the hydrated product, $NaZr_2O5H$. When this material is equilibrated in water, hydrolysis occurs liberating NaOH and since reaction 2a is quantitative, the NaOH which remains associated with the hydrated ZrO_2 could be determined by solution analysis (Table 1). A comparable experiment was done using ZrO_2-1 where varying amounts were equilibrated with aqueous NaOH solutions using the same OH^-:Zr mole ratio (0.5) as the ZrO_2-II material. Comparison of the total amount of OH^- associated with the solid phases shows that the reaction 2a product sorbs an average of 0.033 mole OH^- per mole of ZrO_2 more than ZrO_2-I material, in good agreement with the stoichiometry observed in the non-aqueous TPZ–NaOH reaction (Fig. 3). The results suggest that phenomena occurring in solution prior to hydrolysis can affect the properties of hydrolysis products subjected to moderate drying techniques. Equilibrium quotients for reactions with NaOH in aqueous solution were also found to be different for ZrO_2-I and ZrO_2-II materials(4).

Acid Reactions

Reaction 2b also provides a potential source for differences in zirconia properties. An equilibrium quotient, KEQ, was determined for the reaction between hydrous ZrO_2 and HNO_3 in aqueous solution. The following stoichiometry provided the best fit to the data

$$ZrO_2 + 1/2\ HNO_3 \longrightarrow ZrO_2 \cdot 1/2HNO_3$$

which suggest that hydroxyl bonds[5] are involved in the reaction. A value of 13.7 $\pm$ 4.2 over a concentration range of 0.5 to 3.3 meq HNO_3 per mole ZrO_2 was determined for KEQ defined as

$$KEQ = \frac{\left[ZrO_2 \cdot 1/2\ HNO_3 \right]}{\left[HNO_3 \right]^{1/2} \left[ZrO_2 \right]}$$

where $\left[ZrO_2 \cdot 1/2\ HNO_3 \right]$ = meq NO_3^- sorbed and $\left[ZrO_2 \right]$ = $\frac{mmole\ Zr}{2}$ − meq NO_3 sorbed. In reaction 2b, it is necessary to adjust pH to 2 or lower to achieve quantitative removal of Na^+. At pH = 2, the KEQ predicts a $ZrO_2 \cdot 1/2\ HNO_3 : ZrO_2$ ratio of 1.37 which is equivalent to a NO_3^- content of 12.7% by weight, which was in good agreement with nitrate analyses done by ion chromatography and combustion methods.

ZrO_2-I containing a comparable amount of NO_3^- introduced by adjusting a slurry of a vacuum dried material to a pH of 2 with HNO_3 was prepared and calcined at 500°C to volatilize the nitrate. No change in surface area or pore size distribution were found as compaed to ZrO_2-II calcined at the same temperature.

CONCLUSIONS

Partial PNZT composition can result in changes both in reaction temperatures and microstructure. The ZrO_2 component was found to be an important parameter in determining PNZT microstructure. This investigation suggests that differences in properties at the ZrO_2 precursors were related to the solution chemistry used in their preparation.

REFERENCES

1. Hammetter, W. F., Hankey, D. C., and Dosch, R. G., "Microstructure Development in Niobium − Modified Lead Zirconate − Titanate (PNZT) Ceramics, I. Effect of Raw Material Property Variations on the Calci- nation Reaction," submitted for publication.

2. Harrison, W. B., "Effect of ZrO Source Variation on PZT Piezoelectric Properties," Tech. Report No. 1, to the OffICe of Naval Research, Contract N00014-76-C-0623, NR 032-566, Dec., 1977, Honeywell, Inc., Minneapolis, MN.

3. Keizer, K., Janssen, E. H. J., deVries, J. J., and Burggrat, A. J., Mat. Res. Bull. 8, 533-44, 1973.

4. Dosch, R. G., unpublished results.

5. Muha, G. M., and Vaughan, P. A. J. Chem, Phys., 33, 194(1960).

TABLE I

Comparison of Hydroxide Sorption of $NaZr_2O_5H$ and ZrO_2-I in Aqueous Solution

mmole Zr[1]	$NaZr_2O_5H$	ZrO_2-I	[2]
1.35	0.090	0.056	0.034
2.70	0.111	0.081	0.030
10.78	0.171	0.140	0.031
16.18	0.192	0.158	0.034
26.97	0.218	0.181	0.037
			Ave = 0.033

1) Liquid phase volume was 100ml. Mole ratio of Zr:OH⁻ was 0.5.

2) Meq OH⁻ sorbed/mmole Zr for $NaZr_2O_5$ minus that for ZrO_2-I.

204

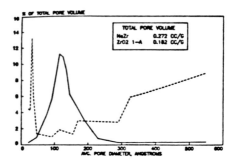

Fig. 1. Incremental Pore Volume Distribution of ZrO$_2$ Precursors
Calcined at 500°C. Dotted Line - Prep 1 ZrO$_2$;
Solid Line - Prep 2 ZrO$_2$.

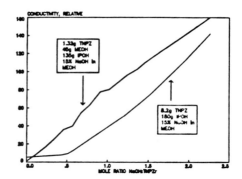

Fig. 2. Conductivity of TPZ-NaOH mixtures in alcohol solutions.

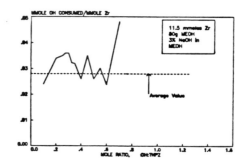

Fig. 3. Stoichiometry of TPZ-NaOH reaction in methanol.

GLASSES AND CERAMICS FROM COLLOIDS

G. W. SCHERER, R&D DIVISION, Corning Glass Works, Corning, NY 14831

ABSTRACT

Flame-generated particles can be suspended in a nonpolar solvent, molded, and gelled. The microstructure obtained after solvent evaporation depends on the characteristics of the colloid. Higher solids loadings lead to denser green bodies and narrow pore size distributions. The mean pore size is proportional to the particle size. Sintering occurs at relatively low temperatures, producing amorphous or crystalline bodies of high purity.

INTRODUCTION

This paper describes a process for making amorphous or crystalline ceramic bodies using particles made by flame oxidation. The particles are suspended in a nonaqueous, usually nonpolar, solvent, then molded and gelled. The characteristics of the colloidal dispersions and of the dried gels are discussed. Materials produced in this way include SiO_2, TiO_2, and Al_2O_3.

PARTICLES

Flame oxidation produces submicron oxide particles from reactions such as:

$$SiCl_4 + O_2 \rightarrow SiO_2 + 2Cl_2 \tag{1}$$

$$TiCl_4 + O_2 \rightarrow TiO_2 + 2Cl_2 \tag{2}$$

$$4AlCl_3 + 3O_2 \rightarrow 2Al_2O_3 + 6Cl_2 \tag{3}$$

$$Si(CH_3)_4 + 8 O_2 \rightarrow SiO_2 + 4CO_2 + 6H_2O \tag{4}$$

The reactants are passed through a gas/oxygen torch and the oxide products, formed in the flame, condense into droplets. This process is used to make optical waveguide fibers (1), because of the extraordinary purity of the particles. Flame oxidation is used commercially to make amorphous SiO_2 and crystalline Al_2O_3 and TiO_2 for use as fillers and pigments.

The oxide particles (often called "soot") are typically 10-300nm in diameter. The size and size distribution can be controlled by adjustment of the concentration and residence time of the reactants in the flame [2]. A "narrow" distribution for this process would include diameters ranging over a factor of ~4. The particles are believed to nucleate rapidly and grow by coalescence [3]. The particle diameters seen in electron micrographs agree with those found by nitrogen adsorption, indicating that the particles are nonporous.

COLLOIDS

It is convenient to disperse the soot in a nonaqueous solvent in order to avoid:

(1) OH contamination (important for optical silica);
(2) dissolution of certain oxides (e.g. P_2O_5);
(3) high capillary stresses resulting from the high surface tension (γ) of water.

The low γ of organic solvents (~20-30 ergs/cm^2 vs 72 ergs/cm^2 for H_2O) facilitates drying of gels without cracking. Agglomeration of the particles

Mat. Res. Soc. Symp. Proc. Vol. 32 (1984) Published by Elsevier Science Publishing Co., Inc.

is prevented by adsorbing a steric barrier of chain-like molecules on their surfaces [4]. Interpenetration of the adsorbed layers exacts a penalty in enthalpy (to displace a layer of solvation around the adsorbed molecules and entropy (to restrict the motion of the chains) [5]; it is not known which of these factors is dominant in these colloids. The closest approach of two particles is limited to about twice the thickness of the adsorbed layer. If, at that range, the energy of attraction is no more than a few kT, agglomeration is prevented and the colloid is stable. For example [4], soot can be suspended in chloroform by adsorbing 1-decanol ($C_{10}H_{21}OH$) on the particles. The hydroxyl group of the alcohol hydrogen bonds to a surface silanol, and the hydrocarbon chain is solvated by the nonpolar solvent. Such a colloid containing 20 vol.% retains a viscosity below 0.1 Pa·s for weeks and does not settle.

As shown in Table 1, the maximum solids content in a stable SiO_2/ chloroform/alcohol colloid increases with particle size and with the size of the adsorbed molecule. These factors also affect the microstructure of the dried gel as we shall see.

TABLE I.   Factors affecting maximum solids content

Surface Area (m^2/g)	Dispersant	Maximum Solids (Vol.%)
130	1-decanol	~ 18
50	1-decanol	~ 30
50	1-propanol	~ 20

The dependence of the viscosity (η) of the colloid on solids content and shear rate is shown in Figure 1 (soot surface area = $50 m^2/g$). Up to ~ 20 vol.%, η is Newtonian and insensitive to concentration, but higher loadings lead to non-Newtonian behavior. Evidently, there is some interparticle attraction (i.e. the Van der Waals forces extent beyond the absorbed layer) which influences the viscosity increasingly as the particles are crowded together. These attractive forces are responsible for the relatively low solids contents indicated in Table 1.

GEL

The colloid gels upon exposure to a base, such as NH_3 vapor or an amine, which deprotonates the silanol groups [4]. The negatively charged particles are drawn together by the positive ammonium ions to produce a semi-rigid gel. Whereas charges promote stability in polar solvents, the tendency to form ion pairs in a non-polar solvent leads to the opposite effect (6,4). As the solvent evaporates, capillary forces cause rearrangement of the particles, so that shrinkage is irreversible. The density of the green body (i.e. dry gel) increases with the solids content of the colloid, as shown in Figure 2. Note that the denser gel in Figure 2 has a narrow pore size distribution, even though the particle size distribution is rather broad. The particles are not packed in an orderly way, but in a random way. The narrow pore size distribution permits uniform sintering.

Figure 3 reveals that a better dispersing agent leads to a denser gel. The pore size of the gel increases with the particle size of the soot, as shown in Figure 4. When the surface area is $200 m^2/g$ (particle diameter ~14 nm), most of the pore volume consists of pores narrower than 10nm (the limit of the Hg porosimeter). Such gels are difficult to dry intact, because of the high capillary forces caused by the small pores.

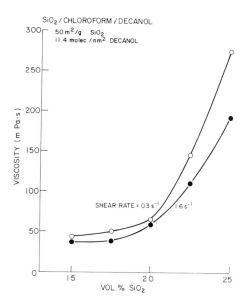

Fig. 1. Viscosity vs. solids content; soot surface area = $50m^2/g$

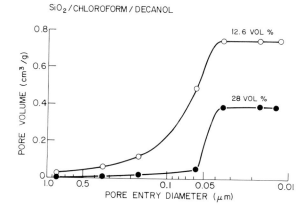

Fig. 2. Pore volume distribution in dried gels; soot surface area = $50m^2/g$

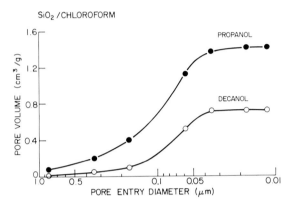

SiO$_2$/CHLOROFORM

Fig. 3. Pore volume distribution in dried gels; soot surface area
= 50m^2/g; solids content 20 vol.%.

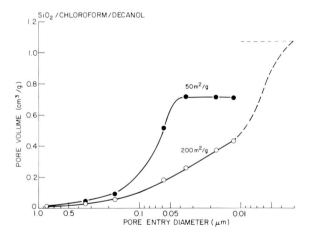

SiO$_2$/CHLOROFORM/DECANOL

Fig. 4. Pore volume distributions in dried gels; solids content
~ 18 vol.%.

TITANIA

TiO$_2$ soot produced according to reaction (2) can be dispersed in chloroform using oleic acid as a dispersant; the colloid gels when exposed to NH$_3$
vapor. Figure 5 shows the pore size distributions of dried gels made from
soots with different surface areas (equivalent particle diameters of 30 and
250nm); the particles are shown in Figure 6. Larger particles permit higher
solids loading in the colloid, and lead to a denser gel. The sintering
behavior of the denser gel, during heating at 5°C/min, is shown in Figure 7.
The shrinkage at ~ 800°C reflects the transformation from anatase to rutile,
and some associated rearrangement of particles. Sintering is rapid between
1100° and 1200°C.

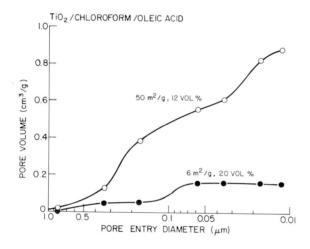

Fig. 5. Pore volume distributions in dried gels.

ALUMINA

Al$_2$O$_3$ soot made according to reaction (3) can be suspended in chloroform using methanol as a dispersant; the colloid is gelled by NH$_3$ vapor.
Using soot with 100m^2/g surface area (~15nm particle diameter) leads to a
gel with the microstructure shown in Figure 8. After firing at 1500°C, all
of the pores $\leq$ 100nm have closed, but the larger ones remain. By starting
with larger soot particles, and possibly using a better dispersant, a
narrow pore size distribution could be obtained in the green body. Then,
complete densification should be possible at ~1500°C.

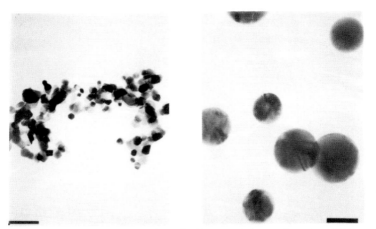

Fig. 6. Electron micrographs of TiO$_2$ particles made using different flame settings (Bar = 0.1μm).

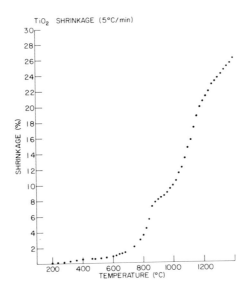

Fig. 7. Sintering of gel made from TiO$_2$ soot (6m^2/g) during heating at 5°C/min.

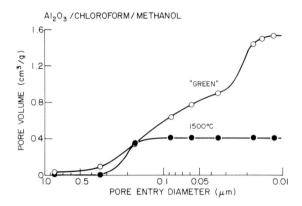

Fig. 8. Pore volume distributions in Al₂O₃ gel before and after firing at 1500°C.

CONCLUSIONS

Colloidal suspensions of flame-generated particles can be used to make amorphous and crystalline ceramic bodies. The particles are small enough to permit sintering at modest temperatures, but large enough to avoid disastrous capillary forces. The green microstructure is controlled by the particle size and solids content, and the quality and quantity of dispersant.

ACKNOWLEDGEMENT

This paper summarizes the work of many people, including John Luong, Dave Crooker, Roger Bartholomew, Gordon Foster, Doris Shaw, and Lina Echeverria.

REFERENCES

1. P. C. Schultz, Proc. IEEE, 68 [10] 1187-1190 (1980).
2. M. Formenti et al, J. Colloid and Interface Science, 39 [1] 79-89 (1972)
3. G. D. Ulrich, Comb. Sci. Tech., 4, 47-57 (1971).
4. G. W. Scherer and J. C. Luong, J. Non-Crystalline Solids, 63, 163-172 (1984).
5. D. H. Napper, J. Colloid and Interface Science, 58 [2] 390 (1977).
6. G. Féat and S. Levine, J. Colloid and Interface Science, 54 ,34, (1976).

AVOIDING CERAMIC PROBLEMS BY THE USE OF CHEMICAL TECHNIQUES

Peter E. D. Morgan
Rockwell International Science Center, Thousand Oaks, California 91360

INTRODUCTION

As the title of this meeting suggests, a paradigm change is underway in the ceramic community; methods of synthesis and use of ceramic powders are being hurriedly reappraised. The imperative for change was all in place by an important meeting that took place in 1977 [1].

POWDER PROBLEMS

Nagging problems with the use of powders for the production of ceramics break down into three broad categories:

Diffusion Problems

Much ceramic research has concerned itself with the way that oxides interact to produce a final polycrystalline product, preferably with high density and strength. It was assumed that the starting materials should be mixed oxides. However, oxides are not generally the large scale primary products of the chemical industry. One of the few that is, rutile TiO_2, is available for the paint industry. The only other really large scale products that I can think of that are at least closely related to an oxide are bayerite and boehmite; the former, is the starting material for the production of aluminum and the latter material a by-product of the Zeigler alcohol process. These two hydroxides produce very active oxides as they decompose; first to convert them into calcined oxides prior to producing ceramics is the approach that has been taken previously.

Primary products of the chemical industry come directly from the extraction and purification of the elements from their ores. It is these chemicals that should be used directly to produce ceramics. Slow diffusion reactions between oxides and the intermediate and/or metastable products that may be produced before the final ceramic form are problems to be avoided. Rather, one should develop new reactive chemical routes utilizing the naturally available large scale inorganic sources for the final ceramic. This will be illustrated later. The use of reactive compounds can avoid various ceramic problems at high temperatures: the loss of volatiles, excessive grain growth, or eutectics leading to disproportionation, and so on.

Apart from large scale bulk chemicals, certain specialty chemicals are becoming much more readily available because of their use in other fields: (for example, alkoxides, e.g. Tyzor™, acetylacetonates for catalysts and interesting stearates used by the paint industry). A present curious feature of the ceramic industry is that the methods used on the relatively large scale are quite similar to methods that would be used by a student studying ceramic forming in the laboratory. The chemical industry on the other hand uses methods almost totally different from the original academic studies. The future ceramic industry must adopt chemical techniques used by the chemical industry on the large scale. This would include highly automated handling of (usually) liquids, control of particle size and shape, and so on, as is already done in the paint industry. If not picked up by the ceramics industry, this will surely be done by the chemical industry.

Mat. Res. Soc. Symp. Proc. Vol. 32 (1984) Published by Elsevier Science Publishing Co., Inc.

We have realized over many years, summarized by Lange [2], that, in the first stages of sintering in a _real_ powder compact, more aggregated or densely packed regions of a compact differentially densify then support local grain growth [3], so that an intermediate porous skeletal network structure of larger, second generation, grains is rapidly formed. (This, as in covalent materials, may be with little overall shrinkage.) Now we must examine the effect of diffusion at the level of these inhomogeneous regions. _Enhanced_ diffusion, leading to densification and/or grain growth within the aggregates, may _diminish_ subsequent sintering and vice versa. Enhanced vaporization-condensation within aggregates can result in the same effect. When such powders are used, the interpretation of results of doping, by measuring the external dimension of compacts as they sinter without separately following the microstructure evolution in detail, cannot lead to any understanding. Without much, if any, densification in the early stages, it was obvious to observers that compacts strengthened so that internal "welding" was occurring. In powders as well as compacts great reduction in measured surface areas and large increases in average grain sizes were often unaccompanied by significant densification at lower temperatures. The characterization of powders and compacts by average surface area, porosity, particle size, etc., did not lead to any predictability in sintering because _geometrical packing arrangement_ including topochemical effects was controlling.

Impurity Problems

The effects of small amounts of impurities on the decomposition of precursors or low temperature aggregate formation or grain growth, has been recognized for many years but hardly studied in any rigorous manner; especially the presence of ubiquitous alkalis and chloride is a problem. A recent paper on the decomposition of dolomites exemplifies this [4]. With the move to higher purity starting materials the effects of impurities are exaggerated; the effects upon decomposition of precursors, aggregate genesis and subsequent sinterability are extremely complex. With active powders these phenomena are substantially controlled by boundary or surface diffusion and _not_ by bulk diffusion. Attempts to relate "active" sintering to bulk diffusional phenomena over many years have been total failures. The relationship of grain growth, which is surely a boundary phenomenon, to impurity effects is also largely unknown.

As mentioned before, we must now review the effects in terms of the evolutionary development of the microstructure. As before the overall average effect does not tell us what has happened in each successive stage. We can imagine many possibilities:

1. An impurity or additive _reduces_ welding in aggregates allowing some rearrangement between aggregates (or easier hot pressing in the early stages), then later stage densification is _enhanced._ This would include the presence of some kind of "fluxing" action (as for boundary films) within aggregates stimulating the rearrangement.

2. An impurity greatly enhances welding in an aggregate but does not support grain growth in the aggregate (possibly an evap-condensation mechanism) or second phase. This can lead to great inhibition of sintering.

3. As 2, but grain growth enhanced; leads to initial slow sintering but later rapid sintering with large grain size and possible discontinuous grain-growth. Grain growth at this stage might be especially deleterious if, as was suggested and demonstrated for β-aluminas [5], it leads to anisotropic platelets or needle-like crystals of the second generation stage [6].

Many other complex situations are possible, which might explain the rate controlled effects occasionally reported [7], including the effects of flash heating (whereby aggregates or units might be able to deform or rearrange before grain growth "stiffens" them).

One of the great difficulties in this area is that of holding all parameters, save one, constant. It would be extremely difficult to produce a range of powders differing only in the impurity level, since this will have already affected the evolution of the powder from its precursor. There is, however, one general guideline that I think has not been properly studied. Many years ago, Pampuch [8] studied the effect of impurities upon the sintering of beryllia and thoria. His conclusions were, however, different from those I now wish to put forward, although he appreciated the overall importance of surface energy effects (which might now be recast at least partially in terms of dihedral angle [2]. Other work carried out in the early 60s supported the general idea that impurities affect sintering and grain growth in some similar fashion to the way in which they affect eutectic temperatures. In the sense that a pure material "knows" where its melting point is, so also, even while below the eutectic temperature, the doped material seems to behave in such a way as to ratio its properties to the eutectic temperature, as the pure material does to its melting point (this very old idea is related to the Tammann homologous temperature). Above the eutectic temperature, it is obvious that very large, usually enhancing, phenomena occur due to the presence of a liquid phase and maybe this has obscured the fact that "sub-eutectic effects" are equally important.

In the case of covalent materials, which were recognized many years ago to perform quite differently [9], boundary effects are quite predominant due to the formation of glassy grain boundary phases; sintering, hot pressing, creep and oxidation of such products are entirely controlled by this.

The relative ease with which the chemical industry has been able to produce ultrapure silicon or ultrapure silica glass for optical fibers implies that, given the opportunity, this industry will have no problem whatsoever in producing high purity chemicals on a scale sufficient to satisfy the ceramic market.

Aggregate Problems

At the inception of the production of very active powders by decompositional approaches, the aggregate problem was immediately realized [10]. If decomposition is carried out at too low a temperature, fine aggregated particles are produced, which often bear a pseudomorphic relationship with the precursor salt. At intermediate calcination temperatures more nearly monosized particles are produced and at higher temperatures "dead burning" (too much crystal growth), which reduces the driving for sintering, occurs. This leads to a maximum in the density vs precalcination temperature [10,11] Ref. [10] clearly shows diagrams to explain the maximum effect. Shortly thereafter Reeve [12] published a paper, which by its very title, indicated differential shrinkage to be the major problem in active or nonuniform systems. Rhodes [13] followed this up by cleverly removing aggregates from a slurry by centrifugation, but had to republish the essential findings many years later [14] before sufficient note was taken. Many observers commented upon the fact that agglomerates lead to fatal flaws in sintering [1].

An attempt was made early on to circumvent the aggregate problem by the use of reactive hot-pressing [15], even as the aggregates formed they might be deformed and pressed closely together to avoid the differential effect. It was implied that hot pressing was a similar phenomena. Differential shrinkage is expected to produce small stresses (~ few hundred psi) between regions shrinking differently [16]. This implies that the application of overpressures of only that magnitude may be sufficient to overcome the differential effect. This may already have been detected in Fig. 13 of Ref. [17]. Suggested is more study of the potential of low pressure "warm pressing."

To illustrate that aggregate particle rearrangement is important, it is found [18] that if shrinkage in sintering or hot pressing is followed (i.e., strain rate versus stress to some exponent $\dot{\varepsilon} = k\sigma^n$) and at a given porosity the stress is suddenly changed, i.e., $\dot{\varepsilon}_1/\dot{\varepsilon}_2 = (\sigma_1/\sigma_2)^n$, then n, calculated from tangents to the curves at the change point lead to very high values of n, e.g., 6-12, if stress is raised, or values near 0 if stress is suddenly lowered. This was explained as a rearrangement mechanism only occurring when the stress is raised. With reduced stress then the structure is essentially fixed until slower mechanisms catch up allowing shrinkage to recommence. The continuum type diffusional hot pressing models of the period (Nabarro-Herring creep, even though it had been shown that ordinary creep in ceramics involved more complex mechanisms! [19]), could not explain the results.

One method of influencing the early aggregate genesis and development would be to include second phase particles within the aggregates. Results on this are conflicting. In the case of decomposition of mixed crystals, which leads to this effect in reactive hot pressing [17], densification seemed to be assisted especially for dolomite, $CaMg(CO_3)_2 \rightarrow CaO + MgO$. However, in work where mixed Al/Zr alkoxides have been used [20] sintering seems to be greatly inhibited. In the former case, sufficient impurities may have been present to "flux" the rearrangement of the aggregates though thin boundary films that may be more frequently present at high temperatures than has been imagined. In the latter case, high purity may lead to an inhibition with the included particles preventing grain growth in the aggregates. The effect may also change with the volume percent of the inclusions, in the former case ~50%, in the latter ~10% of ZrO_2 in $\gamma-Al_2O_3$.

If the effect of grain growth inhibition in the later stages [21] is extended down to the inner-aggregate level some very interesting effects should result.

The earlier idea of using high green pressure [22], presumably crushing soft aggregates together, is being reviewed [23] the achievement of very high green density, with aggregates closely in contact, can lead to full density whereas at lower green pressing pressures this is impossible. The development of monosized particles and studies of their packing is certainly one way to avoid aggregate problems; the suspicion remains, however, that there are many alternative ways of handling this.

SOME USES OF CHEMICAL TECHNIQUES

It is obvious from the foregoing that each step of the sintering process evolves to the next step and that by observing only the average effect over time (e.g., density or overall grain-growth) little can be gleaned of exactly how chemical effects may be operating. In the past, much work has been done on preparing powders by different routes, i.e., from hydroxides, carbonates, oxalates, etc. and on doping them, but analysis was essentially impossible given that only simple models existed for theoretical guidance. An example is shown in Fig. 1 of density results

(with a large scatter!) as a function of temperature for magnesias, pure
and doped by different techniques [24]. With the present environment we
can begin to imagine how to analyze the data (although exacting micro-
graphic confirmation is needed at each step).

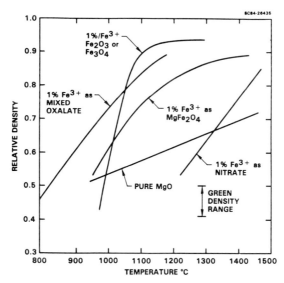

FIG. 1. Density/temperature plots for magnesium
doped in various ways with 1% Fe^{3+}.

General awareness is that iron, at the level of ~1%, stimulates the
average sintering of magnesia. It was long suspected that this was not
due to a simple bulk diffusion vacancy type enhancement because Al^{3+} has
little effect while Cr^{3+} is a great sintering inhibitor even though magnetic measurements show it to go into solid solution. The data shows
that the effect of iron depends greatly on the method of mixing. The
original pure magnesia is fired from synthetic ultrapure magnesite at
900°C, has a particle size, within aggregates, of ~0.07 μm and gives
densities as shown. If Fe^{3+} is added via, an iron nitrate in alcohol
slurry and the powder refired to ~900°C very little effect on agggregates
is seen - the particle size is now ~0.11 μm by TEM. The rough trend is
that average densification is not much affected but is a little slower at
low temperatures and higher at higher temperatures. This might now be
seen as stimulating aggregate welding and strengthening at lower temperatures (e.g., 300°-800°C) with little densification until new effects
appear at higher temperatures.

An unexpected effect was when Fe_2O_3 or Fe_3O_4 was added as a second
phase phase because average sintering was enhanced, this did not fit the
bulk diffusional model at all. The effect may be that, with the iron in
this form, it becomes available only at the higher temperatures, so that
the aggregate welding may be diminished initially; the aggregates may
then be able to rearrange more readily together before enhanced grain
growth, as iron becomes available to boundaries, allows final pores to
shrink [2]. Iron as $MgFe_2O_4$ should be less available (by being nearer to
equilibrium) and this seems to be so. When the iron was added by co-
precipitating a Mg-Fe mixed oxalate and decomposing that at ~900°C then
particle sizes are only ~0.05 μm and the powder sinters well.

Only the most exacting studies are going to clear up the difficulties of interpretation in these cases. When several impurities or doping agents are present, the situation rapidly becomes appallingly difficult. Whereas 2% Ti^{4+} slightly stimulates sintering and grain growth in the early stage of densification of ultrapure alumina (from REMET's dispural®) when Na^+ content is low, in the case of a more ordinary alumina (ALCOA C331) the same addition of Ti^{4+} interacts with the higher level of soda (we believe) to greatly stimulate densification and <u>inhibit</u> grain growth [25]. As the complaint was made years ago that the behavior of Si_3N_4 could not be understood in the absence of phase diagrams, so the Al_2O_3-TiO_2-Na_2O system must be known before more definite conclusions can be reached. However, a eutectic exists in the TiO_2-Na_2O system at 985°C and may lead to both the grain growth inhibition and sintering enhancement in the C331 case. It is tempting to suggest that all sintering and grain growth studies carried out, e.g., in $MoSi_2$ element furnaces, unless special precautions are taken, have extrinsic effects due to the abundant Mo and Si found in these furnace atmospheres.

Fused salt methods are developing for the production of needle or plate-like particles, especially for ferroelectric [26] and ferromagnetic materials. Electric field [27] and magnetic field alignment and appropriate consolidation can lead to very high green density bodies with large preferred orientation. The former has been used with SbSI [27] and the latter for the new barium hexaferrite particles for perpendicular recording tapes. The sol-gel method, long realized to be an alternative to powder techniques, has recently become a large scale industrial process for the production of grinding grain [28] and promises soon, by similar techniques, to be the method of choice for many extremely dense ceramics. Methods use a by-product boehmite available from the industrial Ziegler alcohol process. Sumitomo and DuPont have both brought alumina fiber, made by sol-gel techniques, to the market. Ube Industries of Japan have commercialized a liquid ammonia/silicon tetrachloride reaction at ambient temperature [29] for the production of silicon nitride. In this author's opinion, the liquid ammonia method will supplant gas phase chemistry.

We ourselves are working in several areas where high degree of chemical activation are required. By a hydrothermal technique we have recently been able to produce <u>60Å monoclinic</u> zirconia particles (to be published). This seems to refute an idea [30] that Garvey had <u>very tentatively</u> put forward suggesting that there was a minimum size for the stability of monoclinic zirconia that might lay around 100Å or so, Clearfield [31] having earlier already produced 120Å monoclinic zirconia, i.e., powder as fine as some of the recent hydrothermal work in Japan. Although it could be claimed that fine monoclinic zirconia is stabilized by the high concentration of nitrate ion that is present in this hydrothermal reaction, clearly chemistry has an important role to play in deciding such questions.

In another area we have been engaged in long-term studies of synthesis of many members of the very large magnetoplumbite family and this led us eventually to a new, but related, family of structures having about four times the unit cell volume. One member of this family, $CaTi_3Al_8O_{19}$, CTA, turned out to be a really challenging synthetic problem. Details are elsewhere [32] but briefly the trick, starting with alkoxide and nitrate precursors, was to fire with a very slow temperature ramp up to ~1300°C over a period of three to four days at least. Whereas under normal conditions of fast heating, only the merest traces of this compound would be seen at the level of 1-2%, greater than 95% pure compound could be achieved with the slow heating rate. The fact that work upon the calcium-titanium-aluminum-oxygen system has not identified this compound before suggests how difficult its formation actually is. This compound and the magnetoplumbites turn out to be especially interesting for the 4.55 billion year old so-called CAIs, high calcium aluminum (and titanium) inclusions [33]; in meteorites they are suspected to be amongst the very early grains, either to have condensed in the solar nebular before the formation of the solar system, or to have been produced by some very early heating episode in which other elements were volatilized off.

Another case where high reactivity is required, and which is easily tackled by starting with alkoxide precursors, are new oxynitride crystal structures. Appropriate alkoxides and/or nitrates are decomposed at low temperature, then fired in ammonia gas; a silicon-nitrogen substituted gehlenite series has been achieved, including $Ca_2AlSi_2O_6N$ as well as a similarly coupled substituted $CaAl_4O_7$. Certain early solar system scenarios include quite reducing environments which may have produced these compounds [33]; Si_2N_2O, sinoite, and TiN, osbornite, are both already well known in enstatite chondrites. In systems containing silicon and similar covalently bonded atoms, one must be able to form the oxynitride crystals at temperatures lower than where glasses substantially begin to form.

The chemical preparation of silicon nitride is of continuing interest. The reaction of silicon tetrachloride and ammonia for this purpose has a long history commencing in 1879 with the work of Schutzenberger using originally gas phase reactions [34]. In 1903, Blix and Wirbelaur [35] studied the reaction of silicon tetrachloride with liquid ammonia. In the 1950s in France several workers continued to study these reactions noting that the removal of chlorine from silicon di-imide or amorphous silicon nitride was extremely difficult [36] Liquid phase work continued in the United States both in organic [37] and ammonia [38]. It was shown that the liquid ammonia method was capable of producing silicon di-imide with very low chloride content and, if the material was highly pure, a reactive hot pressing mechanism was possible. Overall density achievement was 85%, with local regions of beta silicon nitride being fully dense. It was shown recently that, when chloride is retained in silicon nitride, it appears in the grain boundary glass [39]. One possible mechanism for the inhibition of sintering in silicon nitride by chloride may [38] be the production of a more ionic chloride containing glass in which silicon nitride is less soluble [40].

In a brief attempt to react silicon tetrachloride, nitrogen and hydrogen at ~1400°C, which thermodynamically is predicted to produce silicon nitride and HCl ($\Delta G = -32.3$ kcal, mole^{-1}), it was rediscovered that silicon tetrachloride attacks alumina or mullite tubes [41] ruling out this method. It was conceived that a move to silicon-sulfur chemistry would avoid this problem. The beneficial use of sulfur as a catalyst [42] during the nitridation of silicon seems to be a result of the much higher vapor pressure of SiS (1 atmosphere at 900°C) as opposed to SiO. Exaggerated vapor phase transport did not lead to changes in alpha-beta ratios except through "seeding" effects demonstrating the importance of substrate reactions [42]. This has recently been nicely confirmed [43]. The use of silicon-sulfur-nitrogen chemistry has much to recommend it over that of the silicon-halogen route. The reaction of silicon disulfide SiS_2 with liquid ammonia has already been studied [35]. Ammonium sulfide is much more soluble than ammonium chloride in liquid ammonia. The potential for intermediate silicon-sulfur-nitrogen compounds as liquid phases or other polymeric species also leads to obvious extensions of the chemistry to produce silicon nitride.

Preliminary tests on this are being carried out. Argon/10% H_2S is passed over finely powdered silicon at 900°C in alumina boats. Beds of long whiskers of silicon disulfide are produced (Fig. 2). Liquid drops are sometimes seen on the ends of the whiskers (Fig. 3). The silicon disulfide product is cooled to room temperature in argon and then argon/ 10% ammonia passed over while slowly heating up to either ~1250°C or ~1450°C; a variety of products have been seen. A spaghetti form of alpha-silicon nitride produced at 1400°C is shown in Fig. 4. We believe this must have been formed by a VLS, vapor-liquid-solid, mechanism. Slightly sintered spheres of amorphous silicon nitride produced at 1200°C are shown in Fig. 5 and the possible mechanism by which they form

FIG. 2. SiS$_2$ whiskers grown at 900°C in A/10% H$_2$S, 100 μm marker.

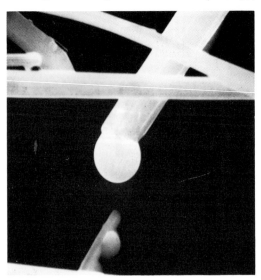

FIG. 3. SiS$_2$ whiskers showing VLS droplets on the ends, 1 μm marker.

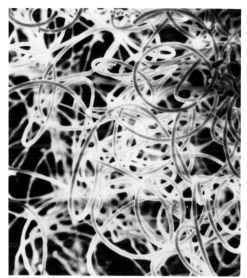

FIG. 4. "Spaghetti" morphology of α-Si_3N_4, from SiS_2 + NH_3, room temperature → 1400°C, 1 μm marker.

FIG. 5. Spherical am-Si_3N_4 from from SiS_2 + NH_3, 900°C → 1250°C

illustrated in Fig. 6. Figure 7 indicates a vermiform type of alpha-silicon nitride that similarly seems to have been produced at up to 1450°C by a VLS mechanism. If ammonia is passed over SiS_2 needles starting at 900°C, a more surprising result is seen as shown in Fig. 8. The top temperature here was about 1250°C and the product by XRD and IR is amorphous. EDAX shows no sulfur in this product and firing to higher

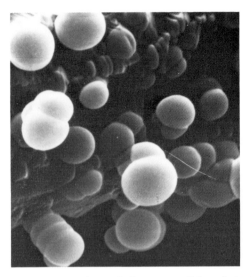

FIG. 6. Growth morphology on a SiS_2 substrate
with NH_3, 900 → 1250°C, 1 μm marker.

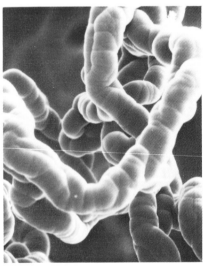

FIG. 7. A "vermiform" $\alpha-Si_3N_4$ at 1400°C,
also VLS mechanism, 1 μm marker.

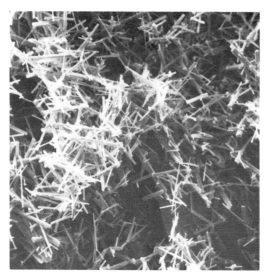

FIG. 8.   am-Si₃N₄ relic (pseudomorphic) "whiskers", 10 μm marker.

temperatures produces alpha-silicon nitride. At larger magnification the
same product is shown in Fig. 9. At 900°C ammonia attacks silicon
sulfide, which has a linear silicon-sulfur chain structure, to produce
relic, or pseudomorphic, amorphous "whisker" silicon nitride. The
largest whisker in Fig. 9 shows markings which suggests how the silicon
sulfide was formed. The markings appear to be previous positions of a

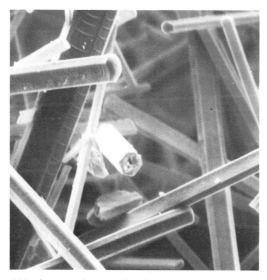

FIG. 9.   am-Si₃N₄ relic (pseudomorphic) "whiskers", 1 μm marker.

liquid droplet at the end of the whisker as can be seen from the shape of one of the whiskers which crosses the whisker in question. Much work remains to be done to elucidate the chemistry of this system, but results tend to confirm the idea that SiS vapor chemistry is important and also indicates that a large Si-S bond polymer and glass chemistry may exist where such polymers and glasses could be treated with ammonia, other gases, etc., to produce silicon nitride forms. We are adapting these techniques to the production of silicon carbide which also works.

It was suggested [42] that the compound Si_2N_2S, analogous to Si_2N_2O, does not exist. This compound still has not been detected in the current work although we do have some evidence for solid solutions. The new SiS_2 chemistry immediately opens up new areas for the forming of powders, pseudomorphic whiskers, fibers, uniform sized particles, etc.

CONCLUSIONS

Some new directions discussed here hopefully exemplify the title of this meeting.

The greater use of chemical techniques is tending to generalize ceramic ideas showing new relationships between ceramic and other sciences. In the past there have been only weak interactions, for example, between ceramists and colloid chemists or between powder formers for ceramics and catalyst experts. This situation is now rapidly changing.

ACKNOWLEDGEMENT

The Si-S-N work was performed for the DoE-OBES, Contract No. DE-AC03-78ER01885, Eloise A. Pugar very ably performed this experimental work. Other work was performed on Rockwell Internal Research and Development funding including the oxynitride study where, again, Eloise A. Pugar carried out the experimental work. Seminal discussions with F. F. Lange are greatly appreciated.

REFERENCES

1. Processing of Crystalline Ceramics, 14th University Conference, Eds. H. Palmour III, R. F. Davis, I. M. Hare, Plenum Press (1978).
2. F. F. Lange, "Sinterability of Agglomerated Powders," J. Am. Ceram. Soc. 67, 83 (1984).
3. C. Greskovitch and K. W. Lay, "Grain Growth in Very Porous Al_2O_3 Compacts," J. Am. Ceram. Soc. 55, 142 (1972).
4. M. Q. Li and G. L. Messing, "Chloride Salt Effects on the Decomposition of Dolomite," Thermochim. Acta. 68, 1 (1983).
5. U. Chowdhry and R. M. Cannon, "Microstructural Evolution during the Processing of Sodium β-alumina," Processing of Crystalline Ceramics, Eds., H. Palmour III. R. F. Davis, and T. M. Hare (Plenum Press) (1978), 443.
6. P. F. Becker, J. H. Sommers, B. A. Bender and B.A. MacFarlane, "Ceramics Sintered Directly From Sol-Gels," ibid, 79.
7. M. L. Huckabee, T. M. Hare and H. Palmour III, "Rate Controlled Sintering as a Processing Method," ibid, 205.
8. (a) R. Pampuch, "Contribution au Problem du Frittage Oxydes Purs avec Additions." Bull. Soc. Franc. Ceram. 46, 3 (1960). (b) R. Pampuch, "Sinterung Reiner Sowie Aktivierter Oxyde in Festen Zustand," Silika-Techn. 10, 69 (1969).

9.   P. E. D. Morgan, "The Sintering of Zinc and Cadmium Sulfides," Proc. Intl. Conf. Sintering and Related Phenomena, Notre Dame, Ind., 543, June (1965) Gordon and Breach, (1967) Ed., G.C. Kuczynski.

10. D. T. Livey, B. M. Wanklin, M. Hewitt and P. Murray, "The Properties of MgO Powders Prepared by the Decomposition of Mg(OH)$_2$," Trans. Brit. Ceram. Soc. 56, 217 (1957).

11. J. F. Quirk, "Factors Affecting Sinterability of Oxide Powders," J. Am. Ceram. Soc. 42, 178 (1959).

12. K. D. Reeve, "Nonuniform Shrinkage in Sintering," Am. Ceram. Soc. Bull., 42, 452 (1963).

13. T. Vasilos and W. Rhodes, "Fine Particulates to Ultrafine-grain Ceramics," 137, in Ultrafine-Grain Ceramics, J. J. Burke, N. L. Reed, and V. Weiss (Eds.), Syracuse Univ. Press (1970).

14. W. H. Rhodes, "Agglomerate and Particle Size Effects on Sintering Yttria-Stabilized Zirconia," J. Amer. Ceram. Soc. 64, 19 (1981).

15. P. E. D. Morgan and E. Scala, "Fully Dense Oxides by the Pressure Calcintering of Hydroxides," Intl. Conf. on Sintering and Related Phenomena, Notre Dame, Ind., June (1965), Gordon and Breach, 861 (1967), Ed., G. C. Kuczynski.

16. B. J. Kellett and F. F. Lange, "Stresses Induced by Differential Sintering in Powder Compacts," in press, J. Am. Ceram. Soc.

17. P. E. D. Morgan, "Superplasticity in Ceramics," Ultrafine Grain Ceramics, Ed., J. J. Burke, N. L. Reed, V. Weiss, Syracuse University Press, 251 (1970).

18. P. E. D. Morgan and N. C. Schaeffer, "Chemically Activated Pressure Sintering of Oxides," Technical Report AFML-TR-66-356, NTIS-AD-815066, Feb. (1967).

19. J. H. Hensler and G. V. Cullen, "Grain Shape Changes During Creep in Magnesium Oxide," J. Am. Ceram. Soc. 50, 584 (1967).

20. Unpublished work in progress.

21. F. F. Lange and B. I. Davis, "Sinterability of ZrO$_2$ and Al$_2$O$_3$ Powders: The Role of Pore Coordination Number Distribution," in press, J. Am. Ceram. Soc.

22. P. E. D. Morgan, "Chemical Processing of Ceramics (and Polymers)," 14th University Conference, Processing of Crystalline Ceramics, Eds. H. Palmour III, R. F. Davis, I. M. Hare, Plenum Press (1978).

23. S. Prochazka, "Optically Translucent Ceramics," U.S. Patent No. 4,427,785, Jan. (1984).

24. P. E. D. Morgan, "Sintering and Grain Growth in Metal Oxide Powder Compacts," Ph.D. Thesis, London University (1963).

25. P. E. D. Morgan and J. E. Flintoff, unpublished work.

26. T. Kimura and T. Yamaguchi, "Morphology of Bi$_2$WO$_6$ Powders Obtained in the Presence of Fused Salts," J. Matl. Sci. 17, 1863 (1982).

27. P. E. D. Morgan, "Preparation and Electric Field Alignment of SbSI Crystals," Comm. Am. Ceram. Soc. C82 (1982).

28. M. A. Leitheiser and H. G. Sowman, "Non-Fused Aluminum Oxide-Based Abrasive Material, U.S. Patent No. 4,314,827 (1982).

29. T. Iwai and T. Kawahito work, "Process for Producing Metallic Nitride Powders," U.S. Patent No. 4,196,178, April 1 (1980).

30. R. C. Garvie, "The Occurrence of Metastable Tetragonal Zirconia as a Crystallite Size Effect," J. Phys. Chem. 69, 1238 (1965).

31. A. Clearfield, "Crystalline Hydrous Zirconia," Inorg. Chem. 3, 146 (1964).

32. P. E. D. Morgan, "Preparing New Extremely Difficult to Form Crystal Structures," Mat. Res. Bull. 19, 369 (1984).

33. P. E. D. Morgan and E. A. Pugar, "New Compounds and Phase Relations, Implications for Refractory Inclusions in Meteorites," Lunar and Planetary Science XV, 566 (1984).

34. M. P. Schutzenberger, Sur l'azoture de silicum, Compte Rendus, Academie des Sciences, Paris, 89, 644 (1879).

35. M. Blix and W. Wirbelauer, "Ueber das Siliciumsulfochlorid $SiSCl_2$, Siliciumimid, $Si(NH)_2$, Siliciumstickstoffimid (Silicam), Si_2NH_3H und den Siliciumstickstoff, Si_3N_4," Ber. 36, 4220 (1903).

36. M. Billy, M. Brossard, J. Desmaison, D. Giraud and P. Goursat, "Synthesis of Si and Ge Nitrides and Si Oxynitride by Ammonolysis of Chlorides", J. Am. Ceram. Soc. 58, 254 (1975).

37. K. S. Mazdiyasni, "Synthesis, Characterization, and Consolidation of Si_3N_4 Obtained from Ammonolysis of $SiCl_4$," J. Am. Ceram. Soc. 56, 628 (1973).

38. (a) P. E. D. Morgan, "Research on Densification, Character, and Properties of Dense Silicon Nitride," Office of Naval Research, AD-757,748, March (1973). (b) P. E. D. Morgan, "Production and formation of Si_3N_4 from Precursor Materials," Office of Naval Research, AD-778,373, Dec. (1973).

39. D. R. Clarke, "Densification of Silicon Nitride: Effect of Chlorine Impurities," J. Am. Ceram. Soc. 65, C21 (1982).

40. R. Raj and P. E. D. Morgan, "Activation Energies for Densification, Creep and Grain Boundary Sliding in Nitrogen Ceramics," J. Am. Ceram. Soc. 64, C143 (1981).

41. A. S. Berezhnoi, Silicon and Its Binary Systems, translated from Russian, Consultants Bureau (1960).

42. P. E. D. Morgan, "The α/β-Si_3N_4 Question," J. Mat. Sci. 15, 791 (1980).

43. H. Inoue, K. Komeya and A. Tsuge, "Synthesis of Silicon Nitride Powder from Silica Reduction," Comm. Am. Ceram. Soc. C205 (1982).

PRECURSOR CHEMISTRY EFFECTS ON DEVELOPMENT OF PARTICULATE MORPHOLOGY DURING
EVAPORATIVE DECOMPOSITION OF SOLUTIONS

TIMOTHY J. GARDNER,* D. W. SPROSON,** AND G. L. MESSING**
*Organization 7472, Sandia National Laboratories, Albuquerque, NM 87185;
** Department of Materials Science and Engineering,
The Pennsylvania State University, University Park, PA 16802

ABSTRACT

Fine-grained, high surface area MgO, NiO and ZnO powders
were synthesized by the evaporative decomposition of solutions (EDS) technique at 1000°C from acetate and nitrate
salt solutions. The powder characteristics were similar in
all cases; however, aggregated powders were obtained from
the nitrate salts and aggregate-free powders were obtained
from the acetate solutions when reacted in air. This difference is attributed to the oxidation of the acetate radical and/or its residue and that this process acts to disag-
gregate the salt droplet/particles that are formed during
EDS.

INTRODUCTION

The reliable fabrication of many electronic and structural ceramic components requires high purity, fine-grained powders. Chemical methods to
synthesize powders with these characteristics include freeze drying [1],
sol gel [2], co-precipitation [3], and hydrolysis of organometallic
compounds [4]. The product of these techniques is usually a salt or hydroxide that requires calcination to obtain the oxide powder. A powder synthesis technique that includes the control over powder characteristics afforded
by chemical synthesis methods and the controlled decomposition of the product salt is the evaporative decomposition of solutions (EDS) technique [5].
EDS refers to a process in which a solution is atomized into a vertical
furnace such that the solution droplets initially dehydrate to yield a
finely divided salt which then subsequently decomposes to form the oxide
powder before exiting the furnace. This process has been successfully utilized for both simple and multicomponent oxide powder production on an industrial scale [6].

It is obvious that the preparation of ceramic powders by EDS requires
careful manipulation of the solution, atomizer and furnace parameters during the process. An often neglected aspect of EDS is the selection of a
salt for solution preparation. Because the primary criterion for salt selection is water solubility, most research on EDS has been "precursor specific" with only the metal nitrate and chloride salts being studied [5,6].
Recently, it was demonstrated [7,8] that the type of salt significantly affects the development of surface area, crystallite size and particle morphology in MgO powders synthesized under identical process conditions.
Specifically, it was shown that magnesium acetate yielded an aggregate-free
powder whereas magnesium nitrate and magnesium chloride resulted in strongly
aggregated powders.

To determine whether this observation was specific to the magnesium
salts, zinc oxide and nickel oxide powders were synthesized by EDS from
their nitrate and acetate salts. The powder characteristics and the decomposition reaction path for each salt were determined for a series of process

conditions to permit the correlation of the precursor chemistry with the development of the powder morphology.

EXPERIMENTAL PROCEDURE

One molar salt solutions were prepared from reagent grade $MgNO_3 \cdot 6H_2O$, $Mg(C_2H_3O_2)_2 \cdot 4H_2O$, $Zn(NO_3)_2 \cdot 6H_2O$, $Zn(C_2H_3O_2)_2 \cdot 2H_2O$, $Ni(NO_3)_2 \cdot 6H_2O$, and $Ni(C_2H_3O_2)_2 \cdot 4H_2O$. The salt solutions were atomized into the EDS furnace (Figure 1) at 4 ml/min and with an atomizer air pressure of 0.14 MPa to yield solution droplets with a median droplet size of 10 micrometers [9]. The powders were separated from the exhaust gas with a standard jet mill bag. Each powder was characterized by X-ray diffraction, BET surface area, X-ray line broadening, and SEM. The decomposition characteristics of the salts were studied by differential scanning calorimetry (DSC) and thermogravimetric analysis (TGA) in either air or nitrogen at heating rates of 10 and 100°C/min.

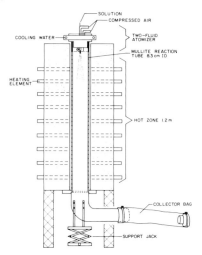

FIG. 1.   Schematic of the EDS Furnace.

RESULTS AND DISCUSSION

Thermal Analysis

The DSC results for the zinc salts heated at 10°C/min are illustrated in Figure 2. If the decomposition reactions of zinc nitrate and acetate in air are compared, it is seen that they are similar except that the acetate undergoes an exothermic reaction. That is, they both show an initial dehydration at <200°C in which the water of hydration is evolved to yield an anhydrous salt. Above 200°C the anhydrous nitrate melts at 220°C, then decomposes by an endothermic process to form ZnO. While the acetate also dehydrates at <200°C and melts at 260°C, it undergoes a series of endothermic and/or exothermic reactions, depending on the reaction atmosphere in which the acetate radical is decomposed. When the acetate is decomposed in nitrogen there is no evidence of the exothermic reactions that were observed when it was decomposed in air. Thus, it can be concluded that part of the decomposition of the zinc acetate involves oxidation of the acetate radical and/

or its residue. Oxidation of a carbonaceous residue is postulated on the basis of observations of incompletely decomposed acetate-derived powders which were gray to black in appearance and the detection of free carbon in these powders [7,8]. Similar DSC results were obtained for both the magnesium and nickel salts.

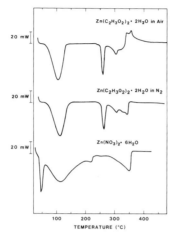

FIG. 2. DSC Results from $Zn(C_2H_3O_2)_2 \cdot 2H_2O$ and $Zn(NO_3)_2 \cdot 6H_2O$ heated at $10°C/min$ in air and/or N_2.

The TGA results were used to interpret the decomposition path of the salts and to determine the temperature at which the sample stops losing weight, which is referred to as the temperature of full decomposition (Table 1). In general, the acetate salts decompose at a lower temperature when reacted in air. This is attributed to the additional heat provided to the sample from the exothermic pyrolysis of the acetate radical and its residue.

TABLE 1. Thermal Analysis of Salt Precursors.

Salt	Heating Rate (°C/min)	Atmosphere	Temperature of Full Decomposition (°C)
$Mg(NO_3)_2$	10	Air	500
$Mg(NO_3)_2$	100	Air	610
$Mg(C_2H_3O_2)_2$*	10	Air	400
$Mg(C_2H_3O_2)_2$	10	N_2	575
$Mg(C_2H_3O_2)_2$*	100	Air	500
$Mg(C_2H_3O_2)_2$	100	N_2	725
$Zn(NO_3)_2$	10	Air	375
$Zn(C_2H_3O_2)_2$*	10	Air	325
$Ni(NO_3)_2$	10	Air	500
$Ni(C_2H_3O_2)_2$*	10	Air	350

*Exothermic decomposition reaction

It is important to note the effect of sample heating rate on the temperature of full decomposition as it is expected that an increase in heating rate would result in an increase in this temperature due to kinetic effects. Indeed, as seen in Table 1 there is an increase of at least 100°C for each

salt with an increase in heating rate from 10 to 100°C/min. Since the estimated heating rate of a salt droplet/particle during EDS is 300°C/sec [8],
the temperature required for full decomposition of these salts during EDS
should be significantly higher than measured by DSC and TGA. This illustrates the importance of using a salt that decomposes at a low temperature
(e.g. <500°C) for EDS powder synthesis.

Powder Characteristics

To ensure the complete decomposition of the salts during EDS, the
powders were synthesized at an EDS furnace temperature of 1000°C. All of
the salts were completely decomposed as confirmed by TGA and were well
crystallized as shown by the sharpness of the X-ray diffraction peaks.
Crystallite sizes ranged from 15 to 34 nm whereas the equivalent spherical
diameter calculated from the BET surface areas ranged from 49 to 139 nm
(Table 2). The difference between these two parameters indicates that the
"particles" consist of multiple oxide crystallites. Interestingly, there
are not large differences in the powder characteristics. This is attributed
to a combination of the narrow range of decomposition temperatures for the
salts and the short transit time (1-2 sec at T max) of the salt droplet/
particle in the EDS furnace. However it is important to note that by adjusting the process conditions a wide range of particle sizes and surface
areas can be obtained as a result of the concurrent processes of grain
coarsening, sintering and decomposition during EDS [7,8].

TABLE 2. Powder Characteristics.

Salt	X-ray Crystal-lite Size (nm)	Surface Area (m^2/g)	Particle Size (nm)
$Mg(NO_3)_2$	15	34.3	49
$Mg(C_2H_3O_2)_2$	16	20.0	84
$Zn(NO_3)_2$	33	7.7	139
$Zn(C_2H_3O_2)_2$	29	18.0	59
$Ni(NO_3)_2$	34	11.5	78
$Ni(C_2H_3O_2)_2$	22	10.8	83

Particle Morphology

Although the powder characteristics of the acetate and nitrate-derived
powders are similar, a significant difference in particle morphology exists.
Figure 3 shows the particle morphology of the nitrate-derived powders. In
general, the "particles" consist of hollow spherical shells or shell fragments. This type of morphology arises due to the rapid evaporation of water
from the surface of the solution droplet and the development of a salt crust
before the droplet is completely dried. With additional drying a positive
pressure develops within the "particle" [10]. Thus, fracture and/or rupture
of the crust occurs and the trapped solution and gas is expelled. Each
"particle" is actually a strongly bonded aggregate of submicron oxide particles which have either coalesced or sintered during EDS.

The morphology of the ZnO powder derived from the acetate is similar to
that observed for acetate-derived MgO powder in that they are aggregate-free
(Figure 4). The powders are composed of individual particles that are between 100 and 300 nm in diameter, which is in reasonable agreement with the
surface area results. The difference in the powder character between the
nitrate and acetate-derived powders is attributed to the exothermic oxidation reaction that occurs during decomposion of the acetate salts in air.

FIG. 3. SEM Photomicrographs of Nitrate-Derived Powders.

FIG. 4. SEM Photomicrographs of Acetate-Derived Powders.

It is believed that the individual salt crystals are precipitated on the surface of the droplet during the drying stage of EDS and that the exothermic decomposition reaction breaks the individual crystals apart from the droplet skeletal structure. In the absence of the exothermic decomposition reaction the skeletal structure is retained as seen for the nitrate-derived powders and observed for acetate-derived powders synthesized at lower temperatures or in N_2 [7,8].

The powders derived from nickel acetate and nickel nitrate are similar in morphology. Unfortunately, preparation of NiO from the acetate was performed with a 0.5M solution and at 950°C. Therefore, the salt droplet/particle did not see the same thermal conditions as the other acetate-derived powders. It is suggested that this is the reason that the droplet relic structure persists in this case. It is possible that the oxidation of the acetate/residue does not possess the prerequisites for disaggregation of the salt crystals during EDS. Alternatively, other chemical differences exist between the nickel and magnesium or zinc species. Additional experiments are in progress to more closely identify these requirements.

SUMMARY

It has been demonstrated that fine-grained, high surface area powders can be synthesized by EDS. However, aggregate-free MgO and ZnO powders were obtained when derived from the metal acetates whereas strongly aggregated powders were obtained from the metal nitrates. This difference is attributed to the exothermic decomposition of the acetate salt which apparently disaggregates the salt droplet structure that is formed during EDS. Whether aggregate-free powders can be obtained from metal acetates or other metal salts with an organic radical, in general, is not yet clear as evidenced by the aggregated structure of the acetate-derived NiO powder.

ACKNOWLEDGMENTS

The authors acknowledge the support of the U. S. Bureau of Mines (Research Grant No. J0225003) and the U. S. Department of Energy (Contract No. DE-AC04-76-DP00789). The technical assistance of C. Casaus, M. Eatough, J. Lanoue, J. Rife and J. Young of Sandia National Laboratories is also greatly appreciated.

REFERENCES

!. F. J. Schnettler, F. R. Monforte, and W. W. Rhodes in: Science of Ceramics, Vol. 4, G. H. Stewart, ed. (Henry Blacklock and Co., Ltd. 1968) pp. 79–90.
2. M. E. A. Hermans, Powder Met. Int. $\underline{5}$(3), pp. 137–140 (1973).
3. W. S. Clabaugh, E. M. Swiggard, and R. Gilchrist, J. Res. NBS $\underline{56}$(5), pp. 289–291 (1956).
4. K. S. Mazdiyasni, Ceram. Int. $\underline{8}$(2), pp. 42–56 (1982).
5. D. M. Roy, R. R. Neurgaonkar, T. P. O'Holleran, and R. Roy, Am. Ceram. Soc. Bull. $\underline{56}$(11), pp. 1023–1024 (1977).
6. M. J. Ruthner in: Ceramics Powders: Preparation, Consolidation, and Sintering, P. Vincenzini, ed. (Elsevier, Amsterdam 1983) pp. 515–531.
7. T. J. Gardner, and G. L. Messing, to be published, J. Am. Ceram. Soc., (1984).
8. T. J. Gardner, M. S. Thesis, The Pennsylvania State University (1983).
9. R. R. Ciminelli, M. S. Thesis, The Pennsylvania State University (1983).
10. D. H. Charlesworth, and W. R. Marshall, Jr., J. Am. Int. Chem. Eng. $\underline{6}$(1), pp. 9–23 (1960).

AN INFRARED STUDY OF METAL ISOPROPOXIDE PRECURSORS FOR SrTiO$_3$*

R. E. Riman**, D. M. Haaland and C.J.M. Northrup, Jr.†, and H. K. Bowen**
and A. Bleier***

***Massachusetts Institute of Technology, Cambridge, MA; Presently at
  Oak Ridge National Laboratories
 **Massachusetts Institute of Technology, Cambridge, MA
 †Sandia National Laboratories, Albuquerque, NM

ABSTRACT

    A Sr/Ti bimetallic isopropoxide complex was synthesized by
two methods. The complex served as a precursor to the production
of homogeneous SrTiO$_3$ powders via alkoxide hydrolysis. Infrared
spectra were obtained for Sr(OPr^i)$_2$, Ti(OPr^i)$_4$, and the product of
the syntheses. In addition, the IR spectra of the solutions of
each of the alkoxides were followed as hydrolysis reactions
proceeded. Detailed analysis of the spectral features support the
existence of a 1:1 Sr/Ti bimetallic alkoxide. The new Sr/Ti
compound exhibits characteristic absorption bands at (1017, 993, 972,
961 cm^{-1}), (844, 838, 827 cm^{-1}) and (620, 596, and 572 cm^{-1}).
A band at 819 cm^{-1} might also be associated with the new Sr/Ti
bimetallic alkoxide. The infrared spectra suggest that the isopropoxide ligands in the bimetallic alkoxide are in at least three
separate local environments. This information offers insight into
possible structures for the complex.

INTRODUCTION

    Uniform and densely packed SrTiO$_3$ powder with a narrow size distri-
bution of equiaxed, pure, homogeneous and unagglomerated particles may
produce optimized SrTiO$_3$ microstructures at a reduced processing temperature [1]. This goal has been achieved in the production of TiO$_2$ compacts
by packing and sintering alkoxide based TiO$_2$ powder [2]. We believe
that SrTiO$_3$ powder can be synthesized using the general synthesis route
pioneered by Mazdiyasni, et al. with the formation of BaTiO$_3$ [3,4].

    In this paper we describe several methods of double alkoxide synthesis
and present infrared spectral evidence for the formation of a Sr/Ti
bimetallic complex which hydrolyzes to form homogeneous SrTiO$_3$[1,3].
In addition, the feasibility of monitoring the complex under different
solution environments and studying hydrolysis reactions is examined.

BACKGROUND

    Smith, et al.[3] were the first to describe titanate formation from
a double alkoxide precursor using 1:1 solutions of Sr(OPr^i)$_2$ and titanium
tertiary amyloxide, Ti(OAm^t)$_4$ [3]. Infrared characterization of Sr(OPr^i)$_2$,
Ti(OAm^t)$_4$ and the mixture of the two alkoxides suggested that the mixture

*A portion of this work was performed at Sandia National Laboratories
supported by the U.S. Department of Energy under contract number
DE-AC04-76DP00789. The M.I.T. work is supported by an industry consortium.

possessed unique spectral features which arise from the bimetallic alkoxide. Bimetallic complexation appears to describe the assemblage of hydrolysis products obtained by Suwa, et al. [6] when different ratios of $Ba(OPr^i)_2:Ti(OPr^i)_4$ were prepared in isopropanol solutions. Turevskaya, et al. [7] synthesized bimetallic alkoxide crystals with two different compositions (1:1 and 1:4 Ba:Ti) in the Ba-Ti ethoxide system. Similarly, Riman[1] synthesized crystals from $Sr(OPr^i)_2-Ti(OPr^i)_4-Pr^iOH$ solutions which can be solubilized and hydrolyzed to form a homogeneous $SrTiO_3$ powder as deduced from x-ray diffraction patterns. Alkoxide solutions prepared with a stoichiometry deviating from 1:1 could be hydrolyzed in two fractions. Hydrolysis of the supernatant produced a powder containing secondary phases while hydrolysis of the solubilized crystals produced single phase $SrTiO_3$.

Spectral assignments for the IR absorption of alcohols and metal alkoxides have received considerable attention [3, 9-17]. In this study the C-O stretching and skeletal vibrations of the isopropoxide group were of interest. Barraclough, et al. assigned the C-O stretch of $Ti(OPr^i)_4$ to 1005 cm^{-1} while Lynch, et al. found the absorption to be at 1000 cm^{-1} [13,14]. Smith, et al. [3] reports that C-O stretch of $Sr(OPr^i)_2$ is found at 970 cm^{-1}. Ziess and Tsutsui [9] found the C-O absorptions for secondary alcohols to fall in the 1085-1125 cm^{-1} region with the C-O stretch of Pr^iOH assigned at 1105 cm^{-1}. Simpson and Sutherland [17] calculated the expected frequencies of three types of isopropyl skeletal vibrations to be at 1170, 1145 and 850 cm^{-1}. Experimentally, the 850 cm^{-1} band was found to shift over a range of ~50 cm^{-1}. Independently, through synthetic structural studies, the isopropyl skeletal vibrations were similarly assigned to the 1163, 1126 and 855 cm^{-1} bands for $Ti(OPr^i)_4$ [13,14] and the 1153 and 1115 cm^{-1} bands for $Sr(OPr^i)_2$ [3]. An isopropyl skeletal vibration at 950 cm^{-1} was found in Pr^iOH and in $Ti(OPr^i)_4$ [13, 14]. The literature cited indicates that the skeletal and C-O regions show significant overlap in the range of 950-1200 cm^{-1}. However, absorptions in the lower 850 cm^{-1} region can be positively identified as isopropyl skeletal vibrations.

EXPERIMENTAL PROCEDURE

The $Sr(OPr^i)_2$ was prepared following the method of Smith et al. [3]. During synthesis, 5 vol% toluene was added to the Pr^iOH to increase the synthesis yield. The $Sr(OPr^i)_2$ dissolved in the parent solution was either assayed for its Sr content colorimetrically and used in solution form or prepared as a solvent free compound. The latter was accomplished by precipitating the compound and replacing the Pr^iOH with methylene chloride using a vacuum distillation-solvent replacement technique. Liquid $Ti(OPr^i)_4$ was purchased from Alfa Products in Danvers, Massachuesetts, assayed for its Ti content gravimetrically, and used with no further purification.

Two methods were attempted to synthesize the bimetallic isopropoxide. Method #1 consisted of mixing a $Sr(OPr^i)_2/Pr^iOH$ solution and $Ti(OPr^i)_4$ which produced a viscous solution that could be crystallized. With method #2, strontium metal was reacted with Pr^iOH and $Ti(OPr^i)_4$ using a slight excess of $Ti(OPr^i)_4$ to complex all of the $Sr(OPr^i)_2$ produced. Following filtration to remove unreacted metal, crystallization occurred. In both methods, after significant crystal growth had taken place, the supernatant was decanted and the remaining crystals were heated in vacuo to remove any Pr^iOH and $Ti(OPr^i)_4$. Method #1 produces a compound which hydrolyzes to a homogeneous $SrTiO_3$ powder as determined by x-ray diffraction. However, compounds synthesized by method #2 hydrolyze to a powder which does not show the same degree of homogeneity.

The alkoxide solutions were examined over the concentration range of 0.5-1.4 M. Nujol mulls were prepared with 5-10 weight percent metal alkoxide. Hydrolysis of the metal alkoxides was facilitated through direct mixing of wet solvents with the precursors or the application of metal organic solutions to salt windows exposed to water vapor. Most of the alcoholic solutions of alkoxides were studied on a Nicolet 7199 FTIR. The pure alkoxides and solutions of the alkoxides in methylene chloride and alcohol were all examined on a Perkin Elmer 1430 ratio recording infrared spectrometer or a Perkin Elmer 283B infrared spectrometer. Samples were examined either in a 0.015 mm path length cell or as a capillary film held between KBr or NaCl windows.

RESULTS AND DISCUSSION

The spectra of each monometallic and bimetallic alkoxide compound in nujol mulls were compared with their corresponding spectra in solution (methylene chloride or isopropanol). No significant differences were found between the spectra of the solutions and that of the compounds in nujol mulls except in the region of 600-300 cm^{-1} for the bimetallic alkoxides. Furthermore, no spectral differences were found between bimetallic alkoxides synthesized with method #1 or method #2. The alkoxide/$Pr^i OH$ solutions were examined with the FTIR spectrometer. Following digital subtraction of the spectral features due to $Pr^i OH$, the peaks associated with the metal alkoxide could be identified and compared with the spectra of the alkoxide compounds in nujol mulls. Shown in Figure 1 and Table 1 are the major spectral peaks for $Sr(OPr^i)_2$, $Ti(OPr^i)_4$ and the double alkoxide after spectral subtration of the $Pr^i OH$ solvent.

The spectra of the single alkoxide solutions correspond to the literature cited previously for metal isopropoxides [3,9-17]. The bands associated with the isopropoxide ligand are found to be environmentally sensitive. The C-O assignments for the $Sr(OPr^i)_2$ at 970 cm^{-1} and $Ti(OPr^i)_4$ at 1002 cm^{-1} were detected. In addition the skeletal vibrations for the isopropyl group were observed at 953 and 852 cm^{-1} for $Ti(OPr^i)_4$ and 819 cm^{-1} for $Sr(OPr^i)_2$. The 962 cm^{-1} absorption reported for $Sr(OPr^i)_2$ could be either a skeletal or C-O stretching vibration but again is clearly associated with the isopropyl group.

Table 1: Spectral Characteristics (1200-450 cm^{-1}) of $Pr^i OH$ and of the $Pr^i OH$ solutions of isopropoxides $Ti(OPr^i)_4$, $Sr(OPr^i)_2$, and $Sr(OPr^i)_2$-$Ti(OPr^i)_4$ after digitally subtracting the $Pr^i OH$ spectrum.

$Pr^i OH$	$Ti(OPr^i)_4$	$Sr(OPr^i)_2$	$Sr(OPr^i)_2$-$Ti(OPr^i)_4$ Solution
1162	1162	1154	1159
1130	1124	1138	1130
1111		1125	
953	1002, 953	970, 962	1017, 993, 972, 961
817	852	819	844, 838, 827, 819(?)
650	620, 560		619, 594, 570
	511	513	
488	472	464	464

236

The 1:1 Sr/Ti double alkoxide solution has a more complex spectrum than its precursors. Figure 1 and Table 1 show that in 3 regions (961-1017, 819-844, and 570-619 cm^{-1}) the spectrum is not a simple superpostion of the spectra of Sr(OPr^i)$_2$ and Ti(OPr^i)$_4$. First, four absorptions were found in the lower wavenumber skeletal region at 844, 838, 827, and 819 cm^{-1}. The band found at 819 cm^{-1} may be due to the 1:1 Sr/Ti alkoxide complex, the Pr^{i}OH group associated with the complex, or simply due to a poor spectral subtraction of the solvent spectrum. Absorptions were also found at 1017, 993, 972 and 961 cm^{-1} and may be attributed to C-O stretching or isopropyl skeletal vibrations. The final group of absorptions found at 619, 594, and 570 cm^{-1} were found in a region often attributed to metal-oxygen stretching frequencies [5, 13].

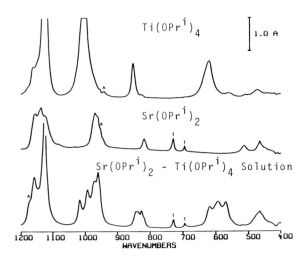

Fig. 1. Absorbance spectra of isopropoxides in Pr^{i}OH after subtracting the Pr^{i}OH solvent spectrum. Peaks labeled "T" are due to toluene incorporated during the Sr(OPr^i)$_2$ synthesis. Peaks labeled "A" are considered artifacts of the spectral subtraction.

Hydrolysis reactions of Sr(OPr^i)$_2$, Ti(OPr^i)$_4$ and the bimetallic complex were performed in methylene chloride and Pr^{i}OH. In both solvent systems, two regions of absorption bands in each case were monitored with increasing degrees of hydrolysis (Ti(OPr^i)$_4$: 1002 cm^{-1} and 852 cm^{-1}; Sr(OPr^i)$_2$: 970, 962 and 819 cm^{-1} and the bimetallic complex: 1017-961 and 844-819 cm^{-1}. In all cases, the intensity of the absorption bands decreased with increasing water concentration. In each system, as hydrolysis proceeded to further degrees, peaks associated with the Pr^{i}OH product emerged at 953, 817, 3200-3500 and 3600 cm^{-1}.

The infrared spectral data coupled with earlier existing data give further insight into the nature of the bimetallic complex. First, the spectra showed that the compounds synthesized were alkoxides since they reacted with water to produce a product alcohol. Second, the homogeneity of the SrTiO$_3$ hydrolysis product showed that the compounds examined had a 1:1 Sr-Ti ratio. The four absorption peaks found in the OPr^i skeletal

region (819–844 cm^{-1}) suggest that a minimum of three near-field environments exist for the OPr^i ligand in with the bimetallic complex. It is not
clear whether the fourth peak is associated with Pr^{i}OH, the complex, or is
a superposition of both. Of course, the potential of overlap of spectral
bands makes more than four near-field environments possible for the OPr^i
ligands.

Many structures can be envisioned for a Sr/Ti bimetallic complex that
have the requirements of covalent bonding, charge neutrality assuming conventional valence states, a Sr/Ti ratio of 1:1, coordination environments
found in titanium alkoxides [18, 19] and three or four distinguishable
near-field sites for the isopropoxide group. Two of the several possible
structures which satisfy these criteria are shown in Figure 2. Further
work (X-ray diffraction, NMR, molecular weight determination, elemental
analysis, etc.) is required to fully characterize the Sr/Ti bimetallic
alkoxide and to determine a unique composition and structure.

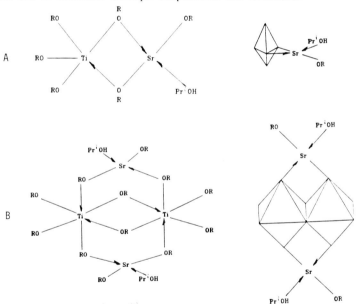

Fig. 2. Possible structures for a Sr/Ti bimetallic alcoholate: a) a complex with Ti pentahedrally coordinated by OPr^i ligands exhibiting
three distinguishable isopropoxide near-field environments; b) a
complex with Ti octahedrally coordinated exhibiting four distinguishable isopropoxide near-field environments.

CONCLUSIONS

Infrared spectroscopy is useful for detecting alkoxide complex formation and monitoring the quality of the alkoxide precursors. The examination
of the complex under different solution environments and studying hydrolysis
reactions is possible. Based on the infrared spectra, hydrolysis reactions
and the literature, there is strong evidence for the existence of a Sr/Ti
bimetallic complex. Furthermore, solutions of the complex in methylene
chloride or Pr^{i}OH did not change the nature of the complex. Finally, the
progress of hydrolysis reactions could be monitored by infrared spectroscopy.

REFERENCES

1. R. E. Riman, Ph.D. Thesis, in progress.

2. E. Barringer and H. K. Bowen, "Formation, Packing and Sintering of Monodisperse TiO_2 Powders," J. Am. Ceram. Soc. Comm. 65 (12), C199-201 (1982).

3. J. S. Smith II, R. T. Dolloff, and K. S. Mazdiyasni, "Preparation and Characterization of Alkoxy-Derived $SrZrO_3$ and $SrTiO_3$," J. Am. Ceram. Soc. 53 (2) 91-95 (1970).

4. K. S. Mazdiyasni, R. T. Dolloff, and J. S. Smith II, "Preparation of High-Purity Submicron Barium Titanate Powders," J. Am. Ceram. Soc. 52 (10) 523-526 (1969).

5. N. Ya. Turova and E. P. Turevskaya, "Primary Hydrolysis Product of Titanium Ethylate" Koord. Khim. 3 (5) 679-684 (1977).

6. Y. Suwa, Y. Sugimoto and S. Naka, "Preparation of Compounds in $BaO-TiO_2$ System by Coprecipitation from Alkoxides," Funtai Oyobi Fummatsu Yakin 25 (5), 20-23 (1978).

7. E. P. Turevskaya, N. Ya. Turova and A. V. Novoselova, "Investigation of the Formation of Bimetallic Alkoxides Reaction of Barium Ethoxide and Titanium Ethoxide," Dok. Akad. Navk SSSR 242 (4), 883-886 (1978).

8. D. C. Bradley, R. C. Mehvotra and W. Wardiaw, "Structural Chemistry of the Alkoxides, Part I. Amyloxides of Silicon, Titanium, and Zirconium," J. Chem. Soc., London 2027-2032 (1952).

9. H. H. Zeiss and M. Tsutsui, "The Carbon-Oxygen Absorption Band in the Infrared Spectra of Alcohols," J. Am. Chem. Soc. 75, 897-900 (1953).

10. Von H. D. Lutz, "IR-spektroskopische Untersuchungen an $Mg(OC_2H_5)_2$, $Ca(OC_2H_5)_2$, $Sr(OC_2H_5)_2$ and $Ba(OC_2H_5)_2$," Z. Anorg. Alleg. Chem. 356, 132-139 (1968).

11. N. Sheppard and D. M. Simpson, "The Infra-Red and Raman Spectra of Hydrocarbons. Part II Paraffins.

12. R. M. Silverstein and C. C. Bassler, Spectroscopic Identification of Organic Compounds, John Wiley and Sons, NY, Second Edition, 1967 p. 84.

13. C. G. Barraclough, D. C. Bradley, J. Lewis and I. M. Thomas, "The Infrared Spectra of Some Metal Alkoxides, Trialkylsilyloxides and Related Silanols," J. Chem. Soc., London 2601-2605 (1961).

14. C. T. Lynch, K. S. Mazdiyasni, J. S. Smith and W. J. Crawford, "Infrared Spectra of Transition Metal Alkoxides," Anal. Chem. 36 (12), pp. 2332-2337 (1964).

15. F. H. Seubold, "Anionic Conjugation: The Infrared Spectra of Sodium Alkoxides," J. Org. Chem. 21, 156-160 (1956).

16. A. I. Grigorev and N. Ya. Turova, "The IR Absorption Spectra of Alcoholates of Berryllium, Magnesium and the Alkaline Earth Metals," Akad. Nauk SSSR Proceedings Chem. Section 162. pp. 424-427 (1965).

17. L. J. Bellamy, The Infrared Spectra of Complex Molecules, 1, J. Wiley & Sons, NY (1975).

18. G. W. Suetich and A. A. Voge, "Crystal Molecule Structure of Tetraphenoxytitanium (IV) Monophenolate, $Ti(OPh)_4 \cdot HOPh$," Chem. Comm. 676-677 (1971).

19. H. Stoeckli-Evans, "Studies of Organometallic Compounds XVI, The Crystal Structure of Di-µ-ethoxy-bis (dibenzlethoxy-titanium (IV)," Hel. Chim. Acta 58 (2) 373-377 (1975).

PREPARATION OF STRONTIUM TITANATE CERAMICS AND INTERNAL BOUNDARY LAYER
CAPACITORS BY THE PECHINI METHOD

K. D. BUDD AND D. A. PAYNE
Department of Ceramic Engineering and Materials Research Laboratory
University of Illinois at Urbana-Champaign
105 South Goodwin Avenue, Urbana IL 61801

ABSTRACT

High purity, fine-grain size, $SrTiO_3$ ceramics were
fabricated from powders which were chemically derived by the
Pechini method. The preparation is based upon a thermally
polymerizable water-ethylene glycol-citric acid system in
which a wide range of metal alkoxides, carbonates, and metal
salts are soluble. The process is a closed system, and
lends itself to the preparation of complex oxides which can
contain numerous additives and dopants. In this study, W-
doped $SrTiO_3$ powders were prepared with various Sr:Ti
ratios, and fine grain internal boundary layer capacitors
were fabricated by controlled segregation in sintering.
Characteristics of the resin, powder, microstructures, and
electrical properties are reported. The advantages and
disadvantages of this method of chemical preparation for
ceramic materials are noted.

INTRODUCTION

Strontium titanate is one of a large number of titanate perovskite
materials which are widely used in the electronic ceramics industry. The
widespread use of titanates is a result of their highly polarizable crystal
structure, and of the wide range of electrical properties possible through
chemical modifications and process control manipulations. The latter is
exploited in internal boundary layer capacitors, base metal electrode
compatible dielectrics, and highly conducting ceramic electrodes. It is in
these newer applications that the degree of process control afforded by
conventional ceramic processing methods has often been inadequate. As a
result, the use of novel methods of chemical preparation has increased
dramatically over the past few years, as in other areas of high technology
ceramics.

In this study, a liquid solution resin-forming technique, known as the
Pechini method [1], was used to prepare high purity W-doped $SrTiO_3$
powders. This method can be used to produce a variety of oxide materials
including titanates, zirconates and niobates. As described in the orignial
patent [1], polybasic acid chelates are formed from alpha hydroxycarboxylic
acids, such as citric acid, with titanium, zirconium, or niobium containing
solutions. When heated with a polyhydroxy alcohol, such as ethylene
glycol, polyesterification and resin formation gradually occurs. Details
of the procedure are given later.

Since a variety of different chemical methods of ceramic powder
preparation are now available, it is useful to note the fundamental
differences between, and compare the advantages and disadvantages of, a
specific method with its various alternatives.

Mat. Res. Soc. Symp. Proc. Vol. 32 (1984) Published by Elsevier Science Publishing Co., Inc.

COMPARISON OF CHEMICAL METHODS

Nearly all chemical powder preparation techniques offer the potential
for finer uniform particle size distributions of reactive powders with
greater purity and compositional homogeneity than conventional mixed oxide
methods. Atomistic or molecular mixing in liquid solutions, and relatively
low temperatures for decomposition into oxides (and other materials) are
primarily responsible. Two general categories of chemical preparation
techniques can be distinguished, though, which have unshared advantages and
disadvantages.

(1) The first category is that involving precipitation or
coprecipitation from solution. These techniques are applicable to a wide
variety of materials, and allow one to wash away soluble impurities before
heating, but they have several weaknesses. Homogeneity is only assured if
a specific unique compound is precipitated, and even so, its stoichiometry
is usually different from that of the original solution, and is sensitive
to precipitation conditions, namely, pH, temperature, concentration, etc.
Doping, in particular, must be carefully and empirically controlled.

(2) Methods in which the atoms are spatially fixed prior to powder
formation are immune to nearly all these problems. These methods include
sol-gel processing, certain types of freeze-drying, and resin forming
methods such as the Pechini method. In these techniques it is essential to
avoid precipitation of any of the components. Assuming that the cations
are not lost in volatilization, the cation composition is identical to that
of the original solution. Atomistic mixing is maintained because the
solution becomes a solid while still clear and homogeneous.

In the Pechini method, a substantial amount of organic material must
be burnt off prior to and during powder formation. This large organic
concentration, however, is probably partially responsible for one of this
method's advantages. A wide range of metal alkoxides, carbonates and even
small amounts of water soluble metal salts remain soluble in this system
throughout resin formation, allowing formation of complex oxides, and the
additions of numerous additives and dopants without changing the process.
Only C, N, O, and H atoms, in addition to the desired cations, are used in
this process, hence anion contamination is also avoided. Because of the
high degree of compositional control afforded by this closed-system
technique, it has been used in a number of studies of high temperature
equilibrium electrical properties. [2,3] Closed-system solution techniques
are very useful for maintaining batch to batch uniformity while studying
the effects of slight changes in composition.

PURPOSE

A systematic study of the effects of composition and processing
conditions (PO_2, sintering, temperature, cooling rate) on the electrical
properties of $\check{W}$-doped $SrTiO_3$ was carried out with the aid of the Pechini
method. The aim was to develop fine grain internal boundary layer
capacitors (IBLCs) from chemically derived powders and a single air firing
process. Today, $SrTiO_3$ IBLCs are typically formed by industrial two step
firing processes in which semiconducting grains are produced by an initial
firing in a reducing atmosphere, and insulating boundaries are later
created during a second firing step in which an acceptor or insulating
phase is infiltrated along the grain boundaries, or partial reoxidation is
allowed to occur on cooling.

This paper reports on the preparation and characterization of

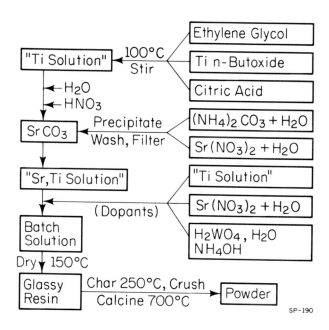

FIG. 1. Flow diagram for the preparation of $SrTiO_3$ based powders by the Pechini method.

chemically derived powders, ceramic microstructures, and dependent electrical properties of certain IBLC-forming compositions.

PROCEDURE

Figure 1 illustrates the basic process used in this study. Three major steps were involved: i) preparation of clear, homogeneous solutions containing the requisite amounts of cations, ii) evaporation and thermal polymerization of the solutions to form viscous resins and eventually glassy polymers, and iii) pyrolysis (calcination) of the glassy resins to form ceramic powders. A large stock solution containing Sr and Ti was prepared and used throughout the study. The compositions of individual batches were adjusted with small additions of dopant solutions (Figure 1). A typical preparation procedure was as follows.

A Ti-containing solution was prepared from reagent grade ethylene glycol, tetraorthobutyltitanate (TBT) and citric acid. 550 grams of TBT were mixed with 1 litre of ethylene glycol at 60°C in a 5 litre pyrex beaker to which were added 875 grams of citric acid, and the mixture was heated to 110°C with vigorous stirring so as to redissolve any hydrated titania. The solution was kept at 110°C for 2 hrs to promote evaporation of butyl alcohol and homogenization of the solution. Gravimetric analysis determined the TiO_2 content by evaporation and combustion of known weights of the solution in a platinum crucible. A value of 0.05642 ± 0.00004 grams of TiO_2 per gram of solution was determined in this study. The master stock solution was prepared with a Ti:Sr stoichiometry ratio of

242

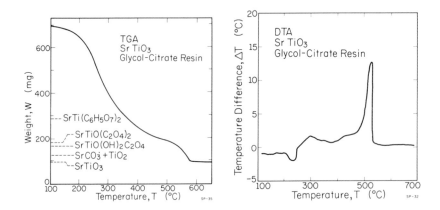

FIG. 2. TGA of resin decomposition. FIG. 3. DTA of resin decomposition

1.01 to 1 by dissolving freshly precipitated $SrCO_3$ into the "Ti-solution" with small additions of water and nitric acid. Several batches were homogenized and combined in a 5 litre nalgene container which was heated and kept at 75°C for 15 hours. Wet chemical analyses of powders obtained from combusted solutions gave a composition of $SrTiO_3$ + 1.04 mole % TiO_2. The final solution contained 4098 grams of the "Ti-solution," 422.96 grams of $SrCO_3$, 2 liters of deionized H_2O, and 100 mls of reagent grade HNO_3. The $SrCO_3$ was precipitated and filtered from an aqueous mixture of $Sr(NO_3)_2$ with $(NH_4)_2CO_3$. Small but precise amounts of the "Ti-solution," or other gravimetrically analyzed solutions containing Sr or W, were used to adjust the concentrations of individual experimental batches.

Powders obtained from this process were mixed with a binder solution (65% H_2O, 25% glycerin, 10% PVA) in a mortar and pestal for 20 minutes and dried at 80°C before granulating through an 80 mesh sieve and pressing at 25,000 to 40,000 psi. Pellets were sintered in an electrically heated tube furnace at 1400° or 1500°C which was equipped with atmosphere control. Reproducibility of the overall thermal processing cycle was effected by a fifty segment microprocessor based programmable control system.

Dielectric properties were measured with an HP 4191A impedence analyzer on polished samples with sputtered gold electrodes. Microstructural analyses were carried out on an I.S.I. DS-130 (SEM), and thermal analyses were carried out on a Dupont 9000 Thermal Analyzer.

RESULTS AND DISCUSSION

Characteristics of the resin and powders are considered first. Batch solutions prepared by the Pechini method are quite stable and completely miscible with water or ethylene glycol. Solutions have been stored over one year without visible precipitation. Upon heating, the solutions lose solvent and become progressively more viscous, with no definite transition point, in contrast to many sol-gel processes. The resins decomposed into a fine white powder at well under 600°C. Figure 2 illustrates a typical thermogravimetric analysis (TGA) of a glassy resin fired in air. The decomposition occurred rather continuously, and for a heating rate of 7°C per minute, was completed by 575°C. No definitive plateaus, indicating the

FIG. 4. Effect of calcination temperature on crystallite size.
Top: T_c = 500°C and T_c = 600°C. Bottom: T_c = 700°C and
T_c = 800°C. 1 cm ≡ = 0.25 microns.

lack of well-defined intermediate decomposition products, were identified
on the TGA. The weight changes for various potential precursors, such as
citrates, oxalates, and carbonates, are identified in Figure 2.

Figure 3 illustrates a DTA for a similar resin, at a heating rate of
5°C per minute. Endothermic volatilization occurred up to near 300°C. The
initial flat region was because the resin had been dried previously at
200°C. From 300°C to 480°C the reactions were only mildly endothermic or
exothermic, corresponding to the region of decreasing slope on the TGA. A
large exothermic peak accompanied carbon burn-off at 525°C, where a sharp
increase in negative slope was previously observed with the TGA.

Figure 4 illustrates powders which were calcined at 500°C, 600°C,
700°C and 800°C. As shown, excellent control of the initial crystallite
size was possible. All of the powders had uniform crystallite size
distributions, with average sizes ranging from less than 300 Å for powders
calcined at 500°C to greater than 1000 Å for those processed at 800°C.
Although the crystallites were small and highly reactive, suggesting the
potential for reduced temperature sintering, extensive aggregation or
initial stage sintering had already occurred by 500°C, reducing the pressed
density and hindering fired densification. Special decomposition
techniques or powder treatments would be necessary if this method were to
be used for low temperature densification. Grain sizes and densities
varied with composition, but densities of 95% of theoretical, with average
grain sizes of a few microns, were typical.

A number of interesting electrical properties were exhibited by W-

TABLE 1. Dielectric properties of $SrTiO_3$ based internal boundary capacitors.

$\frac{Sr}{Ti + W}$ ratio	Tungsten content (mole %)	Average grain size (microns)	Firing temp. (°C)	Cooling rate (°C/Hr.)	K (at 10 KHz)	tan δ
1.015	0.25	0.9	1400	10^3	3,100	0.02
1.015	0.50	0.5	1400	10^3	2,200	0.02
1.015	0.15	2.6	1500	10^2	2,900	0.04
1.015	0.15	2.4	1500	10^3	6,800	0.03
1.015	0.15	2.4	1500	10^4	8,900	0.02
0.985	1.50	6.0	1500	10^2	6,100	0.08
0.985	1.50	5.8	1500	10^3	13,500	0.08
0.985	1.50	5.8	1500	10^4	38,500	0.08

doped $SrTiO_3$, most of which are discussed in a separate report.[5] A brief summary of the processing—composition relationships for dielectric properties are given in Table 1. Note, the intrinsic dielectric constant for $SrTiO_3$ at room temperature is approximately 300. Most of these properties were developed as a direct result of careful processing control made possible by chemical methods of preparation. Specifically, development of internal boundary layer capacitors with fine grain sizes, and manipulation of boundary thicknesses using controlled segregation during sintering and cooling, were possible, as indicated in Table 1. The former is significant because of the potential application to multilayers and film devices; the latter because it facilitates identification of the mechanism of boundary formation, and also allows control of properties.

SUMMARY

The Pechini method is an excellant technique for producing homogeneous powders with closely controlled compositions. It lends itself well to the fabrication of complex multicomponent oxides. $SrTiO_3$ powders with uniform crystallite sizes of less than 300 Å were produced for calcination temperature of less than 600°C. The powders were used to produce air-fired IBLCs with dielectric constants up to 40,000, and submicron average grain-size IBLCs with dielectric constants of over 3000.

ACKNOWLEDGEMENTS

This work was supported by the U. S. Department of Energy, Division of Materials Sciences, under contract DE-AC02-76ER01198. KDB also acknowledges partial support of an ONR-ASEE Fellowship.

REFERENCES

1. M. Pechini, U. S. Patent 3,330,697, July 11, 1967.
2. N. H. Chan, R. K. Sharma, D. M. Smyth, J. Electrochem. Soc. Sol. St. Sci. and Tech. 128 1762 (1981).
3. U. Balachandran and N. G. Eror, J. Sol. St. Chem. 39, 351 (1981).
4. K. D. Budd, Processing and Properties of Tungsten—Doped $SrTiO_3$ Ceramics, M. S. Thesis, University of Illinois, 1983.

PREPARATION AND OPTICAL PROPERTIES OF POLYCRYSTALLINE ALUMINUM GERMANATE

S. Prochazka, G. A. Slack

INTRODUCTION

Aluminum germanate, the germanium analog of mullite, $3 Al_2O_3 \cdot 2 GeO_2$, was first prepared by Gelsdorf, Muller-Hesse and Schweite (Ref. 1), who demonstrated that both partial and complete substitution of SiO_2 by GeO_2 in mullite was possible. They also determined the lattice constants of the solid solutions. The following studies reported some physical properties of aluminum germanate such as density (Ref. 2) optical constants (Ref. 2) crystal structure (Ref. 3) and I. R. absorption (Ref. 4). Phase equilibria in the system Al_2O_3-GeO_2 were investigated by Miller et al. (Ref. 5) and Perez-y-Jorba (Ref. 6). The latter author found several compounds of which germanium mullite was the most stable and showed a relatively wide compositional range. Miller et al. observed only 3 $Al_2O_3 \cdot 2 GeO_2$ with no evidence of solid solutions. The compound was reported to melt incongruently at 1530°C. Recently Yamaguchi et al. (Ref. 7) prepared aluminum germanates of a wide range of stoichiometry from alkyl oxides and found, in addition to germanium mullite, the compound Al_2O_3-$2GeO_2$ which was stable between 1190° and 1310°C. The authors also report infrared absorption spectra.

Germanium mullite has recently attracted interest as an electromagnetic window material because it is expected to possess transmittance further into the infrared than mullite. So far, however, except for i.r. powder data by Muller-Hesse (4) and Yamaguchi (7), no information on optical absorption has been published.

The objective of the present work was to prepare polycrystalline $3 Al_3O_3 \cdot 2 GeO_2$ samples sufficiently dense to make possible infrared transmission measurements on macroscopic specimens. The preparation of aluminum germanate in polycrystalline form is complicated by the volatillity of GeO_2 that precludes the use of high temperatures necessary for its synthesis from oxides in open systems. In addition GeO_2 reduces easily to GeO or Ge at elevated temperatures, thus making the use of hot-pressing impractical. These constraints make the synthesis via amorphous coprecipitated precursors a very attractive approach.

PRECURSOR PREPARATION

Hydrolysis of alkyloxides, a procedure used previously for aluminum silicate compositions (7, 8, 9) and many other compounds, was the selected route. Amorphous GeO_2 is soluble in water and, therefore, a nonaqueous medium was preferred for the hydrolysis. The use of alcohols, however, resulted in the formation of insoluble complexes at room temperature; cyclohexane was finally found to be a convenient solvent.

In the preparation procedure germanium ethoxide (Thiokol Co.,Danvers, Mass.) and redestilled aluminum isopropoxide (Chattem Chem. Inc., Chattanooga, Tenn.) were dissolved separately in cyclohexane. The oxide concentration was determined gravimetrically, and the solutions were mixed to obtain the required stoichiometry. The mixture was diluted with the solvent to contain about 5% oxides and were hydrolyzed at room temperature in a Waring blender using the calculated amount of water to bring about complete hydrolysis. (The water was mixed with tertiary butyl alcohol to prevent liquid separation). The resulting thin gel was aged 24 hrs, filtered, dried, and calcined to remove all hydrocarbons. The amorphous, voluminous powder, when calcined at 1050°C, gave a mullite X-ray diffraction pattern. Table 1 gives the composition and characteristics of the preparations used in this study.

The dry gels retained a substantial amount of hydrocarbon. This was removed by slowly heating them in a thin layer (1 cm) to 600°C in air. The carbon content

Mat. Res. Soc. Symp. Proc. Vol. 32 (1984) Published by Elsevier Science Publishing Co., Inc.

PREPARATION AND OPTICAL PROPERTIES OF POLYCRYSTALLINE ALUMINUM GERMANATE

after 16 hrs. heating at this temperature was about 0.1%. The powders were amorphous to X-rays after heating at $600^{\circ}C$, $750^{\circ}C$ and $850^{\circ}C$. At $940^{\circ}C$ crystallization of germanium mullite occured using a heating rate of $10^{\circ}C$/min. This is shown in Fig. 1a. The germanium mullite was the only crystalline phase that could be detected in the stoichiometry range of the GeO_2-Al_2O_3 system investigated. The X-ray diffraction peaks observed after heating at $950^{\circ}C$ and $1050^{\circ}C$ were weak and somewhat broadened. The d_{hkl} obtained from a Debye-Scherrer pattern of a specimen heated in a closed crucible at $1200^{\circ}C$ agreed well with the data in the Powder Diffraction File and with Ref. 2.

Table 1: Characteristics of GeO_2-Al_2O_3 Powder Compositions

Composition wt. %		$\dfrac{Al_2O_3}{GeO_2}$ mole ratio	Calcination Temperature $^{\circ}C$	Sp.Surf. Area m^2/g	Powder Density[+] g/cc	Structure
Al_2O_3	GeO_2					
60.2	39.8	1.55	600	350	2.91	amorph.
"	"	"	1050	120	3.47	cryst.
61.3	38.7	1.63	600	280	–	amorph.
"	"	"	1050	95	–	cryst.
62.3	37.7	1.70	600	290	–	amorph.
"	"	"	1050	80	3.54	cryst.
63.0	37.0	1.74	1200	38	3.60	cryst.
65.2	34.8	1.92	600	–	–	amorph.
"	"	"	1050	81	–	cryst.

+ He pycnometer

The gel calcined at $600^{\circ}C$ was composed of agglomerates of particles a few nanometers in size that could not be resolved by TEM. The ultimate particles grew, however, on further heating, as was indicated by the decreasing specific surface area, and as is shown in SEM micrographs in Fig. 2. Whereas the mobility was sufficient to bring about growth of the particles no crystallization occured at this temperature, thus indicating a relatively stable amorphous phase.

SINTERING

The calcined powders were ball milled with alumina balls in plastic jars in methanol for 16 hrs and then dried. The processed powders could then be die compacted into thin disks at pressures up to 700 MPa to green densities between 1.65 and 2.25 g/cc. Mercury penetration measurements of the compacts of the amorphous materials revealed pores in three size ranges centered about 70, 10 and 4 nm, thus indicating three levels of pore structure. The largest pores of the distribution were interagglomerate pores. Their volume fraction strongly depended on the applied compaction pressure, and could be nearly eliminated by using the very highest pressures.

Compacts of powders calcined at temperatures $850^{\circ}C$ or below always cracked on heating, even at the slowest heating rates. The cracking was linked to the onset of crystallization, however, and was not simply a consequence of the heat of the reaction, which is quite small. As may be observed on the shrinkage curve shown on Fig. 1b, the crystallization coincides with an abrupt cessation of densification that proceeds at this temperature in the amphous substance most likely by viscous flow. It is believed that the cracking was the result of inhomogeneous nucleation and growth of the crystalline phase in the specimens that locally inhibited shrinkage and generated the stresses that introduced cracks. Calcination of the powders at $1050^{\circ}C$ safely prevented cracking.

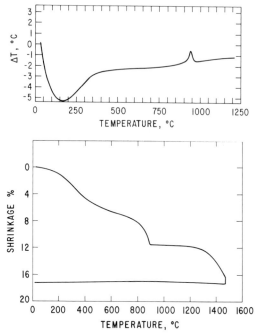

1a. DTA of amorphous aluminum germanate of composition containing 62.3 wt% of Al_2O_3. The exotherm at 940°C is due to crystallization of germanium mullite.

1b. Shrinkage of an aluminum germanate gel compact on heating at 3°C/min. The first wave is due to removal of water and organics, the second wave is probably densification by viscous flow terminated by the crystallization at 900°C. The third wave is due to densification of crystalline germanium mullite.

2. Morphology of gel particles after calcination at 750°C and 950°C SEM, x 100K.

PREPARATION AND OPTICAL PROPERTIES OF POLYCRYSTALLINE ALUMINUM GERMANATE

Sintering of the powder compacts in air gave only limited densification and substantial warpage. Neither the amorphous nor the crystalline powders exceeded 70% of theoretical density. The cessation of densification was caused by the loss of germania, and could be prevented by sintering the specimens in a bed of calcined aluminum germanate in closed alumina crucibles in an oxygen atmosphere. Under these conditions sintering at 1400-1600°C yielded dense bodies. The effect of temperature and some other factors was reported in ref. 10. We note that optimization of the calcination temperature along with the use of very high compacting pressures was essential for removing the last 2% of the pores.

All the specimens sintered at 1500°C and above reached close-to-theoretical density providing the compaction pressure was sufficient, except for the composition with the highest alumina content - 65 wt% Al_2O_3. The sintered specimens were white and translucent in thin sections.

CHARACTERISTICS OF SINTERED ALUMINUM GERMANATE

Microstructure

The microstructure-composition relationship was studied by Prochazka in ref. 10. It was observed that all compositions with a mole ratio less than 1.70 sintered at 1600°C were composed of irregularly shaped grains of germanium mullite up to 150 μm long and pockets of GeO_2 rich glass. In the near surface regions where the specimens were frequently depleted of GeO_2 the microstructures were uniform and fine grained. The composition 1.74 $Al_2O_3.GeO_2$ was single phase and exhibited normal grain growth shown in Fig. 3. All the specimens had a small, variable intragranular porosity, about 0.3%, composed of pores < 2 μm in size. A composition with a mole ratio 1.9 did not achieve comparable high density and showed about 5% α-Al_2O_3 by X-rays.

3. Aluminum germanate 1.74 $Al_2O_3.GeO_2$ sintered at 1600°C. Etched.

When the sintering temperature was increased to 1700°C for one hour the grains grew to 65 μm and severe microcracking resulted from thermal expansion anisotrophy (ref. 11). Such specimens were opaque.

Physical Properties

Measurements of some properties of aluminum germanate were carried out on dense 1.2 and 2.0 cm dia. sintered discs or on segments cut from these discs. The results are reported in Table 2.

PREPARATION AND OPTICAL PROPERTIES OF POLYCRYSTALLINE ALUMINUM GERMANATE

Table 2: Properties of Dense Polycrystalline Germanium Mullite

Property	Unit	Mullite	Technique
Density	g/cc	3.67	X-ray
Density	g/cc	3.66	liq.displacement
Lattice Parameter	Å		X-ray
a		7.662	
b		7.787	
c		2.925	
Micro-hardness	kg/mm^2	1290	Vickers,500 g load
Elastic Modulus	GPa	146	Pulse-echo
Sp. heat	cal/g		
$27^{\circ}C$		0.203	Laser-flash
$800^{\circ}C$		0.243	
Thermal Conductivity	$Wcm^{-1}K^{-1}$		
$27^{\circ}C$		0.024	Laser-flash
$800^{\circ}C$		0.019	
Thermal Expansion $27^{\circ}C-800^{\circ}C$	$K^{-1}.10^{-6}$	4.7	Dilatometer

Optical Transmission and Reflection

An optically polished surface of a polycrystalline ceramic sample of germanium mullite was studied in order to determine its near-normal-incidence reflectivity at room temperature in the infrared region. The incident light beam was at an angle of 15° to the surface normal. The results are shown in Fig. 4. The main lattice vibration bands occur between $200\ cm^{-1} < \bar{\nu} > 1100\ cm^{-1}$ wavenumbers. The photon energies at the reflectivity maxima are listed in Table 3. We have also studied the infrared absorption of fine particle size germanium mullite powder mixed and compressed in a KBr disc. The absorption peaks of this powdered sample correspond closely to the reflection maxima, but are not identical. These are listed in Table 3 also. From these results we expect that germanium mullite will be somewhat transparent in the infrared for $\bar{\nu} > 1000\ cm^{-1}$.

Table 3: Wavenumbers of Optical Bands of Germanium Mullite, in cm^{-1}

BAND ASSIGNMENT	ABSORPTION	REFLECTION
α	-	350 P
β	485 P	500 S
	540 P	533 P
γ	720 S	715 S
	-	790 P
	825 P	850 P
2 β	1035 S	1030 S
	1070 P	1065 P
	1120 P	-
2 γ + 3 β	1505 P	
	1610 S	
3γ	2550 P	
Hydroxyl	3350 P	
	3550 P	

P = peak, S = shoulder

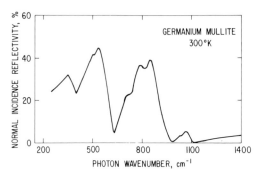

4. Normal-incidence reflectivity of polycrystalline germanium mullite in the infrared region.

The nature of this region of transparency is shown in Fig. 5 where the optical absorption coefficient, α, is plotted versus wavenumber. These data were calculated from percent optical transmission measurements on a number of parallel faced, carefully polished ceramic samples with thicknesses from 6×10^{-3} cm to 2×10^{-1} cm. The intrinsic ultraviolet absorption edge occurs at $\bar{\nu} = 35,000$ cm^{-1} (2860Å wavelength). This edge is the steeply rising portion of Fig. 5 for $\alpha > 100$ cm^{-1}. The intrinsic transparency region is then 1500 cm^{-1} $<\bar{\nu}>$ 35,000 cm^{-1}. Most of the optical opacity in this range, see Fig. 5, is caused by the residual porosity in the ceramic samples. The opacity caused by the residual OH shows up as the double peak near $\nu = 3500$ cm^{-1}. This OH-produced opacity is a true absorption. This water was not removed during consolidation because the sintering was carried out in a closed crucible in order to prevent the loss of germania. The wavenumber region 3700 cm^{-1} $<\bar{\nu}<$ 35,000 cm^{-1} in Fig. 5 shows the equivalent α values determined from in-line percent transmission data. A lot of the beam light is lost by scattering out of beam, and is not a true absorption. The scattering is caused by the residual porosity. An analysis of this curve shows that the maximum scattering occurs for $\bar{\nu} = 15,000$ cm^{-1} or a photon wavelength of 0.67 microns. This is close to the average pore diameter in these ceramics of 1.0 microns. The total pore volume was about 0.3 percent.

Figure 6 shows the infrared-cutoff region from Fig. 5 in more detail. There are multiphonon lattice-vibration peaks at $\bar{\nu} = 1100$ cm^{-1} (actually a double peak), 1500 cm^{-1}, and 2500 cm^{-1}. A common feature of the infrared cutoff-edge of almost all transparent ceramics is the presence of such multiphonon peaks. If we designate the main lattice vibration bands, shown in Fig. 4 as the α, β, δ bands, then the multiphonon bands in Fig. 6 are:

$$\bar{\nu} = 1100 \text{ cm}^{-1} = 2 \ \beta$$

$$\bar{\nu} = 1500 \text{ cm}^{-1} = 3 \ \beta \ + \ 2 \ \delta$$

$$\bar{\nu} = 2500 \text{ cm}^{-1} = 3 \ \delta$$

The multiphonon absorption gets progressively weaker as the number of simultaneously generated phonons increases. Thus in the best possible ceramic with zero porosity one would have absorption coefficients below $\alpha = 1$ cm^{-1} for the wavenumber region 3500 cm^{-1} $\leq \nu \leq$ 30,000 cm^{-1}. The processing will need considerable modification in order to achieve such near-intrinsic behavior, and the concentration of the residual OH will have to be reduced.

PREPARATION AND OPTICAL PROPERTIES OF POLYCRYSTALLINE ALUMINUM GERMANATE

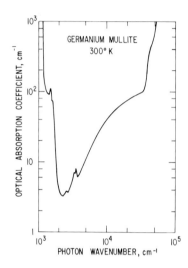
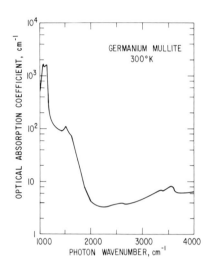

5. Absorption coefficient of polycrystalline germanium mullite, 1.74 $Al_2O_3 \cdot GEO_2$, over a wide range of wavelengths.

6. Optical absorption in the infrared-cutoff region of germanium mullite.

Figure 7 shows the actual percent transmission in the infrared for two different samples of germanium mullite with differing amounts of Al_2O_3. Sample A had 62.3 wt% Al_2O_3 and Sample B had 60.2 wt% (see Table 1). Because these two samples may have had slightly different porosities it is not possible to unambiguously assign the higher transmission to the higher Al_2O_3 content.

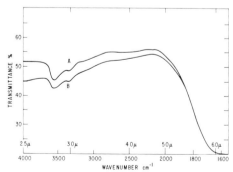

7. Spectral transmittance of germanium mullite - A - 1.70 $Al_2O_3 \cdot GeO_2$, B - 1.55 $Al_2O_3 \cdot GeO_2$. Specimen thickness 0.76 mm.

PREPARATION AND OPTICAL PROPERTIES OF POLYCRYSTALLINE ALUMINUM GERMANATE

The actual band energies shown in Table 3 have been derived from the peaks in the reflectivity or the absorption coefficients. Some but not all of these bands have been reported in previous literature (Ref. 6 and 7) on Al-Ge-O ceramics of slightly different compositions.

DISCUSSION

The hydrolysis of mixed solutions of germanium and aluminum alkyloxides yielded aluminum germanate precursor powders that were sufficiently active to permit sintering to high densities at relatively low temperatures. The main benefit of this approach was, however, the purity, homogeniety and exact stoichiometry that would have been unachievable by synthesis from oxides due to the volatility of GeO_2. The hydrolysis in a nonaqueous medium had the added advantage that the removal of the solvent from the precipitated gel by drying did not bring about strong agglomeration of the material, as is typical with the removal of water. The loose agglomeration was relfected in the high compaction densities, up to 62%, obtained on pressing. In the present work dense bodies of aluminum germanate could be obtained by the sintering of compacts with compositions from 1.55 Al_2O_3.GeO_2 to 1.74 Al_2O_3.GeO_2. Compositions that were richer in alumina contained α-Al_2O_3 and, on sintering, did not yield very high densities. The sintering retarding effect of excess Al_2O_3 was magnified by the consequent loss of germania due to late pore closure. On the other hand there was an indication that very small amounts of α-Al_2O_3 prevented exaggerated growth in specimens of composition 1.74 Al_2O_3.GeO_2, and were thus helpful in reducing intragranular porosity. This was inferred from the observation of a compositional difference between the surface and the interior of the sintered specimens by X-ray fluorescense and the variation of the accompanying microstructure. The near-surface layer in the composition 1.70 Al_2O_3.GeO_2 showed a very regular, uniform grain size, and the as-fired surface showed a depletion of the GeO_2 content and some α-Al_2O_3 was detected by X-ray diffraction. The bodies obtained were sufficiently phase-pure and dense to make possible optical transmission measurements. Substantial further refinements of the process would be required in order to improve on the residual absorbance caused primarily by the remaining pores and, to a lesser extent, by the pockets of glass. or alumina precipitates.

Microstructural and X-ray studies showed only one crystalline phase in the sintered bodies, i.e., germanium mullite. This compound has previously been thought to have a wide range of composition from 1.5 Al_2O_3.GeO_2 to 2.0 Al_2O_3.GeO_2 (4). The present study found that all sintered compositions with an Al_2O_3/GeO_2 ratio less than 1.70 were two phase and contained a GeO_2-rich glass phase in the form of small, intergranular pockets. On the other hand a composition with a mole ratio of 1.92 contained alumina. Thus the stoichiometry of germanium mullite, at least at high temperatures, may be placed between these limits. The absence of any significant difference in the lattice parameters of the sintered specimens suggests that the stoichiometry range of Ge mullite at 1400°C and above is quite small, as was concluded previously by Miller et al. (5). These authors proposed a phase diagram with a peritectic decomposition of germanium mullite at 1530.°C. This could not be confirmed in the present work. Sintering experiments at up to 1700°C showed no decomposition if the GeO_2 loss by vaporization was prevented. Attempts to melt germanium mullite sealed in tungsten tubes resulted in failure of the envelopes due to chemical reaction. Melting was achieved, however, in an oxygen-hydrogen torch, and a full set of mullite diffraction lines, in addition to weak alumina lines, was obtained by X-ray diffraction from the quenched melt. It appears then that germanium mullite melts congruently above 1750°C.

SUMMARY

1. The use of coprecipitated precursors greatly facilitated the preparation of aluminum germanate. This compound crystallized from the amorphous oxides

PREPARATION AND OPTICAL PROPERTIES OF POLYCRYSTALLINE ALUMINUM GERMANATE

at 940°C, i.e., several hundred degrees below the temperature required for the synthesis from mixed oxides.

2. The crystalline powders could be sintered to near theoretical densities at 1400-1600°C in oxygen if loss of germania by evaporation was prevented. The composition of single phase germanium mullite sintered at 1600°C was near 1.75 Al_2O_3.GeO_2 with no indication of a wide stoichiometry range.

3. Melting of 1.74 Al_2O_3.GeO_2 is congruent and occurs at >1750°C.

4. The best sintered specimens were sufficiently translucent to permit optical absorption measurements. The intrinsic transparency region of polycrystalline aluminum germanate is between 6 μm in the infrared and 0.285 μm in the ultraviolet. The minimum apparent absorption coeficient was 3.6 cm^{-1} at 4.65 μm.

ACKNOWLEDGEMENTS

This work was supported by the Defense Advance Research Project Agency. The authors would like to thank their colleagues, G.M. Renlund and F.J. Klug for assistance in thermal analysis and Dr. R.A. Tanzilli for permission to use some of the data in Table 2.

REFERENCES

1. G. Gelsdorf, H. Muller-Hesse, H.E. Schwiete, "Einlagerungsversuche an synthetischem Mullit und Substitutionsversuche mit Gallium und Germaniumoxyd", Arch. F. Eisenhuttenwesen, 29, 513-520 (1958).

2. N.A. Toropov, I.F. Ardreev, V.A. Orlov, S.P. Schmitt-Fogelevich, "Synthesis of Germanium Mullite and Study of its Properties", Zhur. Prikl. Khimii, 43, 2143-47 (1970).

3. S. Durovic, P. Fejdi, "Synthesis and Crystal Structure of Germanium Mullite", Silikaty 20, 97 (1976).

4. H. Muller-Hesse, "Infrarotspektrographische Untersuchungen im System Al_2O_3-SiO_2", Fortschritte Mineralog, 38, 173 (1960).

5. J.L. Miller, G.R. McCormick, S.G. Ampian, "Phase Equilibria in the System GeO_2-Al_2O_3", Jour. Am. Cer. Soc., 50, 268 (1967).

6. M. Perez y Jorba, "Les systemes GeO_2-Al_2O_3 et GeO_2-Fe_2O_3", Silicates Ind., 33, 11-17 (1968).

7. O. Yamaguchi, T. Kanazawa, M. Yokoiga, K. Shimizu, "Formation of Akoxy-Derived 3Al_2O_3.2GeO_2" Ceramics International 9 (1) 18-21 (1983).

8. K.S. Mazdiyasni, L.M. Brown, "Synthesis and Mechanical Properties of Stoichiometric Aluminum Silicate (Mullite), Jour. Am. Ceram. Soc. 55, 548-552 (1972).

9. M. Hoch, K.M. Nair, "Densification Characteristics of Ultrafine Powders", Proc. 3rd CIMTEC Int. Meeting on Modern Ceramic Technologies, May 1976.

10. S. Prochazka, "Sintering and Properties of Dense Aluminum Germanate" in Ceramic Powders, P. Vincenzini ed., 1983, Elsevier Amsterdam.

PREPARATION AND OPTICAL PROPERTIES OF POLYCRYSTALLINE ALUMINUM GERMANATE

11. S. Prochazka, I.C. Huseby, R.P. Goehner, "Thermal Expansion Anisotropy
     and Microcracking in Mullite and Germanium Mullite", G.E.   CRD Report
     82CRD288, November 1982. (Available on request)

CHARACTERIZATION OF GELS AND POWDERS

CHEMICAL ANALYSES OF SOL/GEL SURFACES AND THIN FILMS

Carlo G. Pantano, C. A. Houser and R. K. Brow
Department of Materials Science and Engineering
Pennsylvania State University
University Park, Pennsylvania 16802 USA

ABSTRACT

    The application of surface analysis techniques to the characterization
of sol/gel surfaces and thin films is described. Secondary-ion mass
spectroscopy (SIMS), x-ray photoelectron spectroscopy (XPS) and
sputter-induced photon spectroscopy (SIPS) are used to measure the
composition of multicomponent silicate films, the relative water content of
alumina films, the nitrogen content of ammonia treated silica films, and
the depth profiles for films on black chrome. The determination of
chemical structure using XPS and SIMS is also discussed. Finally, a brief
introduction to temperature-programmed desorption (TPD) and its potential
for studying surface chemical reactions, in situ, is presented.

INTRODUCTION

    Thin films can be prepared by spraying, dip-coating or spinning of an
organometallic solution onto a solid substrate (1-4). The composition and
chemistry of this solution is readily specified and controlled, but the
composition and chemical structure of the resulting film is dependent upon
a variety of factors. If partially hydrolyzed, the solution will contain a
distribution of polymer species. Upon deposition of this solution, the
solvent phase very rapidly evaporates and the extent of additional
polymerization within the film depends upon the drying and firing
atmospheres. Likewise, the composition of the film can be influenced by
volatization of unreacted monomer or adsorption of atmospheric species
during the drying and firing steps. Although these effects are of some
concern in the case of bulk gels, they are greatly exaggerated in the case
of thin films and surfaces. Usually, bulk gels are prepared by gelation
under closed conditions, whereas thin films undergo gelation and solvent
evaporation almost simultaneously and essentially in an open system.
Moreover, the dimensions of these films impose no real kinetic limitations
upon evaporation or adsorption rates. The point is that in the preparation
of sol/gel thin films, only the initial solution composition is known with
any certainty. Thus, a meaningful characterization and understanding of
these films requires extensive chemical analysis.

    In this paper, the utilization of secondary ion mass spectroscopy
(SIMS), x-ray photoelectron spectroscopy (XPS), sputter induced photon
spectroscopy (SIPS), and temperature programmed desorption (TPD) is
described with regard to the characterization of sol/gel surfaces and thin
films. Although these methods have been well-developed for the
characterization of solid surfaces and thin films (5,6), their application
to sol/gel surfaces and thin films is less than routine. The main reason
is that all of these analytical methods are susceptible to matrix effects
which severely complicate both qualitative and quantitative analyses.
Since gelation, drying and firing bring about a continual evolution of the
thin film matrix (e.g. molecular weight, atomic and bulk density,
electronic structure, etc.), the effect of these matrix changes upon the
chemical analyses can be quite bothersome (7). In addition to these matrix
effects, the ultra high vacuum required for all of these surface analysis

Mat. Res. Soc. Symp. Proc. Vol. 32 (1984) Published by Elsevier Science Publishing Co., Inc.

methods is a point of concern. The loss of water and other volatile species in vacuum will not only influence the compositional analyses, but can also lead to changes in the chemical structure of the film. It must be frankly stated that at the present time, the ability to quantitatively characterize these novel thin films – particularly with regard to chemical structure – is limited. It is hoped that this brief review will help to clarify the present capabilities of these methods, and at the same time, highlight those areas where additional development of the methods is needed.

COMPOSITION

The composition and in-depth profiles of multicomponent sol/gel thin films are most readily obtained using secondary ion mass spectrometry (SIMS). In this method, an energetic beam of inert ions is used to continuously erode, or sputter etch, the film. The sputtered species which are ionized (so-called secondary-ions) are mass-analyzed and detected using a quadrupole mass spectrometer. The instrument used in this work (Gatan Model 591C) contains a 7 keV mass-analyzed argon ion beam which is rastered over the film in a square pattern. The detection system is electronically-gated so that only those secondary ions sputtered from the central, flat-bottomed region of the ion-beam crater are accepted. An electron gun can be used to neutralize ion-bombardment induced charging, but this was not a real problem in the analysis of these films. A computerized data acquisition system is used to scan the mass spectra and sequentially measure the count rates for the secondary ions of interest (e.g. $^1H^+$, $^{11}B^+$, $^{16}O^+$, $^{23}Na^+$, $^{27}Al^+$, $^{28}Si^+$, $^{42}SiN^+$, $^{47}SiOH^+$, etc.). The vacuum pressure is typically 1×10^{-9} torr during these analyses.

A series of 4-component and 5-component sol/gel thin films were prepared (8) utilizing solutions whose oxide concentrations were 76 SiO_2, 17.9 B_2O_3, 4.8 Al_2O_3, 1.3 BaO and 71.2 SiO_2, 16.8 B_2O_3, 4.5 Al_2O_3, 1.3 BaO, 6.3 Na_2O, in mole percent respectively. The objective of these analyses was to determine the effect of hydrolysis conditions, solution pH and oxide composition upon the boron concentration in the films. Thus, solutions were prepared at high and low pH and with varying ratios of water to tetraethoxysilane (TEOS) Otherwise, the relative concentrations of the various alkoxides (TEOS, aluminum sec-butoxide, trimethyl borate) and the sodium or barium acetate were held constant to yield the specified oxide compositions.

Figure 1 presents the SIMS depth profiles for a 4-component film on Si prepared from a high pH solution whose water to TEOS ratio was 12. Qualitatively, the profiles in Figure 1 are typical of those obtained for all of the specimens. The presence of a distinct film containing Al, Si, B, O and Na is clearly evident (Ba was detected in the film but the profile is not shown in Figure 1). Sodium is a common impurity that has been found in all the single and multicomponent silicate sol/gel films that we have examined with SIMS. The apparent pile-up of the sodium at the film/substrate interface is probably an artifact of the analysis due to ion-beam enhanced migration of sodium ions. Perhaps the drop in the Si signal upon crossing the film/substrate interface is most surprising since the substrate is pure silicon. This is one of a number of matrix effects associated with SIMS and is due to the oxygen-enhancement of the $^{28}Si^+$ signal in the oxide film (5). Fortunately, this effect does not hamper the analysis of the oxide film itself. It should be noted finally that the sputtering rates used to calibrate the depth scales were obtained independently by measuring the film thickness with ellipsometry.

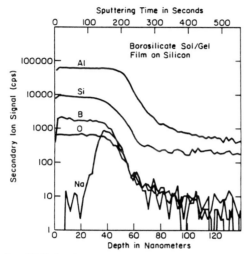

Figure 1.   SIMS depth profiles for an unheated 4-component
sol/gel film ($\cong 24\mu A/cm^2$ Ar$^+$ @ 7KeV).

Of particular interest to us is the boron oxide content of these
films. Thus, a set of bulk glass standards were prepared for a multipoint
calibration of the $^{11}B^+/^{16}O^+$ SIMS signal ratio. Most of these standards
were boroaluminosilicate glasses prepared by melting with varying amounts
of boron oxide; these glasses were subjected to independent chemical
analyses. One of the glass standards, however, was prepared by the sol/gel
method rather than by conventional melting so that any effects of porosity,
residual water or organics upon the SIMS signal could be acounted for.
This bulk gel was prepared from a solution whose oxide composition was 71.2
SiO_2, 16.8 B_2O_3, 4.5 Al_2O_3, 6.3 Na_2O and 1.3 BaO (in mole percent). Upon
gelation, drying and firing at 500°C in N_2, this glass was also
subjected to an independent chemical analysis. The resulting composition
was 78.7 SiO_2, 9.2 B_2O_3, 4.9 Al_2O_3, 6.1 Na_2O and 1.1 BaO (in mole percent).
Thus, it is apparent that boron losses occur during gelation, drying, and
firing of these materials even in bulk. (Note: A white residue was observed
to form on the walls of the container used to dry the gel; it was analyzed
as boric acid and verifies that a significant amount of the boron loss
occurs during drying). Nonetheless, this bulk gel glass, whose composition
was independently obtained, could be used for calibration of the SIMS
signals measured on the films. The results of these quantitative boron
oxide analyses were the same regardless of whether the sol/gel or melted
glass standards were used for calibration; the B/O signal ratio exhibited a
linear dependence upon the boron content over the range 0-12.5% B_2O_3. The
SIMS signal ratios used for these quantitative analyses were those obtained
after the standards and the films had attained the 'equilibrium sputtered
surface composition'; in this way, preferential sputtering effects are
eliminated.

The SIMS analyses for many of the films are summarized in Table I. It
is immediately apparent that up to 90% of the boron in solution can be
evolved from the film during deposition, drying and firing. All of the
heated 4-component films had boron oxide contents of the order .5-2.0 mole
percent irrespective of the pH or extent of hydrolysis in the solution.
The only apparent correlation between the boron content and the solution
chemistry concerned the pH. That is, the acid solutions produced films

whose boron losses were a factor of two greater than for the alkaline solutions. It can also be seen that the drying atmosphere and the oxide composition also influence the boron losses. There is a slightly greater loss of boron when the films are dried at high relative humidity; this is a bit surprising since $B(OH)_3$ is expected to be less volatile than $B(OMe)_3$. More interesting is the observation that sodium additions to the solution attenuate the boron loss. It is likely that the boron and sodium oxides are chemically combined in the gel, and thereby less susceptible to volatization. In general, though, it is clear that the loss of boron oxide is considerably greater for these thin films, than for the bulk gels, primarily because the kinetics are enhanced by the short diffusion distances.

The preparation and characterization of sol/gel thin films on smooth substrates such as silicon single-crystal wafers greatly facilitates research studies, and at the same time, has a practical relevance to the use of sol/gel films as spin-on dopants, interlayer dielectrics and barrier coatings for microelectronic devices. However, there are many situations where the films must be deposited on rough substrates which make subsequent compositional analyses very difficult. Figure 2a, for example, presents the SIMS depth profiles for solar absorbing black-chrome (deposited on a nickel alloy) which has been dip coated with the 4-component boroaluminosilicate solution (9). The sol/gel coated substrate was heated at $500^{\circ}C$ in N_2 for densification and was then subjected to an accelerated oxidation test at $400^{\circ}C$ in O_2. The profiles do not exhibit a sharp interface between the sol/gel constituents (Si, B, Al, etc.) and those of the black chrome (Cr, Al, C, Ni, etc.). The profiles for the uncoated black-chrome are shown in Figure 2b for reference. Of course, the apparent co-existence of the black chrome and sol/gel is due to

TABLE I

Boron Oxide Content[+] of Boroaluminosilicate
Sol/Gel Films (in mole percent)

	$\%B_2O_3$
Dried at 26% R.H., H_2O/TEOS = 12, high pH, 4-component	1.8
Dried at 70% R.H., H_2O/TEOS = 12, high pH, 4-component	1.2
Dried at 26% R.H., H_2O/TEOS = 12, high pH, 5-component	4.9
Dried at 70% R.H., H_2O/TEOS = 12, high pH, 5-component	4.1
Heated to $500^{\circ}C$ in N_2, H_2O/TEOS = 12, high pH, 4-component	1.5
Heated to $500^{\circ}C$ in N_2, H_2O/TEOS = 8, high pH, 4-component	1.5
Heated to $500^{\circ}C$ in N_2, H_2O/TEOS = 5, low pH, 4-component	.9
Bulk gel dried in air, slow-fired at $500^{\circ}C$ in N_2, 5-component	9.2[*]
Solution, 5-component	16.8[*]

* by emission spectroscopy
+ by SIMS

the roughness associated with these substrates. Any attempt to perform a quantitative compositional analysis of the oxide coating with these data would be fruitless. Even attempts to verify the relative concentrations of the sol/gel constituents is hampered because the black chrome, which essentially penetrates the oxide coating, will influence the secondary ion yields. Qualitatively, though, it can be seen that chromium and nickel are detected at the outermost surface. This indicates that the asperities of the black chrome coating are exposed, and hence are not being passivated by the sol/gel coating. The profiles also suggest that nickel has out-diffused through the black-chrome during the thermal treatments.

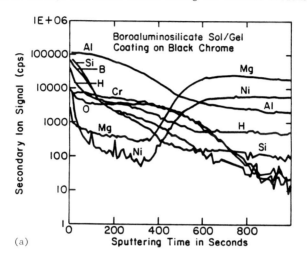

(a)

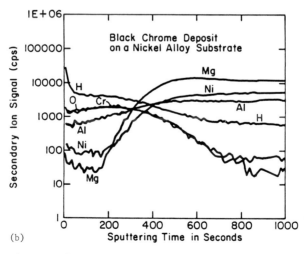

(b)

Figure 2. SIMS depth profiles for (a) sol/gel coated black chrome, (b) black chrome as-deposited on the nickel alloy substrate; obtained with $\cong 20\mu A/Cm^2$ Ar^+ @ 7KeV.

The major problem associated with quantitative analysis by SIMS is due to the matrix effect. This means that the electronic and chemical structure of the calibration standards must approximate those of the sample. Thus, the quantitative analyses already presented utilized oxide glass standards (prepared by melting and by sol/gel) containing known concentrations of boron oxide. Unfortunately, this is not possible for all species of interest. For example, we have an interest in measuring the nitrogen concentration in silica sol/gel films which have been thermally treated in ammonia (3,4). It is not possible to prepare bulk silica glass standards by either melting or sol/gel processing which contain significant concentrations of nitrogen. Although it is possible to prepare multicomponent oxide glasses with nitrogen, these are not ideal SIMS standards for nitrogen analysis in pure silica films. Similarly, stoichiometric silicon oxynitride (Si_2N_2O) and silicon nitride (Si_3N_4) are unacceptable because of the matrix effects introduced by the substantially non-oxide character of these materials. For these reasons, it is necessary to utilize another technique for quantitative nitrogen analysis. We have used x-ray photoelectron spectroscopy (XPS) because it is least susceptible to matrix effects and is essentially non-destructive. Auger electron spectroscopy (AES) can also be used for nitrogen analysis, but electron backscattering and electron beam damage effects cause some additional complications.

In the XPS technique, the film is irradiated with a flux of relatively soft x-rays. The ejected photoelectrons are energy analyzed to provide both elemental and chemical-state identification. The relatively short inelastic free path of these photoelectrons provides for the surface sensitivity of XPS. The intensity of the photoelectron emission for each element and/or chemical-state can be used for quantitative analysis. Figure 3, for example, is a calibration curve obtained using some multicomponent soda-calcia-silica-oxynitride glasses (prepared by melting and independently analyzed for nitrogen content by gas fusion) and both crystalline silicon-oxynitride and crystalline silicon nitride. The linear relationship between the N1s/Si2p photoelectron intensity ratio, and the analyzed N/Si atomic ratio, verifies the insensitivity of XPS to the matrix differences. Thus, the N1s/Si2p intensity ratio can be measured for nitrided silica sol/gel films and this calibration curve can be used to determine their N/Si atomic ratio. One of these analyses is shown on Figure 3; it indicates that ~16 atomic percent N can be introduced by treating a silica film at $1000°C$ in ammonia. A complete study of nitrided silica sol/gel films utilizing XPS and infrared spectroscopy is reported elsewhere in these proceedings (10).

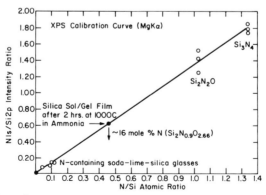

Figure 3. The relationship between the N1s/Si2P photoelectron intensity ratio and the N/Si atomic ratio.

Hydrogen is a ubiquitous constituent whose concentration at gel surfaces and within sol/gel thin films is always of interest. Of course, the propensity for these gel materials to adsorb/desorb water in the ambient atmosphere or in the analytical vacuum chamber makes the interpretation of even the most precise analyses of questionable value. Nonetheless, the concentration and chemical structures associated with hydrogen in gels and gel-derived materials is very important. Here, we describe a method which has proven useful for qualitatively evaluating the relative hydrogen content of sol/gel films with no particular reference to chemical state or structure. The method is called sputter-induced photon spectroscopy (SIPS) (11,12). In this method, the film is eroded with an energetic ion beam and the characteristic photon emissions are analyzed and recorded. These photons arise due to relaxation of the metastable neutrals being sputtered from the film. A monochromator selects the photons of interest and the emission intensity is recorded continuously as the film is eroded. In contrast to SIMS, the SIPS signal is less susceptible to matrix effects. Thus, hydrated and dehydrated oxides can be used as standards for semi-quantitative hydrogen analysis.

Table II presents the result of a SIPS analysis for an alumina sol/gel film deposited on a glass substrate; the analyses for hydrated and dehydrated alumina standards is also shown. The photon emission at 656 nm (due to hydrogen) was referenced to the emission at 396 nm (due to aluminum). The H/Al signal ratio is used to normalize for differences in density and sputtering rate between the films and standards. One of the problems associated with these hydrogen analyses is immediately apparent after comparing the signal ratios for the single crystal sapphire (dry alumina) with the compacted aluminum trihydrate powder. That is, the dynamic range, or sensitivity, is limited to about a factor of 20. However, one can conclude that the water content of the film is less than 3% and is greater than the water content of the sapphire. The non-linear relationship between the H/Al signal ratio and the hydrogen contents of the standards precludes any real quantitative analysis with this limited number of specimens. The non-linearity is probably due to errors in the assumed water content of the standards since dehydration can occur in the SIPS vacuum system. Nonetheless, these semi-quantitative analyses have been useful for ranking the water contents of alumina films which were prepared and processed in different ways.

TABLE II

SIPS Signal Ratios for H(656nm)/Al(396nm)

Single crystal sapphire	$.7 \times 10^{-3}$
Aluminum tri-hydrate ($\cong$30% H_2O)	16.8×10^{-3}
Partially dehydrated aluminum tri-hydrate ($\cong$3% H_2O)	7.7×10^{-3}
Alumina sol/gel film	4.1×10^{-3}

CHEMICAL STRUCTURE

The elemental compositional analyses described in the last section were carried out with no reference to the associated electronic or chemical stucture. In fact, it was pointed out that compositional analyses, particularly in the case of SIMS, are complicated by the bonding and matrix dependence of the ion, electron or photon emission process. In the case of sol/gel materials, the changes in electronic and chemical structure which accompany hydrolysis, polymerization and densification are phenomena of particular interest. Thus, the question arises naturally: Can the bonding and matrix sensitivity of these methods be utilized to provide information about the chemical structure of sol/gel materials?

In the case of XPS, the extraction of structural information is rather straightforward. For example, the Cls photoelectron binding energy is distinguishable for carbon atoms in alkoxy groups (e.g.,-Si-O-C-), saturated hydrocarbons (-C-C-C-) or carbides (-Si-C-). Thus, it is possible to follow the evolution of the carbon in films and at surfaces in terms of their chemistry. Likewise, the coordination structures associated with Si and N in ammonia treated silica gels can be determined by measuring the binding energy of the Si2p and Nls photoelectrons. Since one of the papers in these proceedings describes the use of XPS for chemical structure analysis in nitrided silica gels (10), nothing further will be presented here concerning this approach. However, it is noteworthy that the conventional XPS technique cannot detect hydrogen nor distinguish between oxygen atoms in siloxane or other metal-oxide bridges (e.g. -Si-O-Si-) versus those in terminal hydroxyl groups (e.g. -Si-OH). Thus, there is some limitation in the applicability of XPS to these hydrated oxide gels.

On the other hand, the potential of SIMS for evaluating the chemical structure of these materials is substantial because the mass spectra provide information about molecular clusters (e.g. OH, SiOH, Si_2N, etc.) and molecular weight distributions. Figure 4 shows, for example, the SIMS spectra for an untreated silica sol/gel film. In addition to the elemental secondary ions used for compositional analysis (e.g. $^1H^+$, $^{11}B^+$, $^{16}O^+$, $^{28}Si^+$, etc.) one notes the presence of molecular clusters at higher masses (e.g. $^{44}SiO^+$, $^{56}Si_2^+$, etc.). One would expect the relative abundance of SiO, SiO_2 and Si_2O clusters versus SiOH and $Si(OH)_2$ to increase with polymerization of the film structure. Of course, one must be careful not to confuse $^{45}SiO^+$ (due to the ^{29}Si isotope) with $^{45}SiOH^+$ (due to the ^{28}Si isotope). In this study, the $^{47}SiOH$ signal (due to the ^{30}Si isotope) was used to avoid any interference.

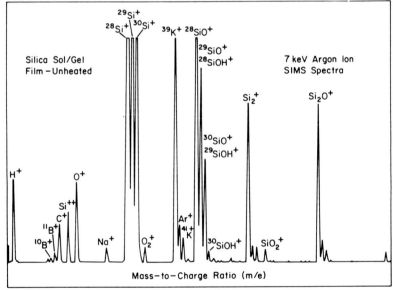

Figure 4. SIMS scan obtained at $\cong 24\mu A/cm^2$ which shows the elemental and molecular secondary ion species.

The molecular secondary ions can also provide insight to the chemical structure of ammonia treated gels. That is, the nitrogen in amine groups (e.g. NH_3, NH_2 etc.) may be distinguishable from metal nitride groups (e.g. SiN, Si_2N, etc.). Unfortunately, an incomplete understanding of the dynamics of sputtering and secondary ion formation make this chemical structure analysis less than a routine undertaking. In particular, it is known that many of the molecular clusters observed in the SIMS spectra do not arise by direct ejection, but rather by recombination of sputtered species above the surface. Nonetheless, it has been shown that many of these recombination clusters are, in fact, related to the parent molecule; in addition, the susceptibility to fragmentation and recombination can be controlled to some degree by varying the primary ion beam energy and current density (13). In an effort to define the extent to which this method could be used to characterize the chemical structure of sol/gel materials, a study was undertaken to fingerprint the SIMS spectra associated with silica gel films which had been hydrolyzed and thermally treated in various ways. The fingerprint spectra for SiO_2, Si_3N_4 and SiC, for example, are already available in the literature (14).

Figure 5 plots the integrated secondary ion intensity ratios for $^{47}SiO^+/^{28}Si^+$ and $^4SiOH^+/^{28}Si^+$ in silica thin films heated at various temperatures in oxygen. Since the silicon content through the film is relatively constant, the $^{28}Si^+$ signal is used to normalize the measured intensities against fluctuations and instabilities in the primary ion beam and detection system. It can be seen that the relative intensity for SiO^+ increases over the temperature range 200°C to 700°C. In contrast, the intensity for $SiOH^+$ is nearly invariant up to 700°C and then drops off sharply at higher temperatures. This may be due to the reversible nature of the dehydroxylation reactions which occur at temperatures less than 800-900°C (15); i.e. the heat-treated films were exposed to the ambient atmosphere before these SIMS analyses. Clearly, additional work is needed to provide a structural interpretation of these data, but it does appear that the relative intensity of these molecular secondary ions does, in fact, vary with the thermal treatment. Likewise, the relative intensity of SiO^+ and $SiOH^+$ in unheated films varied systematically with the extent of hydrolysis in the solution. The ratio $SiO^+/SiOH^+$ in the films was observed to increase as the H_2O/TEOS ratio in the solution was increased.

Figure 6 shows the integrated secondary ion intensity ratio $^{42}SiN^+/^{28}Si^+$ for ammonia treated silica films. The data indicates that the formation of $^{42}SiN^+$ molecular ions is greatly enhanced in films treated at temperatures in excess of 700-800°C. A complementary set of analyses on these films with XPS verified that the $^{42}SiN^+$ molecular ion arises only when three-coordinated nitrogen is present in the film structure. Unfortunately, the resolution of the mass spectrometer used in these studies was insufficient for separating $(NH_2)^+$ (m/e $\approx$ 16.023) from $^{16}O^+$ (m/e $\approx$ 15.999).

Perhaps the most significant effect of the thermal treatment upon the SIMS analysis is in regard to the sputtering rate. It is found that although the films are thinner after a higher temperature treatment, the time required to sputter through them is increased. Thus, the removal of volatiles and the subsequent densification of the film reduces the sputter rate.

264

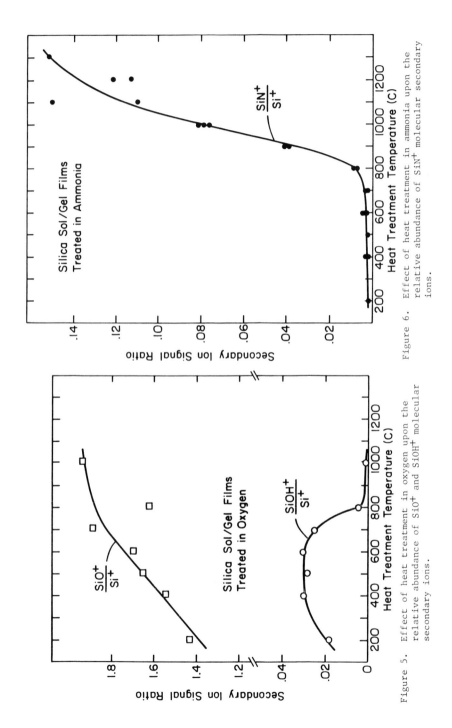

Figure 5. Effect of heat treatment in oxygen upon the relative abundance of SiO^+ and $SiOH^+$ molecular secondary ions.

Figure 6. Effect of heat treatment in ammonia upon the relative abundance of SiN^+ molecular secondary ions.

TEMPERATURE PROGRAMMED ADSORPTION/DESORPTION

Most recently, we attempted to use coordinated measurements of weight change and gas evolution to investigate, in situ, adsorption/desorption and chemical reactions during thermal treatment of sol/gel materials. The experiments were carried out in an all-metal high vacuum system equipped with a microbalance and gas analyzer (16). Figure 7 presents a typical set of data for an acid catalyzed silica gel (made from TEOS). Upon evacuation of the system to 10^{-7} torr, a weight loss of approximately 13.5% was recorded due to desorption of weakly physisorbed water and ethanol. The gel was then heated at a programmed rate to 415°C during which time an additional 4.5% weight loss was reached. The corresponding gas analysis shows a temperature dependent evolution of ethanol and water. It appears that desorption of ethanol may occur by two mechanisms since the spectra show peaks at ~170°C and ~220°C. The water spectrum is more complex, perhaps indicating three peaks in the temperature-programmed desorption (TPD) spectrum. It is also apparent that the gel is still evolving water at 415°C (the maximum operating temperature of this particular apparatus). These TPD's can be carried out at various rates to investigate adsorption-desorption equilibria and to provide kinetic data and activation energies for the various desorption mechanisms.

We have used this approach to examine reactions between silica gel and ammonia. The gel which had been dehydrated at 415°C was cooled to room temperature and then exposed to 3 torr of NH_3. This yielded a weight gain of about 3% due to physical adsorption of NH_3. The subsequent vacuum evacuation to 10^{-7} torr removed most, but not all, of this ammonia. Thus, it is likely that some chemisorption of ammonia (~8% of the total) occurs even at room temperature. The temperature programmed heating of this gel showed water, ammonia, and ethane in the desorption spectra. When the ammonia adsorption is carried out at 125°C, a weight gain is observed initially, but this is followed by a substantial weight loss. This suggests that ammonia reacts with these materials at the relatively low temperature of 125°C. Although more work is needed here to define the reaction pathways, the method is well suited to the study of gel treatment and densification in reactive atmospheres.

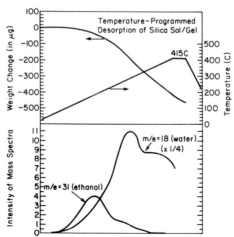

Figure 7. Weight change and mass spectrometric gas analysis during the temperature-programmed heating of a silica gel.

SUMMARY

It should be apparent that no one technique is sufficient to characterize the composition and chemical structure of sol/gel surfaces and thin films. And even with the coordinated use of multiple techniques, a quantitative characterization of composition and chemical structure is difficult to achieve. However, the application of these methods to these chemically-complex sol/gel materials is still in its infancy. The continued use of these methods to analyze sol/gel materials, and the development of a more substantial data base of spectra, will not only improve the quality and precision of the analyses, but should also make them more routine. Nonetheless, it is possible at the present time to obtain meaningful compositional analyses, providing that oxide standards are prepared for signal calibration. And in many cases, the effects of the various process variables can be followed with just the semi-quantitative analyses of composition, hydrogen and chemical structure.

ACKNOWLEDGMENT

The authors gratefully acknowledge the financial support of the National Science Foundation (DMR-8119476) and Sandia National Laboratory.

REFERENCES

1. H. Schroeder, Oxide Layers Deposited from Organic Solutions, Phys. Thin Films, 5, 87 (1969).
2. B. E. Yoldas and T. W. O'Keefe, Antireflective Coatings Applied from Metal-Organic Derived Liquid Precursors, Appl. Opt., 18, 3133 (1979).
3. C. G. Pantano, P. M. Glaser and D. H. Armbrust, Nitridation of Silica Sol/Gel Thin Films in Ultrastructure Processing of Ceramics, Glasses and Composites, Hench and Ulrich, Eds., (John Wiley, New York, 1983).
4. P. M. Glaser and C. G. Pantano, Effect of the H_2O/TEOS Ratio Upon the Preparation and Nitridation of Silica Sol/Gel Films, J. Non-Cryst. Sol., 63, 209 (1984).
5. A. W. Czanderna, Editor, Methods of Surface Analysis (Elsevier 1975).
6. C. G. Pantano, Surface and In-Depth Analysis of Glass and Ceramics, Am. Ceram. Soc. Bull., 60 (11), 1154 (1981).
7. G. Smith and C. G. Pantano, SIMS Studies of Glass Thin Films Prepared with Sol/Gel Techniques, Appl. Surf. Sci., 9 345 (1981).
8. C. J. Brinker and S. P. Mukherjee, Comparison of Sol-Gel-Derived Thin Films with Monoliths in a Multicomponent Silicate Glass System, Thin Solid Films, 77, 141 (1981).
9. R. B. Pettit and C. J. Brinker, Sol-Gel Protective Coatings for Black Chrome Solar Selective Films, SPIE, 324, 176 (1982).
10. R. K. Brow and C. G. Pantano, Composition and Chemical Structure of Nitrided Silica Gel, in this volume.
11. H. Bach, Ion Beam Induced Radiation Applied to Investigations of Thin Surface Layers on Glass Substrates, J. Non-Cryst. Sol., 19, 65 (1975).
12. N. H. Tolk, et al., In Situ Spectrochemical Analysis of Solid Surfaces by Ion Beam Sputtering, Anal. Chem., 49, 16A (1977).
13. R. J. Colton, Molecular Secondary Ion Mass Spectrometry, J. Vac. Sci. Technol., 18(3), 737 (1981).
14. A. Benninghoven, et al., Comparative Study of Si(111), Silicon Oxide, SiC, and Si_3N_4 Surfaces by (SIMS), Thin Solid Films, 28, 59 (1975).
15. M. L. Hair, Hydroxyl Groups on Silica Surfaces, J. Non-Cryst. Sol., 19 (1), 299 (1975).
16. F. Ohuchi and U. Chowdry, Temperature Programmed Reaction and Desorption Studies on Methanol Oxidation Catalysts, in Proc. 12th No. Am. Thermal Analysis Society Meeting, Williamsburg, VA, 1983.

IN SITU FT-IR STUDIES OF OXIDE AND OXYNITRIDE SOL-GEL-DERIVED THIN FILMS*

DAVID M. HAALAND AND C. JEFFREY BRINKER
Sandia National Laboratories, Albuquerque, NM   87185

ABSTRACT

A high-temperature infrared cell was developed to study the gel-to-glass conversion of sol-gel-derived thin films. FT-IR spectra of matched thin-film borosilicate sol-gel samples were taken as the samples were heated at 100°C intervals to 700°C in either air or ammonia. The gels were converted to oxide and oxynitride glasses, respectively, by these heat treatments. The gel-to-glass conversion could be followed and compared for these two treatments by monitoring changes in the vibrational bands present in the spectra. Comparisons between the infrared spectra of NH_3-treated and air-treated films heated above 500°C reveal the appearance of new B-N bonds at the expense of B-O-Si bonds for the NH_3-fired films. These spectra also exhibit changes which may indicate the formation of Si-N bonds. Thus, ammonolysis reactions can result in thin-film oxynitride glass formation at relatively low temperatures.

INTRODUCTION

Oxynitride glasses have been shown to exhibit greater fracture toughness and microhardness, higher glass-transition temperatures, better chemical durability and altered thermal-expansion properties relative to the corresponding oxide glasses [1-6]. We have recently demonstrated that oxynitride glasses can be prepared without melting by heating porous gels in gaseous ammonia [6-8]. Fourier transform infrared spectroscopy (FT-IR) has been shown to be useful in identifying the mechanisms involved in the synthesis of the oxynitride glasses from silicate gels [6-7]. In these earlier studies, however, bulk gels were heated in ammonia and then disperses in a KBr matrix. Thus, the samples were exposed to the ambient atmosphere prior to analysis and only the relative infrared band intensities between samples could be compared.

Using a high-temperature infrared cell designed for in-situ FT-IR studies, the infrared investigation of sol-gel-derived oxide and oxynitride glasses was extended to thin-film glasses. This new cell made possible the infrared study of thin-film glasses as a function of temperature and atmosphere without exposure to the ambient environment. In addition, by studying the various treatments of sol-gel samples of nearly identical initial composition and thickness, it was possible to monitor changes in absolute infrared band intensities with treatment atmosphere and temperature.

EXPERIMENTAL PROCEDURES

Thin-film samples were prepared by spinning multiple coatings of a hydrolyzed solution of the binary composition 1 $SiO_2 \cdot 1$ B_2O_3 (3 wt. % oxides)

*This work performed at Sandia National Laboratories supported by the U.S. Department of Energy under Contract Number DE-AC04-76DP00789.

onto polished 1 mm thick polycrystalline silicon substrates. The solution was prepared from $Si(OC_2H_5)_4$ and $B(OCH_3)_3$ in an ethanol solvent and hydrolyzed with 4 moles of water per mole of alkoxides using an acid catalyst (HCl). Argon bubbled through an aqueous solution of 1 M NH_4OH was used as the deposition atmosphere to help prevent reesterification (and subsequent volatilization) of the hydrolyzed borate species. After applying 4 coats, the film thickness measured by ellipsometry was 199 nm. Infrared spectra of the samples indicated that some boron was lost by volatilization and that the final boron content of the coating was ~10 wt. % B_2O_3.

The infrared samples (12 x 6 x 1 mm) were all cut from the same sol-gel-coated Si wafer (50 mm dia. x 1 mm). The infrared spectra were taken of each sample, and those samples of identical coating composition and thickness (~1% difference) were selected for study. Bare silicon samples were prepared from coated Si samples by chemically stripping the sol-gel coatings in an ammonium bisulfide-fluorosilicic acid solution. These samples were used to obtain the Si substrate spectra at the same temperature after identical treatment as the coated Si samples. One sol-gel-coated and one un-coated Si sample were each placed in the stainless-steel sample holder. The sample holder could be resistively heated with a sheathed nichrome wire heater brazed into the holder. To reduce radiative heat losses, a layer of Pt was sputtered on the sample holder which had been oxidized in air at 800°C. This arrangement made sample temperatures approaching 800°C possible. A type-K sheathed thermocouple was pressed against the sol-gel coated sample for temperature measurement. The holder was then placed into a water-cooled infrared cell with KBr windows. The IR cell was mounted on a computer-controlled positioner (1μm resolution) for accurate and automatic sample positioning. The three sample positions included the sol-gel coated sample, the bare Si sample and an aperture for obtaining spectral backgrounds.

A Nicolet 7199 FT-IR spectrometer equipped with a 4800-400 cm^{-1} range Hg-Cd-Te detector was used to obtain 2 cm^{-1} resolution spectra. Each sol-gel thin-film spectrum is presented after digitally subtracting the spectrum of the simultaneously treated bare Si sample. Because the Si substrate becomes IR opaque above ~ 400°C, spectra of samples treated above 300°C were taken after cooling the samples to 300°C. This procedure also prevented temperature dependent spectral shifts from affecting the spectral subtractions. Spectra were obtained after 15 min treatments at 100°C intervals. Room temperature spectra were also taken of the virgin, 300°C and 700°C treated samples.

RESULTS AND DISCUSSION

Monitoring the infrared spectra of the thin-film gels as a function of temperature yields detailed information about the mechanisms occurring during the gel-to-glass conversion. This is demonstrated for the air-fired borosilicate gel in Fig. 1 in the spectral region from 1600 to 400 cm^{-1}. The virgin gel has infrared bands tabulated in Table I along with their respective assignments. The initial high energy of the B-O stretching vibration (~ 1415 cm^{-1}) suggests it is composed primarily of non-bridging B-O groups. The assignment of the 1305 cm^{-1} vibration is uncertain, and this band is only observed when sol-gel solutions of high boron content are used in making the thin films. It is possible that this band is due to a B-O-B stretching vibration but its frequency is 40 cm^{-1} higher than normally found in vitreous B_2O_3 [9,10]. More likely the 1305 cm^{-1} band is related to BOH vibration found at 1325 cm^{-1} in CVD borosilicate thin films exposed to water vapor [10]. The high-energy shoulder (~ 1200 cm^{-1}) on the intense 1065 cm^{-1} band is due primarily to the longitudinal optical (LO) mode of the asymmetric Si-O stretching vibration [11] since we find the relative

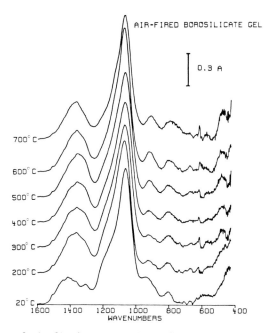

Fig. 1. Spectra of air-fired coating after 15 min hold at the indicated temperatures. Spectrum of the silicon substrate has been subtracted. Spectra at 300°C and above were recorded at 300°C.

Table I. Infrared bands of virgin sol-gel borosilicate

Frequency	Assignment	Ref.
1415	B-O stretch	9,10
1305	BOH (?)	10
1200	Si-O stretch (LO)	11
1065	Si-O stretch (TO)	9-11
942	Si-OH stretch	9-11
915	Si-O-B stretch	9,10
803	O-Si-O bend	9
675	Si-O-B bend	9,10
440	Si-O-Si bend	9,11

intensity of this vibration to be strongly dependent on incident angle and polarization of the infrared beam. The 1065 cm^{-1} band is the transverse optical (TO) mode of the same vibration. It is possible that the 1200 cm^{-1} shoulder and the 942 cm^{-1} band, which is primarily due to SiOH, also have small contributions from ethoxide vibrations [12]. The 915 cm^{-1} vibration of Si-O-B groups is only visible in the virgin sample as a vague shoulder on the 942 cm^{-1} band, but it becomes more pronounced after higher temperature treatments. We attribute the 803 cm^{-1} band to the O-Si-O bending vibration following the assignment of Tenney and Wong [9]. However, this band has been assigned by others as a symmetric Si-O stretching vibration [11] or as due to vibrational modes of ring structures of SiO$_4$ tetrahedra [13]. A very small feature at 675 cm^{-1} is due to a Si-O-B bending vibration [9, 10]. The ~ 440 cm^{-1} band is due to Si-O-Si band bending [9,11]. The spectral regions around 610 and 740 cm^{-1} are quite unreliable due to the subtraction of the very intense Si phonon bands centered at 610 cm^{-1} and 738 cm^{-1}. As the sample is heated in air, the 1305 cm^{-1} band diminishes and shifts to higher energy and the 1415 cm^{-1} band increases in intensity and shifts to lower energy. Thus, the separate BOH groups become incorporated into the borosilicate phase as more B-O-Si bridging bonds are formed. The 942 cm^{-1} band decreases in intensity as non-bridging Si-O terminal groups react to increase the intensity of the 915 cm^{-1} Si-O stretching vibration of Si-O-B bridging groups. The corresponding Si-O-B bending vibration at 675 cm^{-1} also grows stronger at the highest temperatures, and the high energy shoulder on the 1065 cm^{-1} band decreases in intensity. The 803 cm^{-1} band of O-Si-O bending vibrations increases in intensity and broadens as more O-Si-O linkages are formed, and the bond angles become more variable with the greater incorporation of B-O-Si linkages in the glass. The position and intensity of the 440 cm^{-1} band is difficult to accurately monitor due to the poor spectral subtractions of the hot silicon substrate at these low energies. The strong Si-O stretching vibration narrows and shifts 12 cm^{-1} to higher energy as the sample is heated. In addition, we observed the loss of O-H and C-H stretching vibrations (not shown in Fig. 1) as the samples are heated. Therefore, heating in air causes bridging B-O-Si and Si-O-Si bonds to form at the expense of non-bridging SiOH and SiOR groups. A careful examination of Fig. 1 gives the temperatures at which these processes become important.

Fig. 2 presents similar spectra for the sol-gel sample heated in anhydrous NH$_3$. Again we find a loss of the BOH vibration as the sample is heated. Correspondingly, the 1415 cm^{-1} B-O stretching vibration increases in intensity and shifts to lower energy and the 942 cm^{-1} vibration disappears. The 915 cm^{-1} B-O-Si stretching band initially forms but is lost at the two highest temperatures studied. This band disappears at the same temperature that yields the formation of a new vibration at ~1490 cm^{-1} and a decrease in intensity of the shifted B-O stretching vibration at 1360 cm^{-1}. We attribute the 1490 cm^{-1} vibration to the formation of B-N bonding which occurs primarily at the expense of B-O-Si bonds. As described previously [6], this result is expected due to the adsorption of ammonia on the Lewis acid surface sites (primarily boron) followed by dissociation to yield BNH$_2$ + SiOH as shown below:

$$\underset{\underset{M\vdots O\text{-}M}{\overline{\phantom{M\vdots O\text{-}M}}}}{\overset{H\vdots}{H\vdots N\vdots H}} \longrightarrow \underset{\underset{M\quad M}{\overline{}}}{\overset{H\diagdown\diagup H\quad H}{N\qquad O}} \qquad\qquad 1$$

In Eq. 1, the horizontal line represents the gel surface. Further heating results in chemically dissolved nitrogen (N in 3-fold coordination with M) as evidenced by the extremely altered physical properties of the bulk gels heated in the same manner [6]. Unfortunately, our sensitivity in the N-H stretching region is poor for the thin films, so examination of the N-H

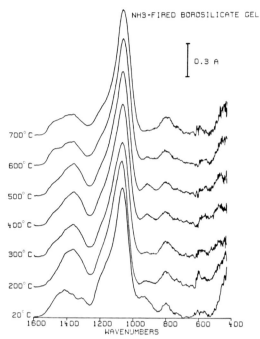

Fig. 2. Spectra of NH₃-fired coating after 15 min hold at the indicated
temperature. Spectrum of the silicon substrate has been subtracted. Spectra at 300°C and above were recorded at 300°C.

bands has been inconclusive. At temperatures of 200°C and above, we observe
very weak bands at 3440 and 3350 cm^{-1}. Although the assignment of these
bands is not consistent in the literature, they might be assigned to N-H
stretching vibrations in which nitrogen is attached to Si [14] and B [15],
respectively. After treatment of the sample at 600°C where reaction 1 has
occurred, we see no evidence of SiO-H vibrations or significant changes in
the 3440 and 3350 cm^{-1} vibrations. However, since the products of reaction
1 can undergo condensation at 600°C to yield water vapor and M-NH-M during
the 15 min temperature hold before before spectra are obtained, significant
changes in the N-H bands and appearance of the SiO-H vibration might not be
expected.

Differences between spectra taken at each temperature for a sample
treated in a given gas environment can more clearly identify the bonding and
chemical changes occurring, but these difference spectra will not be presented here. In addition, the difference spectra at any temperature between
the samples heated in NH₃ and air can yield useful information about the
chemical changes occurring in the two gas environments. For example, the
spectra of the sol-gel samples heated in NH₃ and air to 700°C are presented
in Fig. 3 along with their difference. The difference spectrum clearly
shows the new ~1490 B-N band, the loss of both B-O stretch (~ 1360 cm^{-1}) and
the Si-O-B stretch (915 cm^{-1}) for the NH₃ heated sample relative to the air
heated sample. Also apparent is a slight intensification of the O-Si-O bend
(803 cm^{-1}). The most prominent feature of this difference spectrum is the

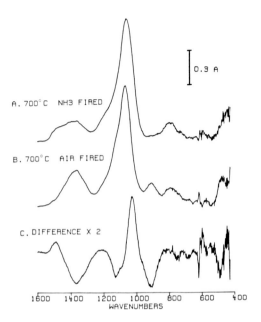

Fig. 3. Spectra are after 15 min hold in each environment at 700°C.
Spectrum of the silicon substrate has been subtracted. C is A
minus B scale expanded by a factor of 2.

change in the Si-O stretching band, which is difficult to see in the spectra
of the individual glasses. There is a shift in intensity from the high-
energy side of the band to the low-energy side. This shift may be simply a
reflection of bond-angle and force constant changes in the Si-O bonds with
the incorporation of nitrogen, or it may be a result of the appearance of a
new silicon oxynitride vibration. This vibration has been observed previously at ~ 980 cm^{-1} [6,16] although its frequency is somewhat dependent on
O/N atom ratio. The difference spectrum places the possible new band at
1030 cm^{-1}. The large difference in frequency between the bulk and thin-film
glasses makes the assignment of this band as a new Si oxynitride vibration
uncertain in the thin film. Previous studies of infrared spectra of bulk
sol-gel materials containing Si, B, Al, Ba, and Na heated in air identified
the presence of Si-N vibrations of the oxynitride [6,7]. If the band at
1030 cm^{-1} is simply a shift in frequency, then the formation of a silicon
oxynitride might require the presence of one or more of the other components.
Further work must be done to clarify this issue.

CONCLUSIONS

     Relative to our earlier IR studies on bulk sol-gel samples [6], the
capability of obtaining in situ, high-temperature spectra of matched thin-
film samples in several gas environments greatly increases the reliability
and information content of the spectra as well as increasing the efficiency
of data collection. Spectra can be obtained as a function of temperature
without exposing the sample to the ambient environment, and therefore, the

samples are not subject to adsorption of atmospheric moisture or other contaminants. Accurate sample positioning and simultaneous treatment of the sample and Si substrate improve the reliability of spectral subtractions. The treatment of matched samples in different environments makes possible the direct comparison of treatments without the necessity of subjective scaling of spectral data. Therefore, absolute rather than relative spectral differences can be noted.

The results presented here help to understand the mechanisms occurring during the gel-to-glass conversions for both air and NH_3-fired borosilicate gels. In addition, the differences in bonding and reaction between the air and NH_3 treatments can be more readily identified and characterized when identical samples are heated in situ in the two environments. The infrared spectra show that borosilicate sol-gel thin films heated in NH_3 form oxynitride glasses at 600°C and above.

ACKNOWLEDGMENTS

C. S. Ashley prepared the sol-gel samples and K. L. Higgins aided in collection of the FT-IR spectra.

REFERENCES

1. R. E. Loehman, J. Am. Ceram. Soc. 62, 491 (1979).
2. T. H. Elmer and M. E. Nordberg, J. Am. Ceram. Soc. 50, 275 (1967).
3. T. H. Elmer and M. E. Nordberg, Proc. VII International Congress on Glass, Brussels, Belgium, Paper No. 30 (1965).
4. P. E. Jankowski and S. H. Risbud, J. Am. Ceram. Soc. 63, 350 (1980).
5. R. E. Loehman, J. Non-Cryst. Solids 42, 433 (1980).
6. C. J. Brinker and D. M. Haaland, J. Am. Ceram. Soc. 66, 758 (1983).
7. C. J. Brinker, D. M. Haaland, and R. E. Loehman, J. Non-Cryst. Solids 56, 179 (1983).
8. C. J. Brinker, J. Am. Ceram. Soc. 65, C-4 (1982).
9. A. S. Tenney and J. Wong, J. Chem. Phys. 56, 5516 (1972).
10. E. A. Taft, J. Electrochem. Soc. 118, 1985 (1971).
11. A. Bertoluzza, C. Fagnano, M. A. Morelli, V. Gottardi, and M. Guglielmi, J. Non-Cryst. Solids 48, 117 (1982).
12. A. L. Smith, Spectrochim. Acta 16, 87 (1960).
13. M. Decottignics, J. Phalippou and J. Zarzycki, J. Mat. Sci. 13, 2605 (1978).
14. J. B. Peri, J. Phys. Chem. 70, 2937 (1966).
15. N. W. Cant and L. H. Little, Can. J. Chem. 42, 802 (1964).
16. F.H.P.M. Habraken, A.E.T. Kuiper, and Y. Tamminga, J. Appl. Phys. 53, 6996 (1982).

STRUCTURE DATA FROM LIGHT SCATTERING STUDIES OF AEROGEL

Arlon J. Hunt and Paul Berdahl
Lawrence Berkeley Laboratory
University of California
Berkeley, California 94720

ABSTRACT

This paper reports a recent advance in the understanding of the structure of microporous optical materials such as aerogel through the interpretation of light scattering data. The Fourier transform of the density-
density correlation function is used to relate measurements of the angular
dependence of scattered light to material structure parameters. The results
of the approach fit the unusual dependence of the intensity of scattered
light as a function of angle for two polarizations. The fit shows that light
scattering from aerogels may be interpreted as having two origins; one from
the small scale structure of linked particles that comprise the material,
and the second due to weak fluctuations in the average density of the microporous structure over distances significantly larger than the pore size.

INTRODUCTION

Low density, microporous optical materials have been prepared that are
transparent and are excellent thermal insulators. A program is underway at
Lawrence Berkeley Laboratory (LBL) to develop these materials for insulating
window applications [1,2]. The current method of preparation utilizes the
sol-gel process followed by supercritical drying to remove the fluid from
the gel. The resulting material (called aerogel) is a low density network
of interlinked particles of extremely small dimensions. Aerogels are transparent because the characteristic size of the microstructure is much
smaller than the wavelength of light, allowing the material to transmit
rather than scatter light. These materials are excellent thermal insulators
because of their low density and the very small pore size.

Microporous optical materials hold considerable promise as new materials for insulating windows and a variety of other applications. It appears
that silica aerogel can be made with a thermal conductivity of 0.022 $Wm^{-1}K^{-1}$
(R7 per inch) and possesses sufficient clarity for use as a view glazing.
Large energy and economic impacts would result from reducing the thermal
losses from buildings which arise from heat flows through the windows. However, before this material becomes commercially viable, its optical and physical properties must be improved and lower cost means of production must be
developed. The work reported here focuses on the relationship between the
microstructure of the material and the scattering of light. In particular,
the technique of polar nephelometry (the study of the angular distribution
and polarization of scattered light) is used [3] to probe the microstructure
characteristics of aerogel and its precursors.

The preparation and properties of silica aerogels are described in the
next section as well as the light scattering technique. The theoretical
relationship between the microstructure of the material and the scattered
light is described in the following section. The final section discusses
the comparison between the experimental and theoretical results.

Mat. Res. Soc. Symp. Proc. Vol. 32 (1984) Published by Elsevier Science Publishing Co., Inc.

EXPERIMENTAL PROGRAM

A program to study the preparation and properties of microporous silica at LBL using sol-gel processing methods was initiated in 1981. The work has three main components; preparation of aerogel, measurements of the properties of aerogel and its precursers, and a theoretical effort to interpret the structure of porous optical materials from light scattering data.

Aerogel Preparation

The aerogel preparation begins with the formation of a colloidal suspension of minute silica particles produced by the hydrolysis of an ester of silicic acid. Monosilicic acid dissociates to form particles of silicon dioxide. The reaction is carried out in a mutual solvent (alcohol) because water and the silica ester are immiscible. Aerogels were made using base catalyzed tetramethylorthosilicate (TMOS)[3] and acid catalyzed tetraethylorthosilicate (TEOS)[4].

The colloidal particles grow in size and link together to form a gel with a density that is determined by the amount of alcohol. The gels are cast into cylinders of diameter 3 cm and 5 cm high. After aging the gels are removed from the mold and placed in an autoclave for supercritical extraction of the alcohol. The pressure and temperature are raised well above the critical point of the alcohol (110 bars and 270° C) and slowly vented. After cooling the gels are removed from the autoclave (subsequent heating above 500° C may be used to remove any remaining hydrocarbons).

Physical Properties

The resulting samples are transparent cylinders similar in size to the gels and generally without cracks. Some shrinkage may occur during formation of the gel or during the drying process. The shrinking is slightly asymmetric, and is often accompanied by an increase in cloudiness. The density of aerogels produced in our laboratory varied from a few percent up to about 10%. Typical particle and pore diameters are of the order of 50-100 and 200-400 Angstroms respectively. The material has an open structure with a relatively small solid cross sectional area through a given plane and as a result is relatively weak compared to bulk silica. Because of its open pores and high surface area, aerogel adsorbs water and must be protected from outside environment. Sealing methods are being developed to provide strength and protection for the aerogel.

Optical Properties

Samples prepared by supercritical drying are very transparent and objects may be clearly viewed through them. The index of refraction is in the range of 1.02 to 1.05 resulting a very low reflectivity. Significant polarization effects that we attributed to strain birefringence were reported earlier and will not discussed further here. Due to the character of the wavelength dependent scattering aerogel has a pale blue cast when observed against a dark background and objects viewed through it appear slightly reddened. Measurements of this wavelength dependence[1,2] produced results consistent with the assumption that the materials are Rayleigh scatterers.

Scattering Measurements

The scattered light intensity was studied as a function of angle and polarization using the polar nephelometer illustrated in FIG. 1. Lasers with wavelengths of 425 and 633 nanometers provided the source of light. The incoming beam was polarized at an angle of 45° to the scattering plane.

The rotating detector arm carries collimation optics, a photomultiplier, and polarizers oriented either parallel or perpendicular to the scattering plane. The intensity for both polarizations is plotted as a function of angle.

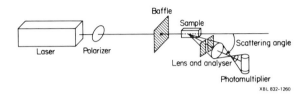

FIG. 1. Experimental apparatus for light scattering measurements.

Scattering measurements were performed on aerogels produced in our laboratory and on samples provided by Schmitt [5] and van Dardel [4]. In addition to measuring the properties of the aerogel, the nephelometer was used to study the growth of the sol and monitor the maturation of the gel. In earlier work [3] we reported the unusual character of the angular dependence of intensity of the scattered light exhibited by samples from Schmitt (the solid lines in FIG. 2) and the more Rayleigh-like scattering from the samples from Sweden. Measurements on samples produced in our laboratory showed a variety of behaviors similar to that found in the earlier measurements. We had postulated that the unusual character of the scattering may be due to density fluctuations in the material on a scale larger than the pore size. The next section describes the predicted scattering from inhomogeneous two-phase materials of the type postulated earlier.

SCATTERING EQUATIONS

Consider a physical mixture of two media, medium A and medium B. Each has a corresponding real bulk dielectric constant ϵ_A and ϵ_B. If these constants are not too different from unity, and the mixture is fairly uniform and statistically isotropic, we expect the resulting effective dielectric constant ϵ to be given by

$$\frac{\epsilon-1}{\epsilon+2} = f_A \frac{\epsilon_A-1}{\epsilon_A+2} + f_B \frac{\epsilon_B-1}{\epsilon_B+2} \quad , \qquad (1)$$

where f_A is the volume fraction of type A material and f_B is the volume fraction of type B material. (We will take $f_B = 1 - f_A$.) This equation is obviously exact to first order in ϵ_A-1 and ϵ_B-1, but in some cases it is accurate in higher order: it is based on the Clausius-Mossotti equation which takes the non-linear dependence of polarization upon atomic polarizabilities into account. Eq. (1) is exact in some special cases, for example, for widely separated spheres in a vacuum. However, it is not difficult to find counterexamples (e.g., aligned rods in a matrix) which show that it is not exact in general.

Following Debye et al. [6,7] and Born and Wolf [8] the scattering amplitudes per unit volume of scatterer (silica, medium A) are given by

$$E_{\parallel} = \frac{3k^2}{4\pi} \left[\frac{\epsilon_R-1}{\epsilon_R+2}\right] \cos\phi \, \cos\theta \, \frac{e^{ikr}}{r} \qquad (2a)$$

and

$$E_{\perp} = \frac{-3k^2}{4\pi} \left[\frac{\epsilon_R-1}{\epsilon_R+2}\right] \sin\phi \, \frac{e^{ikr}}{r} \quad , \qquad (2b)$$

where the scattering angle is θ , the incident wave has unit amplitude and is polarized with azimuth angle ϕ of the electric field relative to the scattering plane. The relative dielectric constant ϵ_R is just ϵ_A/ϵ and k is the magnitude of the wavevector in the medium ($2\pi\epsilon^{1/2}/\lambda$ where λ_0 is the vacuum wavelength). Distance from the scattering center is designated by r. In the x-ray portion of the spectrum $\epsilon_R \cong 1$ and this approach leads to essentially exact equations, whereas in light scattering the results are approximate because ϵ_R is not close to unity. For example, for $\epsilon_A=2.13$ (silica), $\epsilon_B=1$, and $f_a=0.1$, $\epsilon_R=1.96$. Eqs. (2) are accurate only to leading order in (ϵ_R-1).

Squaring Eq. (2a), introducing $\rho(\underline{r})$ as a function equal to unity in phase A (silica) and zero otherwise, after the usual mathematical manipulations, we obtain for the scattered intensity

$$|E_{||}|^2 = \frac{9k^4}{16\pi^2} \left[\frac{\epsilon_R-1}{\epsilon_R+2}\right]^2 \cos^2\phi\cos^2\theta \, \frac{1}{r^2} f_A \, (1-f_A)V \int d^3r \, \gamma(r) e^{ik\underline{s}\cdot\underline{r}} \quad , \quad (3a)$$

where $|\underline{s}| = s = 2\sin(\theta/2)$, V is the illuminated sample volume and $\gamma(r)$ is a correlation function, equal to unity at $r = 0$, defined by

$$\gamma(r_{12}) \, f_A(1-f_A) = \left\{(\rho(\underline{r}_1)-f_A)(\rho(\underline{r}_2)-f_A)\right\} \quad , \quad (4)$$

where the outside brackets indicate an average over the sample volume. Reflection from the sample surface, if significant, must be included separately. In a like manner we have for $|E_{\perp}|^2$:

$$|E_{\perp}|^2 = \frac{9k^4}{16\pi^2} \left[\frac{\epsilon_R-1}{\epsilon_R+2}\right]^2 \sin^2\phi \, \frac{1}{r^2} f_A \, (1-f_A)V \int d^3r \, \gamma(r) e^{ik\underline{s}\cdot\underline{r}} \quad . \quad (3b)$$

The presence of the correlation function $\gamma(r)$ in the scattered intensity equations permits one to use light scattering measurements to determine density-density correlations in the medium. If the scattering medium is statistically uniform and isotropic, γ will depend only upon the magnitude of $\underline{r}$, and will vanish for sufficiently large r.

Short range correlations: Rayleigh Scattering

If the medium contains only short range correlations in comparison with the wavelength of light, then $\gamma(r)$ vanishes as r increases before the exponential factor in Eqs. (3) departs significantly from unity. In this case the angular and wavelength dependence of the scattering is that of simple Rayleigh scattering. The overall strength of the scattering depends upon the details of the correlations in the medium. For the special case of widely spaced small spheres with radius a in a vacuum, γ can be calculated explicitly [9]:

$$\gamma(r) = (1/16) \, (r/a+4) \, (r/a-2)^2, \text{ for } r<2a,$$

$$\gamma(r) = 0, \text{ for } r\geq 2a,$$

and our equations reduce exactly to the standard equations [8] for Rayleigh scattering.

Long range correlations: departures from Rayleigh scattering

If $\gamma(r)$ does not vanish for values of r comparable with the wavelength, equations (3) show that both the angular and wavelength dependence will show new structure. In particular, the scattered intensity will no longer be symmetric in the forward and backward hemispheres (equal intensities at θ and $\bar{\pi}-\theta$). We will choose a convenient parametric form for $\gamma(r)$ identical to that used in Ref. [7] to treat small angle x-ray scattering. We assume that the correlation consists of a short range exponential part and a long-range exponential part:

$$\gamma(r) = (1-w)\exp(-r/a_1) + w \exp(-r^2/a_2^2), \qquad (5)$$

where $a_1 \ll k^{-1} = \lambda_0/(2\pi\epsilon^{1/2})$. Substitution of (5) into (3a) and performing the Fourier transform, one has

$$|E_{||}|^2 = \frac{9k^4}{16\pi^2} \left[\frac{\epsilon_R - 1}{\epsilon_R + 2}\right]^2 \cos^2\phi \, \cos^2\theta \, \frac{1}{r^2} f_A(1-f_A)V \qquad (6a)$$

$$x\left[8\pi(1-w) \, a_1^3 + \pi^{3/2}w \, a_2^3 \, \exp(-k^2 a_2^2 \sin^2 \theta/2)\right] \quad .$$

The expression for $|E_\perp|^2$ is identical except for the replacement of $\cos^2\phi\cos^2\theta$ by $\sin^2\phi$.

RESULTS

Experimental data was fit by equation 6a and its companion by variation of the parameters a_1, a_2, and w. The results given in FIG. 2 are direct experimental data and therefore include the variation in scattering volume intrinsic to light scattering (but not Small Angle x-ray Scattering, SAXS) measurements. The character of the curves at angles above 90° is similar to Rayleigh scattering. The dip in intensity at low angles is very unusual in light scattering measurements but can be fit by the assumed correlation function. Several observations regarding the microstructure of the material may be made from the values used to fit the curve. First, w $\ll$ 1 and is negative. Therefore the correlations causing the scattering are weak and of the type that larger than average concentrations of silica tend to be sur-rounded by lower than average concentrations. Second, the value of a_1 is not known accurately because we measured the relative cross section but is probably of the order of the small particle radius i.e., 50 Angstroms. Third, The value of a_2 is rather large (about 2000 Angstroms) and thus, even though w $\ll$ 1, the ratio of a_2^3/a_1^3 is so large that the weak long-range inhomogeneities cause significant deviations from Rayleigh scattering. This type of long range correlation produces an important contribution to the light scattering. In some materials it is probably the main cause of resi-dual scattering. Such long range correlations would not appear in SAXS measurements.

The correlation functions assumed in this preliminary analysis were chosen for analytical convenience and produced moderately good fits to the experi-mental data. In the future we plan to refine our numerical technquies to provide more accurate fits to the scattering data, and by direct Fourier transformation, more accurate representations of the correlation function $\gamma(r)$.

280

SCATTERING INTENSITY
Two Polarizations

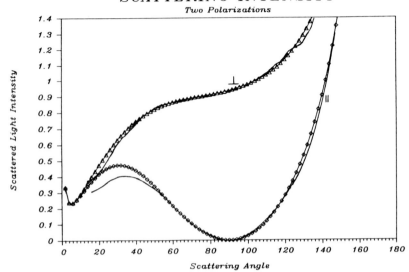

FIG. 2. Intensity of light vs.angle from a TEOS sample for scattered light with polarization parallel || and perpendicular ⊥ to scattering plane. Solid lines give experimental results.

ACKNOWLEDGEMENTS

This work was sponsored by the Assistant Secretary for Conservation and Renewable Energy, Office of Solar Heat Technologies, Passive and Hybrid Solar Energy Division of the U. S. Department of Energy. The authors gratefully acknowledge the contributions of Dr. Richard Russo in preparing and measuring aerogel samples at LBL.

REFERENCES

1. A. J. Hunt, "Microporous Transparent Materials for Insulating Windows and Building Applications, Assessment Report," Lawrence Berkeley Lab.report LBL-15306 (1982).
2. M. Rubin and C. Lambert, "Silica Aerogel for Transparent Insulation" Solar Energy Materials, 7 393-400, (1983).
3. "Light Scattering Studies of Silica Aerogel" A. J. Hunt, Proceedings of the Intern. Conf. on Ultrastructure Processing of Ceramics, Glasses, and Composites, Gainesville, FL, Feb 13-17, 1983.
4. S. Henning and L. Svensson, "Production of Silica Aerogel" Physics Scripta 23 703,707 (1981).
5. W. J. Schmitt, R. A. Greiger-Block and T. W. Chapman "The Preparation and Properties of Acid-catalyzed Silica Aerogel," Presented at the Annual Meeting of the AIChE, New Orleans (1981).
6. P. Debye and A.M. Bueche, Scattering by an inhomogeneous solid, J. Appl. Phys. 20, 518-525 (1949).
7. P. Debye, H. R. Anderson, Jr., and H. Brumberger, "Scattering by an Inhomogeneous Solid. II. The Correlation Function and its Application," J. Appl. Phys. 20 679-683 (1957).
8. M. Born and E. Wolf, Principles of Optics, 4th Ed., (Pergamon Press, Oxford 1970), Section 13.5.2.
9. M. Kerker, The scattering of light, (Academic Press, N.Y. 1969) p. 475.

CHARACTERISATION OF CONCENTRATED COLLOIDAL SUSPENSIONS FOR CERAMIC PROCESSING

R F STEWART AND D SUTTON
ICI Corporate Colloid Science Group, P O Box 11,
The Heath, Runcorn, Cheshire, England.

ABSTRACT

Characterisation and control of the structure and properties of particulate solids suspensions is of widespread importance with respect to many operations, such as separation, dispersion and densification in the ceramics and related industries. In this paper we describe techniques which have been used to study the effects of changes in basic parameters, for example solids content, particle size, electrolyte concentration and added polymer on the aggregated suspension morphology and characteristics. Suspension morphology was examined by a number of methods, notably freeze-etch microscopy. The mechanical properties of suspensions such as their compressional modulus and rheological behaviour was also examined.

INTRODUCTION

In the formation of ceramics from suspensions of fine particles the properties of the material are not only dependent upon the process selected but they are also very much determined by the properties of the suspension itself [1,2]. In many cases the latter are responsible for placing limits on the quality of the final product. Problems involving the manipulation of colloidal suspensions are also prevalent in the chemical industry. There is a need to improve the efficiency with which such systems are handled and this requires the ability to control suspension structure and its derivative properties such as densification, filtration and rheological behaviour.

To achieve an improved understanding of the behaviour of such systems we have investigated how the structure and properties of standard suspensions of flocculated colloidal materials vary with basic parameters, for example solid loading, particle size, electrolyte concentration and added polymer type. In the main, well characterised polymer latices have been used for the work in order to allow unambiguous examination of one variable at a time, though a number of inorganic materials such as Al_2O_3 and TiO_2 have also been studied. Suspension structure has been probed directly using a range of microscopic techniques, freeze-etch microscopy being extensively used for the examination of concentrated suspensions as it avoids the need to perturbate the system by, for example, dilution or drying. Parallel investigations into the properties of the samples such as their consolidation behaviour, shear moduli and rheology have also been conducted.

From the experimental data we map out some general principles concerning the influence of changes in some basic parameters on the morphology and properties of colloidal systems.

Mat. Res. Soc. Symp. Proc. Vol. 32 (1984) © Elsevier Science Publishing Co., Inc.

MATERIALS

The principal dispersions used for the work were surfactant-free, anionic charge stabilised polystyrene latices prepared by methods following those suggested by Goodwin et al. [3]. Monodisperse species with diameters ranging from 0.2-4.0 microns were produced by appropriate modification of reaction conditions and the particles characterised in terms of their size, electrophoretic mobility, surface charge and stability to added electrolyte. For the majority of the studies reported these materials were coagulated by the addition of electrolyte, though a number of investigations were performed with various polymeric flocculants.

STRUCTURAL CHARACTERISATION

For dilute colloidal dispersions a wide range of techniques are available for examining the structure of the aggregates. Procedures used include light scattering, optical microscopy and sedimentation methods. Often different methods measure different characteristics of the aggregates, certain procedures (such as optical microscopy) provide estimates of the envelope occluded by the floc whilst others (such as the "Coulter" counter) provide a measure of the solids in the flocs. Appropriate combinations of techniques can thus be used to obtain estimates both of floc size distribution and of average floc porosity [4]. For colloidal species smaller than ~1μm scanning or transmission electron microscope, however, necessitates drying the sample which may result in a change in its morphology.

Unfortunately, conventional experimental techniques are unsuitable for solids contents of greater than a few percent. Yet it is in this regime that it is most important to be able to characterise the structure of an aggregate or network of colloidal particles if the behaviour of a dispersion in a centrifuge, or the response of a particulate network to an applied pressure is to be predicted.

The structure of concentrated suspensions can be examined in two ways, either directly using specialised electron microscope methods, or indirectly by rheological techniques.

The technique of freeze-etch microscopy provides an excellent means for characterising the morphology of the vast majority of suspensions in which the continuous phase is water or a volatile non-aqueous solvent. The method consists of a number of stages as follows:

1) Crash-freezing of a sample of the material in a suitable cryogen to liquid N_2 temperatures. In the present work freezing rates of between 10^3 and 10^4 °Ks^{-1} were achieved

2) Fracture (or cutting) of the frozen material to yield a suitable section plane

3) Etching of the exposed surface by sublimation of the frozen suspension medium under reasonably high vacuum conditions

4) Shadowing and replication of the exposed surface with evaporated Pt and C

5) Recovery of the replica and its examination in the transmission electron microscope

The principal advantage of freeze-etch microscopy is that the technique allows direct examination of the structure in colloidal suspensions at solids loadings of >50% by volume without dilution, drying or other gross perturbation. Further description of the method, as developed in the ICI laboratories at Runcorn has been provided in a recent publication [5].

The first extensive use of this technique for the quantitative imaging of industrial colloidal suspensions was by Menold [6] and, as can be seen from figure 1, the procedure provides an immediate visual impression of the structure of a flocculated material.

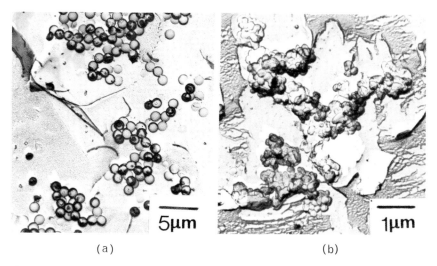

(a)   (b)

Figure 1.   (a)   The structure of electrolyte coagulated polymer latex as revealed by freeze-etch microscopy. Note the characteristic texture of floccule cores with 'tentacles'. Here the medium is deeply etched to reveal several layers of particles. (b) Freeze-etch micrograph of TiO_2 aggregated by adjusting pH to the isoelectric point.

By applying the mathematical techniques of stereology to freeze-etch micrographs it is possible to perform a quantitative analysis of the structure in the sample [7]. This procedure enables quantities such as porosity, particle coordination numbers, surface areas and contiguity to be measured directly. In the present work a "Magiscan" image analyser was used to permit statistically-significant quantities of data to be rapidly collected.

The rheological methods deployed to probe aggregate structure in this work included:

Continuous Shear Rheological Properties which were determined using a Deer PDR81 rheometer fitted with double concentric cylinder platens. For more concentrated samples a Haake RV3 viscometer was employed.

Shear Modulus Measurements which were made with a Rank Brothers Limited Pulse-Shearometer.

<u>Centrifugation Characteristics</u> which were measured by use of MSE "Chilspin" and HS25 laboratory centrifuges. In each case from the graph of equilibrium sediment height versus centrifuge speed, a compressive, or "network", modulus ($K = V\frac{dP}{dV}$, where V = volume of material, P = applied pressure) could be estimated (see Buscall [8]).

It has been shown for a variety of suspensions that the network modulus defined above and the shear modulus are essentially identical numerically.

RESULTS AND DISCUSSION

Before considering the structure and properties of aggregated suspensions in detail, it is useful to contrast some of their general characteristics with those observed for stable colloidal dispersions.

Broadly speaking, at volume fractions of solids <50%, stable dispersions of near spherical particles have rather simple rheological properties. Such systems are usually near Newtonian (ie the viscosity varies only weakly as a function of shear rate), and the viscosity of the suspension becomes markedly larger than that of the host medium at only the highest solid contents. In contrast, flocculated dispersions tend to be much more viscous than stable dispersions of the same particle size and solids content and usually exhibit a decrease in viscosity with increasing shear rate. This is demonstrated in figure 2 in which the behaviour of viscosity as a function of solids content is shown for stable and disperse 1 micron polymer latices.

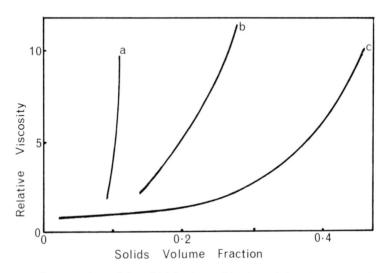

FIGURE 2. Low shear (a) and high shear (b) viscosities for a coagulated 1μm latex compared with results for a stable latex (c).

Sedimentation rates and equilibrium sediment volumes are also strongly affected by aggregation of the suspension. Stable dispersions slowly settle to form highly concentrated, close packed beds. In contrast, flocculation greatly enhances the rate at which a dispersion at low solids sediments. However at slightly higher solids contents the aggregates pack together to form an open, porous network. Not surprisingly it is found that the concentration at which settling slows greatly, or ceases, in a flocculated suspension is similar to that at which characteristics such as the zero-shear rate viscosity diverge.

In principle there is a large range of variables which can influence the structure and properties of flocs [9]. In practice only a few (solids loading, particle size and shape and added polymer (if any) type) are important. The effect of solids content and particle size are well illustrated by network modulus data for simple coagulated latex suspensions, in which aggregation is the consequence of electrical double layer compression (figure 3).

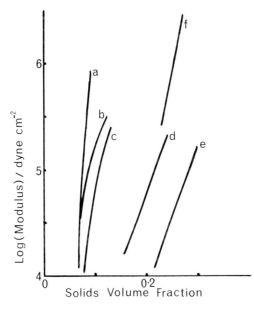

FIGURE 3. Network modulus vs solids content for various sized latices in 0.1M $BaCl_2$ (a = 0.17, b = 0.33, c = 0.52, d = 0.96, e = 2.06 micron). Also shown is modulus for 0.3 micron TiO_2 at its isoelectric point (f).

Three points are immediately obvious from these results:

1) There is a striking particle size effect with curves for larger species being translated to higher solids contents. This effect is due to the rapid increase in the number of particle-particle contacts/unit volume of sediment outweighing the decrease in strength of each individual contact as the size is reduced.

2) For each particle size there is a well defined concentration at which the suspension starts to show solid-like behaviour as the aggregates begin to interact to form a space-filling network. As particle size decreases there is a convergence to a limiting gel point of ~8% v/v solids, a result in harmony with those obtained theoretically from simulation of the build up of a randomly aggregated structure [10].

3) There is an extremely rapid rise in modulus with solids content. Experimental studies (using freeze-etch microscopy) show that this can not be explained in terms of an increase in the average coordination number as solids loading increases [9]. What seems to matter is the progressive entanglement of aggregate "tentacles", with rapid

elimination of weak, low coordination linkages, as the solids rise. Above the gel points compaction becomes rapidly more difficult as even modest changes in network morphology require steadily more cooperation between structural elements. Other suspension properties tend to move in step with the behaviour of the network modulus; for example, sediments become more voluminous as the primary particle size increases.

When polymeric materials are added to suspensions, the differing flocculation mechanisms afford the possibility of differing aggregate structures. In practice, polymers of low to medium molecular weight tend to give structures which are only modestly perturbed from those observed when the same system is aggregated by compression of the electrical double layer [9]. For very high molecular weight polymers changes can be observed in both floc structure and properties, as the particles may now be joined by flexible polymer linkages with the elimination of direct contact between particle surfaces [9]. Systems of this type often display rapid sedimentation to quite a close packed bed.

CONCLUDING REMARKS

Design and operation of processes involving colloidal suspensions are helped by a better understanding of the relationships between process conditions, structure and properties. Qualitative trends assist in focussing quickly on key variables in cases where particular problems have arisen. At high solids the properties of a suspension become much more sensitive to particle size and colloidal stability. In these circumstances a small change in fines content or aggregation can change a free-flowing slurry to a near solid with disastrous consequences.

Quantitative prediction of consolidation behaviour can be facilitated by analysis of easily measured rheological and sedimentation experiments. It has been shown [8] that deconvolution of network or shear modulus data to give pressure-solids content correlations, allows accurate prediction of a limit to the solids content in a pressing or centrifugation operation. This allows the effects of different treatments or chemical additives to be quantitatively screened in the laboratory.

The range of techniques now available to probe the structure of concentrated colloidal dispersions are facilitating elucidation of the links between process conditions, floc structure and suspension properties. Such advances are starting to enable much more careful design and control of operations involving colloidal materials.

REFERENCES

1   H K Bowen, Mat Sci Eng 44, 1, (1980).
2   W H Rhodes, J Am Ceram Soc 64, 19, (1981).
3   J W Goodwin, J Hearn, C C Ho and R H Ottewill, Coll Polym Sci 252, 464, (1974).
4   B Koglin, Ger Chem Eng 1, 252, (1978).
5   P F Luckham, B Vincent, J McMahon and T F Tadros, Colloids Surf 6, 83, (1983).
6   R Menold, B Luttge and W Kaiser, Adv Coll Int Sci 5, 281, (1976).
7   E E Underwood, "Quantitative Stereology", London: Addison-Wesley,(1970).
8   R Buscall, Colloids Surf 5, 269, (1982).
9   R F Stewart and D Sutton, SCI: Proc Int Symp: "Advances in Solid-Liquid Separation" (London, 19-21 September 1983), (1983).
10   M J Vold, J Coll Sci 14, 168, (1959).

THE PYROLYTIC DECOMPOSITION OF OWENS-ILLINOIS RESIN GR650, AN ORGANOSILICON COMPOUND

B. G. BAGLEY,* P. K. GALLAGHER,** W. E. QUINN,* AND L. J. AMOS***
*Bell Communications Research, Inc., Murray Hill, NJ 07974; Work done while at Bell Laboratories; **AT&T Bell Laboratories, Murray Hill, NJ 07974; ***Department of Chemistry, Princeton University, Princeton, NJ 08540; Work done while at Bell Laboratories.

ABSTRACT

The pyrolytic conversion of an organosilsesquioxane (Owens-Illinois resin GR650) to SiO_2 is characterized by ir spectroscopy, thermogravimetry and evolved gas analysis (line-of-sight mass spectroscopy). Scanning calorimetry, ramping at $10°C/min$, on the as-received (room temperature annealed) resin indicates a glass transition temperature of $67°C$ which decreases to $58°C$ for an unrelaxed sample. The ir spectra have bands which can be assigned to $Si-CH_3$ and $Si-O-Si$ modes. For 30 minute isothermal anneals at temperatures above $420°C$ there is a continuous decrease in the bands associated with the $Si-CH_3$ groups such that after 30 minutes at $650°C$ the ir spectrum has evolved to that for SiO_2. Evolved gas analysis indicates that there are four major components evolving. Over the temperature range (ramping at $10°C/min$) ~180 to $\sim500°C$ we observe C_2H_5OH and H_2O, both of which are condensation reaction products from the curing reaction. Methane is a major evolving species over the temperature range ~500 to $\sim800°C$ and the thermal spectrum is double peaked which we attribute to CH_3^+ bound to the inside and outside of the polymer cage structures. The final major component detected was H_2, over the temperature range ~600 to $\sim1100°C$, which was attributed to pyrolysis of the organic components, both trapped and evolving. The features of the weight loss curve can be accounted for by the measured evolving species spectra.

INTRODUCTION

There is much current interest in lowering ceramic fabrication temperatures. One applicable technique for doing this is via organometallic precursors which are subsequently converted to the inorganic ceramic. Some precursors are particularly attractive because their viscosities can be easily adjusted by temperature (at and above the glass transition) and/or by the use of solvents, thus making for easy forming; and the conversion to the inorganic ceramic then can be done conveniently by pyrolysis. Emphasis, thus far, on the use of this process for bulk material preparation has been centered on the carbides and nitrides [1-6] which are generally difficult to fabricate. Organosilicon compounds are currently of interest to the electronics industry where, again because of the ease of forming, they are being used for thin film (e.g. as dopant sources) and coating (encapsulant) applications. These materials are part of a group sometimes called collectively "spin-on glasses." In these electronic applications the material is used in its organometallic state or its conversion to the inorganic is incidental to its intended use (e.g. as when used in dopant sources).

Owens-Illinois, one of several spin-on glass manufacturers, supplies a series (termed Glass Resins) of organometallic polymers for which, in the current applications, the fabrication and processing temperatures are kept low enough such that the organometallic nature is maintained. The polymer that we consider a prototype for study because of its relative chemical simplicity is Resin GR650, an organosilsesquioxane prepared by the hydrolysis and condensation polymerization of methyltriethoxysilane [7]. In this note we consider this resin (as with the others in the GR series) as an SiO_2 precursor with potential low temperature ceramic processing advantages and characterize its conversion by pyrolysis. An ancillary result is a better understanding of the low temperature thermal behavior of the organosilicon itself.

EXPERIMENTAL

Samples were prepared and studied in three different forms: 1) The bulk resin as received; 2) thin films, 8000Å thick, prepared by spinning (after adjusting the viscosity with solvent) onto 7.5 cm diameter polished single crystal silicon substrates and 3) a low density solid prepared by evaporation of a solvent (ethanol) from ∼1 cc of bulk liquid, which produces a bulk sample that mimics the spun-on thin film material.

Thermal measurements were made using a commercial Perkin-Elmer Model DSC-1B scanning calorimeter. Melting of stearic acid was used to calibrate the temperature scale in the region of the resin's glass transition and measurements were made at a temperature scanning rate of 10°C/min. Infrared optical data were taken with a Perkin-Elmer Model 580 Spectrometer, a double beam instrument, with a matching silicon wafer in the reference beam. Our thermogravimetric results were obtained using a Cahn 1000 balance together with a Perkin-Elmer System 4 TGA Furnace and Controller: Temperature ramping rates were 10°C/min and the gas flow was 40 sccm.

The evolved gas analysis was done with a system described in detail elsewhere [8]. In this system, which has a turbomolecular pump and a base pressure of 10^{-7} torr, the sample resides in a heated crucible with a 1 cm line of sight distance to a mass spectrometer (UTI Model 100C). Associated circuitry, under computer control, ramps and controls the crucible temperature and measures the masses of the species evolved. The chamber total pressure is also constantly monitored. Some measurements were made under isothermal conditions but the majority were made with a heating rate of 10°C/min. Initially, survey scans were made for all evolving species (masses 1 to 100) as a function of temperature (ramping at 10°C/min). Subsequently, with fresh samples, attention was focused, and more detailed data were taken, on those masses which dominated the evolution spectra.

RESULTS AND DISCUSSION

In Fig. 1 are shown the results of our thermal measurements in the region of the glass transition. Shown solid are the results on the as-received resin where we observe a glass transition at $67 \pm 1.5°C$ and a large relaxation endotherm due to a volume relaxation occuring ostensibly at room temperature. We define T_g in this note as the inflection point in the ΔCp versus T curve which corresponds (within a degree) to the temperature at ½ (Cp liq. − Cp sol.) in the absence of relaxation effects. Shown dashed is the heat capacity that we observe after heating the resin to 125°C at 10°C/min, immediately cooling at 10°C/min to 25°C and then reheating at 10°C/min through the glass transition. In this case there is no relaxation endotherm observed and, being void of relaxation effects and characterized by the 10°C/min heating rate, T_g is observed to be 58°C. We also note that the enthalpy of the relaxation endotherm equals, within our experimental accuracy, the value for $\int \Delta Cp\Delta T$ between the two curves indicating that this is essentially a volume relaxation with no major structural rearrangements.

The ir spectra we obtained from the thin films are shown in Fig. 2. Shown are the effects of 30 minute heat treatments in dry nitrogen at various temperatures. The spectrum for the film treated for 30 minutes at 100°C we take as that for a fully dried film in which all physically absorbed solvent and water have been eliminated. We note that without a stringent drying procedure the residual solvent (ethanol) can be particularly troublesome in characterizing spectral changes. All bands observed we attribute to Si-CH$_3$ and Si-O-Si modes. The Si-CH$_3$ associated modes are observed at 2990 cm^{-1} (stretch), 1275 cm^{-1} (bend), 780 cm^{-1} (bend) and 410 cm^{-1} (bend). The band observed at 780 cm^{-1} may be a doublet at 770 and 780 cm^{-1} and the band at 410 cm^{-1} also likely contains a component due to an Si-O-Si bend. We attribute both the bands we observe at 1130 and 1035 cm^{-1} to Si-O-Si chain stretching modes. No bands associated with silanol groups were detected. After 30 minutes at 420°C the sample is cured and the only obvious change we detect is the disappearance of a very small peak at 2920 cm^{-1} which has an ambiguous assignment, either CH_3 (Si-CH_3) or CH_2 (in an ethoxy group). The principle bands for the ethoxy groups, which would be expected to decrease upon curing, are buried in the Si-O-Si stretching absorptions and small changes in intensity would have gone undetected.

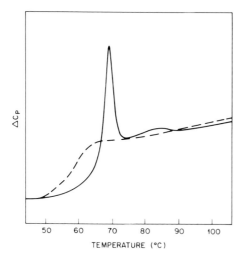

Fig. 1. The temperature dependence of the specific heat in the region of the glass transition for Owens-Illinois resin GR650. Shown solid are the data for the as-received resin for which $T_g = 67°C$. Shown dashed are the results after cooling from 125°C at 10°C/min for which $T_g = 58°C$. Data were obtained while scanning at +10°C/min in a nitrogen ambient.

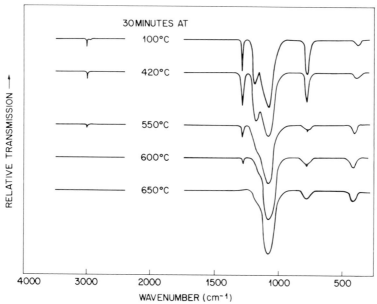

Fig. 2. Transmission infrared spectra obtained from thin films (8000Å thick) of Owens-Illinois resin GR650 after heat treatments for 30 min in dry nitrogen at the temperatures indicated.

Thereafter, upon heat treating for 30 minutes at temperatures above 420°C there is a monotonic decrease in the Si-CH$_3$ bands and an increase in the bands associated with SiO$_2$. After 30 minutes at 650°C we observe only bands associated with Si-O-Si [9] at 1080 cm^{-1} (stretch), 800 cm^{-1} (bend) and 450 cm^{-1} (bend). The band we observe at 1130 cm^{-1} in the polymer and which we attribute to a linear Si-O-Si stretching mode evolves into the higher energy shoulder (~1180 cm^{-1}) of the three dimensional SiO$_2$ network. At temperatures above 420°C, and as the methane evolves, we observe a shift to 1280 cm^{-1} of the 1275 cm^{-1} peak, the 780 cm^{-1} organic band is superimposed on the broad 800 cm^{-1} Si-O-Si band, and the low frequency band shifts monotonically from 410 to 420 to 430 to 440 and finally 450 cm^{-1} for each of the heat-treating temperatures from 100 to 650°C. After 30 minutes in oxygen at 1100°C the refractive index (n) of the thin film, as measured by ellipsometry at a wavelength of 6328Å, was found to be 1.454. For a heat treatment of 30 minutes in nitrogen at 1100°C we measure n = 1.490, which exceeds that for SiO$_2$ indicating either (or both) that there is retained carbon or that the film is off stoichiometry and slightly silicon rich.

The results of our thermogravimetric measurements (weight loss) are shown in Fig. 3. Each sample was a small platelet 1.2 mm thick and weighing about 16 mg. The sample heated in nitrogen pulverized explosively at 670°C as a result of an internal gas pressure increase (due to the pyrolysis). As would be expected, this result is a kinetic effect such that other similar-sized samples (10-20 mg) would pulverize at different temperatures (≥600°C) and small samples (e.g. 4 mg) at slower heating rates (e.g. 2.5°C/min) do not fragment at all. The sample heated in oxygen ignited at 480°C and burned with a concomitant loss in weight until the volatile components were removed. This ignition temperature is not a characteristic temperature, occuring in different samples at different temperatures (around 500°C), but ignition always accompanies copious CH$_4$ evolution. The presence of oxygen in the ambient is clearly advantageous, however, as it enhances the weight loss (i.e. the conversion process) at lower temperatures. The sample heated in air neither ignited nor pulverized explosively. By adjusting the O$_2$/N$_2$ ratio a controlled conversion rate can be achieved, although clearly air is the most convenient.

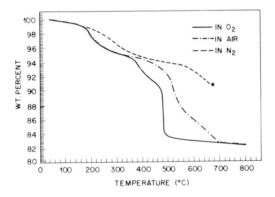

Fig. 3. Weight loss versus temperature (scanning at +10°C/min) of Owens-Illinois resin GR650 in three different ambients (as indicated). Sample heated in nitrogen pulverized explosively at 670°C.

Information on the evolving species leading to the weight loss is provided by evolved gas analysis. We use line-of-sight mass spectroscopy which, of necessity, is done in a high vacuum, a nonreactive yet different ambient from that of the other measurements we report here. We also note that what we measure is what leaves the surface, reflecting (but not determining exactly) the interior and surface chemistries. The total pressure rise in the system as a function of temperature (ramping at 10°C/min) is shown in Fig. 4. The major organic species observed are shown in Fig. 5 and are identified as C$_2$H$_5$O$^+$ (m/e = 45) and CH$_3^+$ (m/e = 15) with assumed parents C$_2$H$_5$OH and CH$_4$. Although C$_2$H$_5$O° and CH$_3$°, after ionization by the mass spectrometer, would appear at m/e = 45 and 15 respectively, we observe hydrogen in the cracking pattern the intensity of which mirrors the organic component, thereby favoring the assumed parents. The onset of ethanol evolution is at the resin glass transition and increases markedly at 180°C, corresponding to the onset of sensible curing (at the 10°C/min heating rate) as evidenced in the weight loss and total pressure curves (Figs. 3 and 4). We believe the ethanol detected below 180°C is residual, physically absorbed, solvent which becomes mobile when the resin becomes fluid at T_g; although we cannot, at this point, rule out subtle curing reactions. The double peaked structure in the CH_3^+ we attribute to methyl bound on the interior and exterior of caged polymer structures. Methyl groups bound to the cage interior require a higher temperature to evolve whereas the exterior units evolve at a lower temperature. The major inorganic components observed in the mass spectrometer, Fig. 6, are H_2O^+ ($m/e = 18$) and H_2^+ ($m/e = 2$). All the features of the total system pressure shown in Fig. 4 can be accounted for by the four components shown in Figs. 5 and 6. The H_2O and C_2H_5OH are condensation products from the curing reaction which occurs over the temperature range 180 to $\approx 500°C$. The structure in the temperature dependence of the ethanol evolution observed above $\approx 450°C$, which is seen in all samples, may reflect a pyrolysis of unreacted ethoxy groups which may be bound on interior or exterior polymer cage surfaces as with the CH_3^+. Why silanol groups are not detected in the uncured, but dried (30 min at 100°C), sample by ir spectroscopy (Fig. 2) and H_2O is such a plentiful reaction product (Fig. 6) is, at this point, an enigma. Most of the hydrogen

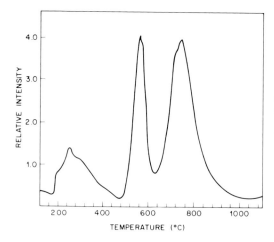

Fig. 4. Total pressure change in evolved gas analysis vacuum system as a function of temperature (scanning at +10°C/min) for Owens-Illinois resin GR650.

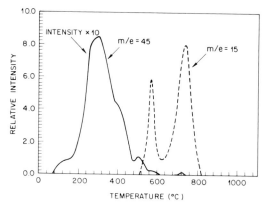

Fig. 5. Mass spectroscopic results for the two major organic components evolving from Owens-Illinois resin GR650 as a function of temperature while ramping at +10°C/min. The observed $m/e = 45$ is attributed to the parent C_2H_5OH which is a condensation product from the curing reaction, and the $m/e = 15$ is attributed to the parent CH_4, a pyrolysis product of the methyl component in the resin.

292

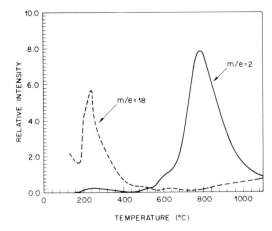

Fig. 6. Mass spectroscopic results for the two major inorganic components evolving from Owens-Illinois resin GR650 as a function of temperature while ramping at +10°C/min. The observed m/e = 18 is attributed to H_2O which is a condensation product from the curing reaction, and the m/e = 2 is attributed to H_2 which, above ~600°C, is due to pyrolysis of the CH_4 (both trapped and evolving).

detected above 600°C is due to pyrolysis of the methane, both trapped and evolving. The pyrolysis of any organic component trapped in the sample will leave behind a residual carbon impurity as the more mobile hydrogen diffuses out. Most of the hydrogen detected below 600°C and a small amount of the hydrogen detected between 600 and 800°C is due to the spectrometer produced organic component cracking pattern. All aspects of the evolving species observed in the bulk resin (Figs. 4-6) are also observed in the low density solid, except that in the low density solid the spectra are shifted by 50°C to higher temperatures, due, we believe, to a poorer thermal contact between the sample and the crucible which results in a 5 minute thermal lag.

ACKNOWLEDGEMENT

The contribution of one of us (LJA) was made while participating in Bell Laboratories Summer Research Program, the support of which is gratefully acknowledged.

REFERENCES

1. R. W. Rice, Am. Ceram. Soc. Bull. 62, 889 (1983).
2. S. Yajima, Am. Ceram. Soc. Bull. 62, 893 (1983).
3. R. West, L. D. David, P. I. Djurovich, and H. Yu, Am. Ceram. Soc. Bull. 62, 899 (1983).
4. R. R. Wills, R. A. Markle, and S. P. Mukherjee, Am. Ceram. Soc. Bull. 62, 904 (1983).
5. C. L. Schilling, Jr., J. P. Wesson, and T. C. Williams, Am. Ceram. Soc. Bull. 62, 912 (1983).
6. B. E. Walker, Jr., R. W. Rice, P. F. Becker, B. A. Bender, and W. S. Coblenz, Am. Ceram. Soc. Bull. 62, 916 (1983).
7. W. J. Crosby, Owens-Illinois, Toledo, OH 43666, Private Communication.
8. P. Gallagher, Thermochim. Acta 26, 175 (1978).
9. P. H. Gaskell, and D. W. Johnson, J. Non-Cryst. Solids 20, 171 (1976).

MICRO-RAMAN SPECTROSCOPY OF FRESH AND AGED SILICA GELS[†]

L. C. KLEIN,* C. NELSON AND K. L. HIGGINS**
*Rutgers Ceramics Department, Piscataway, NJ;
**Sandia National Laboratories, Albuquerque, NM 87185

ABSTRACT

Raman spectra were obtained from fresh and aged silica gels using micro-Raman techniques. Comparison of the spectra from wet and dried gels reveals the stages of reaction in the ethanol-TEOS-water system and the gradual development of the silica structure. The steady disappearance of bands assigned to ethanol and the appearance of bands assigned to silatious species can be traced during the drying process. Increases in either acid concentration or water level increase reaction rates to give gels in a more advanced stage of polymerization. Gels tend to develop fluorescence upon aging for several months. The source of this fluorescent could not be determined, but it's development can be prevented by heating freshly dried gels to about 155°C.

INTRODUCTION

Previous Raman spectroscopy studies (1,2,3,4) report spectra for silica gels heated to relatively high temperatures such as 800 C (1) and fairly dense states (2). The effect of water to TEOS ratio on the sol-gel transition was evaluated, using thermal analysis and Raman spectroscopy, by comparing gels prepared with a ratio greater than six to gels prepared with a ratio less than six. By observing the Raman band at 980 cm^{-1}, it was found that the condensation of hydroxyl groups to give crosslinking occurred above 600 C in the high ratio gels, while pyrolysis below 400 C in the low ratio gels produced more Si-O-Si bonds (3). Regardless of the preparation method, by 800 C all of the silica gels have been largely converted to a silica network. The spectra for these materials show Raman bands at 490 and 600 cm^{-1}, which were attributed to defects in the silica random network (5). Separately, it has been demonstrated that gels of very high surface area show reversible adsorption of water when exposed to air (4). The Raman spectra show indications of bulk hydroxyl, surface hydroxyl and adsorbed water in partially densified gels.

Far more studies on gels using infrared spectroscopy have been reported (e.g. 6, 7, 8). This technique has been used to point out possible differences between sol-gel derived silica and conventional fused silica. Attention has been paid to the 1100 cm^{-1} band associated with bridging oxygens and the 800 cm^{-1} band related to tetrahedral distortion. Based on a comparison of melted glass and hot pressed sol-gel glass (8), it was suggested that sol-gel glasses have Si-O-Si rings of larger size or chains of longer length between crosslinks. Yet, in dense glasses produced using sol-gel processing many bulk properties, such as index of refraction, microhardness and bulk density, are equivalent to those in melted glass (6). Structural relaxation may take place during the densification of the sol-gel glass, as polymerization continues to change the structure (9). The net result in the densified glass is a physical duplicate of melted glass (10).

[†] This work performed at Sandia National Laboratories supported by the
   U. S. Department of Energy under Contract Number DE-AC04-76DP00789.

In this study, Micro-Raman spectroscopy was used to examine the kinetics in the ethanol-tetraethyl orthosilicate (TEOS)-water system during the sol-gel process. The reactants water and TEOS in the solvent ethanol undergo hydrolyzation and polymerization. The time it takes for these reactions to produce a gel depends on the ratio of water to TEOS, the concentration of acid catalyst and temperature. The rigid gel containing solvent, the dessicated gel and the heat treated gel were examined with Raman spectroscopy to reveal the extent of silica polymerization.

Rather than study the high frequency part of the spectra between 3000 and 4000 cm^{-1}, this study looks primarily at low frequencies, between 0 and 1600 cm^{-1}. In the low frequency range, it is possible to observe the disappearance of bands due to solvent and observe the appearance of bands related to the structure of the developing gel.

EXPERIMENTAL TECHNIQUE

Five gel compositions (Table 1) were prepared using a one-step process by mixing at 60°C, TEOS. ethanol, water and HCl in proportions which gave a series IVA(1 mole water)-VA(2 moles water)-VIA(3 moles water) with constant acid addition and a series IIA(.003 M HCl)-VA(.03 M HCl)-VIIIA(.3 M HCl) with constant water/TEOS ratio. The approximate molar ratio TEOS/ethanol is 1:3. The solutions were maintained at 40°C until gelled, which usually took 7 to 10 days. After gelation, the samples were dried over a period of five months.

TABLE I. Compositions investigated with micro-Raman spectroscopy

Sample	Condition	moles H_2O/TEOS	Solvent	HCl(M)	Sapphire Window	Fluorescence Fresh	Aged
IIA	wet	2.0	ethanol	0.003	Yes	No	–
	dried	2.0	ethanol	0.003	No	No	Yes
	heated	2.0	ethanol	0.003	No	No	No
	irradiated	2.0	ethanol	0.003	No	–	Yes
IVA	wet	1.0	ethanol	0.03	Yes	No	–
VA	wet	2.0	ethanol	0.03	Yes	No	–
VIA	wet	3.0	ethanol	0.03	Yes	No	–
VIIIA	wet	2.0	ethanol	0.3	Yes	No	–
	dried	2.0	ethanol	0.3	Yes	No	–

– = No data collected.

Samples were studied both immediately after gelling (fresh samples) and after being stored for five months in capped plastic containers (aged). For both aging conditions, samples were examined by Micro-Raman in a "wet" condtion, where solvent had not been allowed to evaporate; "dried", where solvent had evaporated; and in some cases "heat treated", annealed in air, at 155 C or 255C. All samples were not prepared at the same time.

The sample mount for placing gel samples on the microscope stage was a bored aluminum cylinder. Wet samples were covered with a sapphire window and sealed with vacuum grease. In most cases, dried samples were viewed without the sapphire window. Raman spectra were collected from all samples listed in Table 1.

The optical geometry of the micro-Raman facilitated sample handling and was not used for spatial resolution considerations. The 514.5 nm line from an Ar laser (Coherent CR-4) was passed through a line filter and into the incident illumination port of a Zeiss microscope which had been modified for

use in micro-Raman spectroscopy. Scattered light collected by the microscope was dispersed by a double monochromator (Spex 1404) equipped with
holographic gratings. Dispersed photons were detected by a photomultiplier
tube (RCA C31034), counted by a photometer, and stored in a digital computer. The most commonly used objective was a 10x (0.30 N.A.) with a 4 mm
working distance (13). Experimental conditions resulted in a resolution of
5 cm^{-1}. Spectra were obtained without a polarization analyzer in the collection optics.

RESULTS

Assignments for Raman bands observed in the ethanol-TEOS-water system
are listed in Table 2. The Raman spectra for the first series of gels prepared with constant acid addition are shown in Figure 1. The spectra for
samples IVA and VA have sharp lines at 890, 1057 and 1103 cm^{-1} assigned to
ethanol, as well as the lines at 1090, 1290 and 1460 cm^{-1} due to TEOS. A

TABLE II. Observed peak positions in the Raman Spectra for Ethanol-TEOS-
Water System

cm^{-1}	Assignment
383	Sapphire window
422	Sapphire window
440	Silica glass, ethanol
495-500	Hydrated silicate (probably chain structure)
650-660	TEOS
810-825	Hydrated silicate monomer
885-895	Ethanol
950-985	Hydrated silicate chain
1055-1060	Ethanol
1090-1110	Ethanol, TEOS
1280-1300	Ethanol, TEOS
1455-1480	Ethanol, TEOS

band at 650 cm^{-1} also indicates the Si(OR)$_4$ type linking in TEOS. The sharp
bands in the low frequency range are due to the sapphire window (380 and
420 cm^{-1}).

A marked change in the spectrum occurs between the spectra for samples
VA and VIA. These samples represent just enough water for hydrolysis in VA
and an excess of water for hydrolysis in VIA. With the same amount of catalyst and the same reaction time, the evidence for hydrated silicate species
is more apparent in VIA than VA. The band at 823 cm^{-1} is assigned to the
breathing mode of silicate monomer while the 977 cm^{-1} band is assigned to a
similar mode of silicate structures having an average of two non-bridging
oxygens per tetrahedra (i.e. chains). The band at 497 cm^{-1} is assigned to
the rocking/stretching mode of the Si-O-Si bridging bond in chain structures.
The low signal-to-noise ratio of the spectrum for VIA makes positive identification of bands due to more polymerized silicate structures such as sheets
(1100 cm^{-1}) or three-dimensional network (440 cm^{-1}) impossible. Structure
on the plateau on the low frequency side of the 497 cm^{-1} band is due to the
sapphire window used during data collection.

The Raman spectra for the second series of gels prepared with stoichiometric water for hydrolysis are shown in Figure 2. The acid addition is increased ten times from sample IIA to VA and one hundred times from sample
IIA to VIIIA. The same trends seen with increasing water are seen with

296

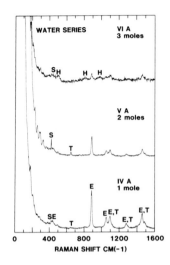

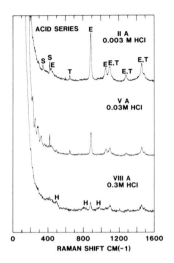

FIG. 1. Raman spectra of gels
with constant acid ad-
dition and ratio of
water to TEOS 1, 2 and
3, soon after gelling.
(S = sapphire window,
T = TEOS, E = ethanol,
H = hydrated silicate).

FIG. 2. Raman spectra of gels
with constant water
addition and concentra-
tion of HCl 0.003, 0.03
and 0.3M, soon after
gelling. (S = sapphire
window, T = TEOS, E =
ethanol, H = hydrated
silicate).

increasing acid catalyst. There is more evidence for reactant TEOS in samples IIA and VA than VIIIA. More evidence for silicate structures also occurs in sample VIIIA than IIA or VA with the appearance again of bands at
813 cm^{-1} (hydrated silicate monomer), and 500 cm^{-1} and 985 cm^{-1} (hydrated
silicate chains).

After drying, Sample IIA was heated to 155C and 255C. The Raman spectra
for the room temperature, 155 and 255°C samples are shown in Figure 3. A
broad asymmetric band at low frequencies (400-500 cm^{-1}) is observed in the
spectra for gels heat treated at 155 C and 255 C. These spectra were collected without the sapphire window, whose Raman bands can complicate
the interpretation of the low frequency region (see Figures 1 and 2). In
the absence of the sapphire bands, the low frequency bands of Figure 3 can
more definitely be assigned to the presence of highly polymerized silicate
structures (i.e. sheets and regions of three dimensional structure) (15).
The broadening at the base of the 1100 cm^{-1} ethanol/TEOS band may be an indication of the presence of silicate sheet structure in the gel. The effect
of heat treatment in acid catalyzed gels should be to facilitate polymerization of surface hydroxyls and possibly to allow for some structural relaxation. This trend toward polymerization of the silica network becomes more
apparent in Raman spectra from gels heated above 600°C (1,2).

When Sample IIA was heat treated soon after drying, the spectra in Figure 3 were measured. These spectra were reproduced at 1.5, 3 and 5 month

intervals and no fluorescence was observed. However, after 5 months of aging, originally wet samples of IIA which showed structure in the fresh condition, developed similar, albeit less intense, fluorescence than that observed in previous studies (11,12). The spectra for the fresh sample and the aged sample are compared in Figure 4. The occurrence of fluorescence is also noted in Table 1.

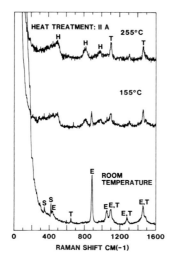

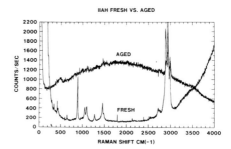

FIG. 3. Raman spectra of gels with 2 moles water per mole TEOS soon after gelling, after heating to 155°C and after heating to 255°C. (S = sapphire window, T = TEOS, E = ethanol, H = hydrated silicate).

FIG. 4. Raman spectra of gels with 2 moles water per mole TEOS before aging and after aging for 5 months at room temperature.

DISCUSSION

The trends observed in the spectra shown in Figures 1, 2 and 3 are consistent with results reported using small angle X-ray scattering and physical property measurements (11), as well as gas chromatography and thermal analyses (12). These trends are that increasing water levels in acid catalyzed gels improve the efficiency of conversion to oxide (14) and that increasing acid concentrations in gels with enough water for hydrolysis increase reaction rates. Higher water levels or higher acid concentrations lead to shorter times for gelling. Since the samples in the water series and acid series were allowed to react for the same length of time, samples VIA and VIIIA are in a more advanced stage of polymerization. The same is true in the temperature series with samples IIA. The evidence for silicate monomers and chain structure increases with increasing temperature from room temperature to 255°C. More polymerized structures such as silicate sheets or regions of three dimensional network may also be present in the gels. However, the spectral regions in which evidence is expected for these structures are in one case obscured by a strong TEOS band. In the other case, the assignment of peak positions is difficult due to the breadth of the plateau on the low frequency side of the Si-O-Si bridging band of silicate chains. At 255 C, the gel should be in a condition where physically trapped water in pores is removed. This temperature generally corresponds to the temperature where the gel has a maximum BET gas adsorption surface area (9). The Raman spectra (Figure 3) clearly show the loss of ethanol from the gel matrix on heating. The 650 cm^{-1}, Si(OR)$_4$ band, also disappears with heating to 255 C. However, prominent Raman bands at 1100 cm^{-1}, 1300 cm^{-1} and 1450 cm^{-1}, which indicate the presence of alkoxy species, persist even in the gel heated to 255 C. This spectroscopic evidence suggests that significant amounts of organic remain in the gel matrix, probably as silico-alkoxide groups, even after heating to 255 C.

The source of fluorescence in the aged gels is not known. However, samples tested soon after drying did not show fluorescence, while samples dried for several months did. The two types of species considered as possible sources are organic impurities and structural defects. In previous Raman studies where fluorescence was observed (1,2) the cause was attributed to organic impurities. In this study, samples being continuously irradiated at laser intensities exhibited photobleaching (i.e. decreasing intensities with time). This phenomenon is commonly observed in laser induced fluorescence studies of organic chromophores, lending credence to the organic impurities explanation.

A heat treated, non-fluorescent sample of IIA was irradiated for 30 mins. at 1 MW in the Sandia ACRR reactor in order to introduce defects into the gel structure. The fluorescence noted in the spectrum of the irradiated sample is significantly shifted in frequency from the fluorescence observed in aged gels (1500 cm^{-1} vs. 1800 cm^{-1}). These results suggest that the fluorescence in the aged gels is not due to structural defects in the silicate network.

With this kind of reactor irradiation, fused silica turns purple due to the creation of color centers in the glass. Although the gel sample showed no visible color change, the Raman spectrum of the irradiated samples showed fluorescence, probably due to the production of structural defects by the radiation. The lack of color change in the sol-gel sample upon irradiation indicates that the gel is better able to "self-heal" the radiation damage than a fused glass (16).

Regardless of the source of the fluorescence, the phenomenon can be prevented. Spectra of samples heat-treated at 155°C and 255°C obtained over a period of several months did not show fluorescence. Therefore, early relatively low temperature heat treatment is sufficient to prevent the development of fluorescence in gels.

CONCLUSIONS

The micro-Raman technique is useful for examining fresh and aged gels in wet, dried and heat-treated conditions. Comparisons of the spectra reveal the stages of reaction in the ethanol-TEOS-water system and the gradual development of silicate structures. The steady disappearance of bands assigned to ethanol can be traced during the drying process. The effect of water level and acid concentration on the reaction rates is that an increase in either one increases the reaction rates so that the gels are in a more advanced state of polymerization.

The cause of the fluorescence in dried gels, which was observed in aged gels but not in fresh gels, could not be determined from these studies. However, the fluorescence can be prevented by heat treating the fresh gel.

ACKNOWLEDGMENT

The technical assistance of C. S. Ashley, C. Scherer and T. A. Gallo is
greatly appreciated, as is the financial support of the National Science
Foundation Division of Materials Research DMR80-12902 (T.A.G., L.C.K.).
Part of the work was performed during the Sandia Summer Visiting Faculty
Program, Summer 1983. C. J. Brinker and D. R. Tallant are thanked for their
helpful advice.

REFERENCES

1. A. Bertoluzza et al. Raman and infrared spectra on silica gel evolving
   toward glass. J. Non-Crystal. Solids 48 (1982) 117-128.
2. D. M. Krol and J. G. vanLierop. The densification of monolithic gels.
   J. Non-Crystal. Solids (1983) to be published.
3. V. Gottardi, M. Guglielmi et al. Further investigations on Raman spectra of silica gel evolving toward glass. J. Non-Crystal. Solids (1983)
   to be published.
4. D. M. Krol and J. G. vanLierop. Raman study of the water adsorption on
   monolithic silica gels. Submitted to J. Non-Crystal. Solids (1983).
5. A. G. Revesz and G. E. Walrafen. Structural interpretations for some
   Raman lines from vitreous silica. J. Non-Crystal. Solids 54 (1983)
   323-333.
6. M. Nogami and Y. Moriya. Glass formation through hydrolysis of
   $Si(OC_2H_5)_4$ with NH_4OH and HCl solution. J. Non-Crystal. Solids 37
   (1980) 191-201.
7. K. Kamiya, S. Sakka and M. Mizutani. Preparation of silica glass fibers
   and transparent silica glass from silicon tetraethoxide. Yogyo-Kyokai-
   Shi 86 (1978) 553-559.
8. M. Decottingnies, J. Phalippou and J. Zarzycki. Synthesis of glasses by
   hot-pressing of gels. J. Mat. Sci. 13 (1978) 2605-2618.
9. T. A. Gallo et al. The role of water in densificaton of gels. This
   volume (1984).
10. L. C. Klein, T. A. Gallo and G. J. Garvey. Densification of monolithic
    silica gels below 1000°C. J. Non-Crystal. Solids (1983) to be published.
11. C. J. Brinker et al. Sol-gel transition in simple silicates. J. Non-
    Crystal. Solids 48 (1982) 47-64.
12. C. J. Brinker et al. Sol-gel transition in simple silicates II. J.
    Non-Crystal. Solids (1983) to be published.
13. D. R. Tallant, K. L. Higgins and C. L. Stein. Raman spectroscopy
    through transparent materials, in Microbeam Analysis-1983, ed. R.
    Gooley, San Francisco, pp. 297-300.
14. L. C. Klein and G. J. Garvey. Effect of water on acid- and base-
    catalyzed hydrolysis and tetraethyl orthosilicate (TEOS) This volume
    (1984).
15. S. A. Brawer and William B. White. Raman Spectroscopic Investigation of
    the Structure of Silicate Glasses. J. Chem. Physics. Vol. 63 No. 6,
    Sept. 15, 1975.
16. A. A. Wolf et al. Radiation-induced defects in glasses with high water
    content. J. Non-Crystal. Solids 56 (1983) 349-354.

^{1}H NMR STUDIES OF THE SOL-GEL TRANSITION*

ROGER A. ASSINK AND BRUCE D. KAY
Sandia National Laboratories
Albuquerque, NM 87185, USA

ABSTRACT

High resolution ^{1}H NMR spectroscopy has been employed to study the dynamics of the sol-gel transition in simple silicates. High magnetic fields (360 MHz) were used to detect and identify the various chemical species present in the $Si(OCH_2CH_3)_4:C_2H_5OH:H_2O$ sol-gel system. Using these techniques, the time evolution of the reactant and product species were monitored. The results of these studies have shown that acid and base catalyzed systems react along very different chemical pathways and that the elementary hydrolysis and condensation reactions occur on widely different time scales.

INTRODUCTION

In previous publications [1,2] Brinker and colleagues reported on polymer growth during acid and base catalyzed hydrolysis of tetraethylorthosilicate (TEOS) in alcoholic solutions and proposed that the form of the resulting polymers was governed by the relative rates of hydrolysis and condensation. We have extended our investigation by employing proton nuclear magnetic resonance (^{1}H NMR) spectroscopy to probe both the temporal and chemical nature of the sol-gel reactions. As will be shown, considerable new kinetic and chemical information has been gained.

EXPERIMENTAL

A two-step process carried out at 25°C was used to prepare all the solutions [1,2]. In the first step TEOS, CH_3CH_2OH(EtOH), H_2O, and HCℓ were mixed in the mole percents displayed in Table I. This amount of water is one-half that required to form anhydrous silica via the stoichometric reaction

$$Si(OEt)_4 + 2H_2O \rightarrow SiO_2 + 4EtOH. \qquad (1)$$

After 90 minutes, additional water plus HCℓ or NH_4OH were added comprising the second step. The concentrations studied are shown in Table I and were chosen to match those previously examined [1,2].

^{1}H NMR spectra were recorded at 360 MHz at the Colorado State University NSF Regional NMR Facility. All spectra were recorded at 25°C using 4 pulses at a .5 Hz repetition rate. All resonances were referenced to tetramethyl silane (TMS) via the ethanol methylene group whose chemical shift is reported as 3.70 ppm relative to TMS [3].

*This work supported by the U. S. Department of Energy under Contract No. DE-AC04-76DP00789.

TABLE I. Compositions (mole %).

STEP	EtOH	H_2O	TEOS	HCl	NH_4OH
1st Acid	65.4	17.9	16.7	0.012	---
2nd Acid	22.7	71.6	5.8	0.004	---
2nd Base	45.5	42.9	11.6	0.005	0.014

RESULTS AND DISCUSSION

General Findings

Figure 1a displays the entire proton spectrum for the initial hydrolysis after the reaction has proceeded 12 min. Figs. 1b and c show an expansion of the 3.5 to 4.1 ppm spectral region for short (12 min) and long (90 min) times during the initial stage. The quartet centered at 3.90 ppm is the resonance of methylene (-CH$_2$-) protons associated with a silicon while that centered at 3.70 ppm is associated with ethanol. Two changes are clearly observed with time (Figs. 1b and c); the Si related quartet is becoming a more complex superposition of many quartets as the chemical environment about the Si becomes more varied and the intensity of this superposition decreases relative to the ethanol quartet. The loss of TEOS and production of EtOH due to reaction is easily quantified by integration of the appropriate peaks centered at 3.70 and 3.90 ppm.

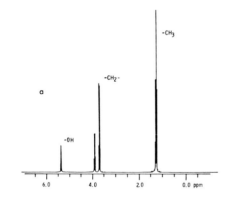

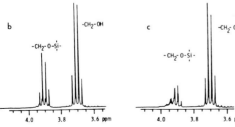

FIGURE CAPTIONS

FIG. 1. ^1H NMR spectra of the initial hydrolysis for a) entire proton spectrum at short time (12 min), and expansions of the methylene quartets at b) short (12 min), and c) long (90 min) times.

Figure 2 shows the fraction of CH_2 groups associated with an Si relative to the total number of CH_2 groups for both 1st and 2nd stages of hydrolysis as a function of time. For the initial stage, the results indicate that complete hydrolysis is achieved after 12 min while the condensation reaction attains stoichometric completion only after long time (~ 24 hrs). The second stage depends strongly upon whether the system is acid or base catalyzed. As shown in Fig. 2 the acid system undergoes complete hydrolysis long before gelation while the base system undergoes gelation and remains incompletely hydrolyzed in the gel state.

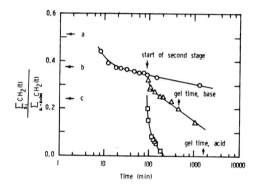

FIG. 2. Fraction of CH_2 groups associated with a Si relative to the total number of CH_2 groups for 1st (o) and 2nd (□ acid, Δ base) stages as a function of time. For the 1st stage, a = starting composition, b = complete hydrolysis, no condensation, and c = complete hydrolysis and condensation.

Figure 3 shows the dramatic differences between the acid and base catalyzed second stages. In the acidic case (Figs. 3a and b) the superposition of quartets associated with silicon decrease rapidly with time and at comparable rates. This is contrasted with the base system (Figs. 3c and d) in which the superposition quartets decrease at vastly different rates and at long times (Fig. 3d) the system has a well resolved single quartet which corresponds to unreacted TEOS. This may indicate a microscopic phase separation and possibly accounts for the cloudy nature of the macroscopic base catalyzed gels [1].

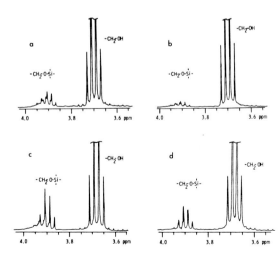

FIG. 3. ^1H NMR spectra of second stage hydrolysis for a) acid catalyzed 2 min; b) acid catalyzed 17 min; c) base catalyzed 6 min; and, d) base catalyzed 16-1/2 hr.

Spectral Deconvolution and Reaction Kinetics

In an attempt to reduce the complex chemical nature of the sol-gel reaction to a tractable level and still retain the salient features we propose the following. Consider the possible reaction types,

Hydrolysis

$$(RO)_4Si + H_2O \rightarrow (RO)_3Si-OH + ROH, \qquad (2)$$

and

Condensation

$$2(RO)_3SiOH \rightarrow (RO)_3Si -O- Si-(RO)_3 + ROH. \qquad (3)$$

Since Si retains a coordination number of 4 and its only possible next to nearest neighbors are (1) $-CH_2-$, (2) $-H$, and/or (3) $-Si$, there are only 15 possible combinations (local chemical environments) at the next to nearest neighbor level. These are shown in the form of a matrix in Fig. 4 and display all species from starting material (TEOS(400)) to orthosilic acid $(Si(OH)_4(040))$ to anhydrous silica $(SiO_2(004))$. Figure 5a shows an expanded region of the Si related quartets in Fig. 1c; as readily seen it is a complex superposition of quartets. Via aid of the sol-gel matrix (Fig. 4) we have simulated this spectra by assigning chemical shifts to the various species possible. This simulation is shown in Fig. 5b and there is remarkable agreement. The specific details will be reported in a forthcoming publication [4].

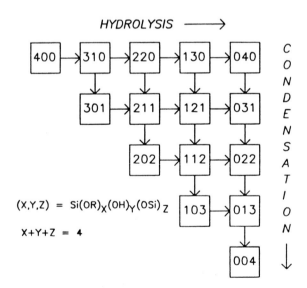

FIG. 4. Possible chemical speciation at the next to nearest neighbor level displayed in matrix form.

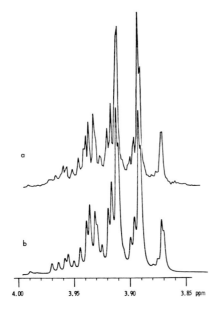

FIG. 5. ^{1}H NMR spectra of the Si related $-CH_2-$ quartets for the initial hydrolysis stage after 90 min of reaction; a) observed spectra, b) theoretically simulated spectra.

Even at the next to nearest neighbor level, modeling the reaction kinetics is an extremely difficult endeavor. From the sol-gel matrix (Fig. 4) it can be seen that there are 65 rate constants necessary to describe the system (10 for hydrolysis and 55 for condensation). The present data are not adequate to determine each of these coefficients. In an attempt to model this complex reactive system with the present data the following simplifying assumption is made. We assume that the hydrolyis and condensation rates are dependent only on the functional group participating in the reaction. The two possible functional reactions are,

Hydrolysis

$$-\overset{|}{\underset{|}{Si}}-OEt + H_2O \overset{k_H}{\to} -\overset{|}{\underset{|}{Si}}-OH + EtOH \qquad (4)$$

and

Condensation

$$2 -\overset{|}{\underset{|}{Si}}-OH \overset{k_c}{\to} -\overset{|}{\underset{|}{Si}}-O-\overset{|}{\underset{|}{Si}}- + H_2O. \qquad (5)$$

and the rate equations governing the concentrations of the various functional groups are

$$d[Si-OEt]/dt = -k_H[Si-OEt][H_2O] \qquad (6)$$

$$d[H_2O]/dt = k_c[SiOH]^2 - k_H[SiOEt][H_2O] \qquad (7)$$

$$d[EtOH]/dt = k_H[SiOEt][H_2O] \qquad (8)$$

$$d[Si-O-Si]/dt = k_c[Si-OH]^2 \qquad (9)$$

$$d[Si-OH]/dt = k_H[SiOEt][H_2O] - k_c[SiOH]^2 . \qquad (10)$$

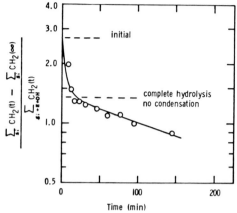

FIG. 6. Temporal behavior of the initial hydrolysis stage, (circles = data, solid line = model).

By numerically integrating these coupled differential equations we have attempted to model the first stage hydrolysis time dependence shown in Fig. 2. In Fig. 6 the measured data has been recast on a scale better suited for kinetic analysis. As clearly shown, there are two distinct temporal domains, a fast initial step, followed by a slower second step. The solid line represents the best fit to the data within the framework of the stated model. The rate coefficients are .025 and .002(ℓ/mole-min) for the hydrolysis and condensation reactions, respectively. The hydrolysis rate is roughly 12 times greater than that of condensation, thus indicating that in the 1st stage condensation is rate limiting. It should be emphasized that these rate coefficients are expected to be strongly dependent on pH, and are thus only valid for the particular system studied. A more detailed explanation will be presented in a future article [4].

CONCLUSION

^{1}H NMR spectroscopy has been demonstrated to yield detailed chemical information about the sol-gel reactions. Using this technique we have obtained quantitative kinetic data on this complicated reactive system. This technique will undoubtedly complement structural scattering data [1,2] in elucidating the physiochemical processes involved in the sol-gel transition.

ACKNOWLEDGEMENT

We would like to thank Drs. C. J. Brinker and K. D. Keefer for bringing sol-gel chemistry to our attention. The support of the Colorado State University Regional NMR Center, funded by National Science Grant No. CHE-8208821, is gratefully acknowledged.

REFERENCES

[1] C. J. Brinker, K. D. Keefer, D. W. Schaefer, and C. S. Ashley, J. Non-Crystalline Solids 48 (1982) p. 47.
[2] C. J. Brinker, K. D. Keefer, D. W. Schaefer, R. A. Assink, B. D. Kay, and C. S. Ashley, J. Non-Crystalline Solids (in press).
[3] N. S. Bhacca, L. F. Johnson, and J. N. Shoolery, NMR Spectra Catalog (Varian Associates, Palo Alto, CA, 1962).
[4] B. D. Kay and R. A. Assink, manuscript in preparation.

EFFECT OF AGING ON ELECTROPHORETIC BEHAVIOR OF SOL-GEL DERIVED ALUMINAS

Burtrand I. Lee and Larry L. Hench
Ceramics Division, Department of Materials Science and Engineering,
University of Florida, Gainesville, Florida 32611

ABSTRACT

Sample alumina powder prepared with the procedures described by Yoldas [1,2] was suspended in 10^{-3}M NaCl and contained in precleaned polypropylene cups with air tight lids. Electrophoretic mobilities of the samples were measured at various pH's using a microelectrophoretic technique. The points of zero charge (PZC) were determined from plots of electrophoretic mobility vs pH of the sample suspensions. The initial values of PZC are 10.6 for γ-Al_2O_3 and 8.3 for α-Al_2O_3. The corresponding BET surface areas are 206 m^2/g and 3m^2/g respectively. Upon aging, the PZC decreased in 2 days and then increased by reaching maximum values of 10.4 for γ-Al_2O_3 and 8.8 for α-Al_2O_3 in 20 days. Aging of sol-gel derived aluminas and commercial alumina derived by precipitation are compared and differences are discussed.

INTRODUCTION

Sol-gel methods have been developed [1-4] for the preparation of various phases of alumina which are otherwise difficult to obtain using traditional methods. The alumina sol-gel method is different from traditional methods in several important aspects which may affect the structure of alumina; 1) homogeneity at the molecular level, 2) a possibility of ultimately pure alumina, and 3) low temperature chemical reactions and mixing. Because of these differences, the electrokinetic properties of sol-gel derived alumina may also be different from the properties of alumina prepared by traditional methods.

The electrokinetic properties of sol-gel derived aluminas yield important information concerning the surface characteristics of the alumina in a liquid environment. A knowledge of the latter is necessary for scientific control of alumina in dispersions and in producing ceramic composite microstructures. Catone and Matijevic [5] prepared spherical particles of monodispersed aluminum hydrous oxide sol by hydrolysis of $Al(OC_4H_9)_3$ in aqueous sulfate solution. They showed that the electrophoretic mobility (EM) at a given SO_4^{-2} concentration increased in a positive direction with aging time in 20 hrs., then the EM decreased up to 30 hrs. of aging at 99°C. No explanations were given for these fluctuations of EM but they suggested that the particle composition may be different than that of sols made from inorganic salt solutions.

It is well known that the PZC of metal oxides varies depending on degree of surface hydration, [6,7,14-17] and the method of preparation. Consequently, one of our objectives is to compare the effects of sol-gel vs precipitation methods of preparing alumina on the PZC of the powders.

There has been much effort to understand the factors affecting the time dependent variations of PZC of metal oxides, the so called aging effect [7-19]. These studies have shown that aging phenomena of metal oxides can be divided into three general categories: 1) phenomena involving impuries, 2) surface hydration and 3) exchange reactions between the surface and the surrounding medium.

Sol-gel derived aluminas have a higher reactivity and surface area than aluminas from precipitates, and a submicron particle size. Each of these factors may also affect the electrophoretic properties and render unique characteristics of sol-gel derived alumina.

EXPERIMENTAL

The alumina gel was prepared from aluminum sec-butoxide using the procedure described by Yoldas [1,2]. Excess water in the mole ratio of 100 of deionized water to reagent grade of Al $(OC_4H_9)_3$ (without further purification) was used to form the hydrolyzed alkoxide which was followed by peptization at 90°C with 0.07 moles of hydrochloric acid per one mole of Al $(OC_4H_9)_3$. The peptized sol was cast over mercury and dried in air at room temperature.

The dried gel was ground using an alumina mortar and pestle. Portions of the gel powder were calcined at 800°C and 1400°C in a microprocessor controlled furnance for 4 hours after preheating at 400°C and 600°C for 3 hours respectively. Phase analysis of the calcined powder is described by Clark and Lannutti [3]. The mean particle size estimated from the SEM micrographs was approximately 1 μm. Other physical characteristics of the powders are shown in Table 1 along with commercially available alumina powders [21,22].

Table 1. Characteristics of sol-gel derived alumina powders calcined at different temperatures compared with commercial aluminas

Sample Number	Calcin. Temp. °C	Mean Particle Size, μm	BET Surface Area, m^2/g	Phase[a]	Initial PZC
A-025	25	<0.5	285	δ-AlO(OH)	9.4
A-800	800	1	206	γ	10.6
A-1400	1400	1	3	α	8.3
Commercial Al_2O_3					
Baikalox CR6[b]	as received	<1	6	α	8.8
A-17[c,e]	as received	1.5	3.02[e]	α	8.8
C-30DB[d,e]	500		108[c]	γ	8.3[e]

a) from Lannuti et al. [3], b) 99.99% pure alum derived by Baikowski, c) Bayer processed high purity alumina by Alcoa, d) Bayer processed high purity alumina by Alcan, e) from Horn [21].

Alumina suspensions for the microelectrophoresis experiments were prepared by dispersing 0.04g of the alumina powders in 40ml of 10^{-3}M NaCl solution. All solution preparations and experiments were carried out with doubly purified water which had been deionized and distilled.

Polypropylene cups and lids to contain the alumina suspensions were prewashed succesively with hot 2M NaOH, 2M HNO_3 for 24 hours followed by 12 hours soaking in hot water and air drying. Aging of the sample suspensions in the containers was carried out at room temperature until each electro-phoretic measurement was made.

Electrophoretic mobilities of the samples were measured in duplicate using a microelectrophoresis apparatus with a rotating prism. Rotating prism measures an average velocity of thousands of particles in both directions.* The pH of the alumina suspensions was adjusted using dilute HCl or NaOH. Four or more measurements using the rotating prism were made for each sample at each time. The mean EM for each sample was used to calculate the zeta potential(ζ) following the procedure that Pask et al. [19] used for their aging study of aluminas.

Alum derived high purity commercial alumina Baikalox CR6 which contained 94% α-phase was obtained from Baikowski International Corporation for use as a reference in the study.

The surface areas of the various alumina samples were determined by using a Quantachrome Nitrogen Adsorption BET Apparatus.

RESULTS

The values of PZC for each alumina sample after each aging time interval were determined by plotting the ζ (mV) versus pH and locating the pH value where the curve intersected ζ at zero mV. Aging plots for samples A-800 and A-1400 are given in Figs. 1 and 2. The values of PZC after each aging time interval are tabulated and given in Table 2 along with the BET surface areas before aging. Duplicate PZC values were in agreement within ±0.1 pH unit.

Table 2. PZC values of aluminas after each aging time interval

Sample	Aging Time (days)					BET Surface Area m^2/g
	0	2	10	20	30-32	
Gel Derived Al_2O_3		PZC Values				
A-025	9.4	9.2	9.0	9.3	9.0	285
A-800	10.6	10.0	10.0	10.4	8.9	206
A-1400	8.3	8.0	8.7	8.8	7.2	2.5
Commerical Al_2O_3						
Baikalox CR6	8.8	9.4	9.6	9.8	8.5	6.0
A-17[a]	8.8	9.7	9.8	9.8	9.8	3.0
		(8 dys)	(16 dys)	(33 dys)		
C-30 DB[a]	8.3	8.5	8.8	9.6	10.2	108
		(1 dy)	(8 dys)	(16 dys)	(33 dys)	

[a] From Horn [21]

*Mark II, Rank Brothers, U.K.

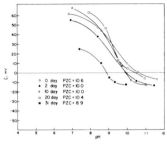

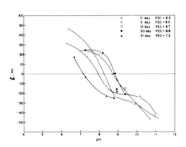

Fig. 1 Zeta potential curves for
A-800 after each interval
of aging. The PZC is the
pH at $\zeta = 0$ mV.

Fig. 2 Zeta potential curves of
A-1400 after each interval
of aging.

DISCUSSION

The sample aluminas were derived from an organometallic compound with
an alcohol as by-product in the sol-gel reaction. This reaction is written
in the equation below.

$$Al\ (O-\underset{\underset{CH_3}{|}}{\overset{\overset{H}{|}}{C}}-CH_2CH_3)_3 + \text{excess } H_2O \xrightarrow{HCl} Al(OH)_3 + 3CH_3-\underset{\underset{H}{|}}{\overset{\overset{OH}{|}}{C}}-CH_2CH_3$$

The alcohol residue should remain in the pores of the alumina at lower
temperatures as weakly physisorbed alcohol [23-25]. When the alumina is
placed in water, the alcohol should desorb into the solution. Thus the
alcohol itself would have no effect on the PZC of the alumina A-025.

In Figs. 3 and 4, all commercial α-Al$_2$O$_3$ (Baikalox CR-6, A-17, C-30
DB) show increasing PZC's with aging. After 10 days of aging, PZC's of all
commercial aluminas continue to increase except Baikalox CR-6 which shows a
decreasing trend. This is thought to be contributed by an effect of impurities from the sample container [26]. This should also apply to samples A-
025 and A-800. An initial decrease in PZC's causing minima within 2 days
of aging is common to all gel derived aluminas, however no minima are shown
for any of commercial aluminas. A greater effect of aging for A-1400 and
A-17 is shown in Fig. 3. A somewhat smaller effect is shown for Baikalox.

In Fig. 4, the commercial γ-Al$_2$O$_3$ C-30 DB shows a greater effect of
aging than that of gel derived γ-Al$_2$O$_3$ A-800 within the same aging
period. In fact, each of the commercial aluminas gave a greater effect of
aging within ca. 30 days. This is difficult to explain by surface hydration factors alone. Other factors such as impurities and the kinetics of
exchange reactions between the surface of alumina and the aqueous medium
need to be considered [7-19].

There are large differences in the initial PZC's for the γ-Al$_2$O$_3$
samples A-800 (10.6) and C-30 DB (8.3). This could be due to differences
in the degree of surface hydration and surface areas. Both are a γ-Al$_2$O$_3$
phase but the surface area of A-800 (206 m^2/g) is nearly twice larger than
that of C-30 DB (108 m^2/g). Although the alumina samples were stored in a
desiccator until used, slow hydration may have occurred during storage
before the powders were placed in a desiccator. Thus the degree of hydra-

tion would be greater for A-800 than that of C-30 DB since the extent of hydration is a function of surface area.

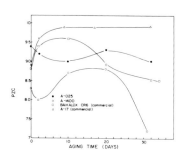

Fig. 3 Variation of PZC with time for α-Al$_2$O$_3$. A-025 is an δ-AlO(OH) with residual impurities.

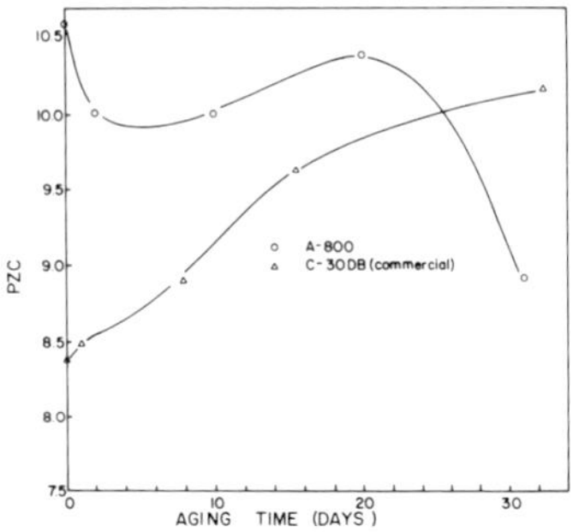

Fig. 4 Variation of PZC with time for γ-Al$_2$O$_3$.

The maximum PZC values of 10.4 for A-800 and 10.2 for C-30 DB are close. However C-30 DB took much longer times (~10 days) to reach its maximum during aging (Fig. 4).

Large differences in curve shapes are shown between gel derived α-Al$_2$O$_3$ (A-1400) and commercial α-Al$_2$O$_3$ (A-17 and Baikalox CR6). A-1400 exhibited a much lower PZC throughout the entire range of aging time. Commercial α-aluminas took shorter times to reach the maximum PZC than those of gel derived aluminas. The shape of the aging curves resemble each other for commercial aluminas and gel derived aluminas i.e. A-025 and A-1400.

Horn [21] prepared the alumina sample C-30 DB by calcining at 500°C to convert the as-received trihydrate to γ-Al$_2$O$_3$. Although he did not specify what containers were used for his aging study, polypropylene bottles were mentioned for measurements of equilibrium pH of an alumina after aging.

CONCLUSION

The electrokinetics of sol-gel derived aluminas in an aqueous medium contained in polypropylene containers are characterized by an initial decrease in PZC in 2 days and an increase thereafter until impurities from the container begin to act on the surface of alumina. Commercial α-alumina, on the other hand, does not exhibit the initial decrease of PZC but exhibits a greater effect of aging than that of gel-derived aluminas.

There are major differences in the electrophoretic behavior of gel derived α-alumina and commercially available alum derived and Bayer processed α-aluminas. The minima shown for all of gel-derived alumina in ~2 days of aging may be related to the variations of EM in the work of Catone, et. al [5]. For most gel-derived aluminas, it takes at least 20 days of aging to reach an equilibrium PZC.

Finally, the cause for the distinctive differences between commercial α-alumina and gel-derived alumina is presently unclear. However, α-Al$_2$O$_3$ is the least stable in water by exhibiting the most aging. In contrast, the uncalcined alumina gel AlO(OH) shows the least aging. Consequently, the difference between sol-gel and commercial aluminas may be

312

thermodynamically related. Further study using contaminant free containers and longer aging times to obtain true behavior of high purity alumina powder derived by a sol-gel method is needed to establish the specific origins for the differences in aging behavior.

ACKNOWLEDGMENTS

The authors are grateful for the financial support of AFOSR (Contract # F49620-83-C-0072). They also appreciate especially, among many others, J. Lannutti for alumina sample preparation and A. Izadbakhsh for a portion of the BET surface area measurement.

REFERENCES

1. B. E. Yoldas, Am. Ceram. Soc. Bull., 54, 289-290 (1975).
2. B. E. Yoldas, Am. Ceram. Soc. Bull., 54, 286-288 (1975).
3. D. E. Clark and J. J. Lannutti, Proc. Internatinal Conf. on Ultrastructure Process of Ceramic, Glasses and Composites, Feb. 13-17, 1983, Gainesville, FL.
4. Y. Ozaki and M. Hideshima, Engineering Report, Seikei University, Japan No. 23, 1611 (1976).
5. D. L. Catone and E. Matijevic, J. Colloid and Interfacial Sci., 48, 291 (1974).
6. G. A. Parks, Chem. Rev., 65, 177-198 (1965).
7. G. Y. Onoda, "The Mechanism of Proton Adsorption at the Ferric Oxide-Aqueous Solution Interface," Ph.D. Thesis, M.I.T. (1966).
8. D. J. O'Connor, P. G. Johansen and A. S. Buchanan, Trans. Faraday Soc., 52, 229 (1956).
9. P. G. Johansen and A. S. Buchanan, Aust. J. Chem., 10, 398 (1959).
10. Y. G. Berube and P. L. DeBruyn, J. Colloid and Interfacial Sci., 27, 305 (1968).
11. H. J. Modi and D. W. Fuenstenau, J. Phys. Chem., 61, 640-43 (1957).
12. Y. G. Berube, G. Y. Onoda, and P. L. DeBruyn, Surface Sci., 7, 448 (1967).
13. Y. G. Berube and P. L. DeBruyn, J. Colloid and Interfacial Sci., 28, 92 (1968).
14. L. Block and P. L. DeBruyn, J. Colloid and Interfacial Sci., 32, 529 (1970).
15. H. F. Trimbos and H. N. Stein, J. Colloid and Interfacial Sci., 77, 386 (1980).
16. H. F. Trimbos and H. N. Stein, J. Colloid and Interfacial Sci., 79, 399 (1980).
17. J. W. Murray, J. Colloid and Interfacial Sci., 46, 359 (1974).
18. D. M. Furlong, P. A. Freeman, and A. C. Lau, J. Colloid and Interfacial Sci., 80, 20 (1981).
19. J. S. Moya, J. Rubio, and J. A. Pask, Am. Cer. Soc. Bull., 59, 1198 (1980).
20. D. N. Furlong and G. D. Parfitt, J. Colloid and Interfacial Sci., 65, 548 (1978).
21. J. M. Horn, Jr., "Electrokinetic Properties of Silica, Alumina, and Montmorillonite," Ph.D. Thesis, University of Florida, 1978.
22. W. M. Flock, "Bayer Processed Alumina" in "Ceramic Processing Before Firing," G. Y. Onoda, Jr. and L. L. Hench, eds., John WIley and Sons, New York, 1978.
23. K. O. Kagel, J. Phys. Chem., 71, 844 (1969).
24. A. V. DeO and I. G. Dalla Lana, J. Phys. Chem., 73, 716 (1969).
25. A. V. Kiselev and V. I. Lygin, "Intrared Spectra of Surface Compounds", p. 254, John Wiley and Sons, Inc., New York, 1975.
26. B. I. Lee and L. L. Hench, J. Colloid and Interfacial Soc., (to be published).

NONAQUEOUS SUSPENSION PROPERTIES OF Al_2O_3 AND SILICATE GLASS POWDERS

M.D. SACKS AND M.I. ALAM
Department of Materials Science and Engineering
University of Florida
Gainesville, Florida 32611

INTRODUCTION

Colloidal suspensions prepared with nonaqueous liquids are important in ceramic processing operations such as tape casting, slip casting, etc. [1,2]. However, investigations of the properties of such suspensions are limited. In the present work, the suspension behavior of Al_2O_3 and silicate glass powders was studied using two nonaqueous liquids - methanol (MEOH) and methyl isobutyl ketone (MIBK). Rheological, sedimentation, and electrokinetic properties were determined for various mixed liquid ratios. The effect of polyvinyl butyral resin, a polymeric binder, on silicate glass suspension properties was also investigated.

EXPERIMENTAL

Experiments were carried out with α-Al_2O_3 (C-75, Alcan Aluminum Co.) and glass (SiO_2-Al_2O_3-MgO-CaO, Pemco Products) powders. The Al_2O_3 had a median equivalent Stokes diameter of ~2µm and a specific surface area of ~2 m^2/g, while the corresponding values for the silicate glass were ~3µm and ~2.5 m^2/g, respectively. Additional information on these powders is provided elsewhere [3,4]. Suspension liquids were methanol (MEOH, duPont Co.) and methyl isobutyl ketone (MIBK, Ashland Chemical Co.). The polymer used in this study was a polyvinyl butyral resin (B-98, Monsanto Co.). All suspensions were prepared with 30 vol% solids, 65-70 vol% liquid, and 0-5 vol% polymer.

Suspension rheological flow curves (shear stress, τ, vs. shear rate, $\dot{\gamma}$, behavior) were determined using a narrow-gap, concentric cylinder viscometer (Haake, Inc.). Viscosity values, η, were obtained from the flow curves using the relationship:

$$\eta = \tau/\dot{\gamma}$$

Sedimentation density was determined by pouring the suspension (of known weight) into a 50 mL graduated cylinder, covering the cylinder to prevent liquid evaporation, and allowing particles to settle until the sediment height no longer changed with time. The measured bulk density was divided by the powder true density in order to obtain the relative density. Electrical conductivity measurements were made on the "pure" liquids (various MEOH/MIBK ratios) and on the supernatant liquids of extremely well-centrifuged suspensions. A microelectrophoresis apparatus (Rank Brothers Co.) was used to measure the electrokinetic mobility of particles. Zeta potentials were determined from the electrophoretic mobilities using the Helmholtz-Smoluchowski equation [5]. Polymer adsorption isotherms were determined gravimetrically.

RESULTS AND DISCUSSION

Effect of MIBK/MEOH Ratio on Suspension Properties

The effect of liquid ratio (MIBK/MEOH) on suspension properties was

Mat. Res. Soc. Symp. Proc. Vol. 32 (1984) © Elsevier Science Publishing Co., Inc.

investigated for Al_2O_3/liquid and glass/liquid suspensions. Plots of viscosity (at shear rates of 3, 30, and 300 s^{-1}) vs. vol% MIBK (in the liquid phase) are shown in Fig. 1. Several observations are noted:

(1) All suspensions show shear thinning behavior, i.e. the viscosity decreases as the shear rate increases. This behavior is characteristic of flocculated suspensions. Under low shear rate conditions, liquid is immobilized in the interparticulate pore channels of flocs and floc networks. The viscosity is increased (relative to dispersed suspensions) due to the increased "effective" solids loading. Shear thinning flow results from the breakdown of the flocculated structure (and the release of entrapped liquid) as the shear rate is increased.

(2) In Al_2O_3 suspensions, the viscosity (at a given shear rate) remains relatively constant over the range ~0-75 vol% MIBK, but increases significantly at higher MIBK contents. The viscosity increases continuously with increasing MIBK content in glass/liquid suspensions. Both results indicate that more extensive flocculation occurs at high MIBK/MEOH ratios.

Sedimentation behavior (Fig. 2) for Al_2O_3/liquid and glass/liquid suspensions also indicates that increased flocculation occurs at higher MIBK contents. Highly flocculated suspensions tend to form sediments with low relative density due to the large amounts of interparticulate porosity associated with flocs and floc networks. For Al_2O_3 suspensions, the sedimentation density remains relatively constant over the range ~0-75 vol% MIBK, but decreases significantly at higher MIBK contents. The sediment density also decreases to very low values with increasing MIBK content in glass/liquid suspensions.

The state of particulate dispersion in suspension is governed by the electrostatic repulsion and London-van der Waals attraction forces that operate between particles [5]. The latter forces are expected to have only a weak dependence on the MIBK/MEOH ratio [3,6]. Therefore, the decrease in stability (against flocculation) at higher MIBK contents is attributed to decreased electrostatic repulsive forces between particles. This is

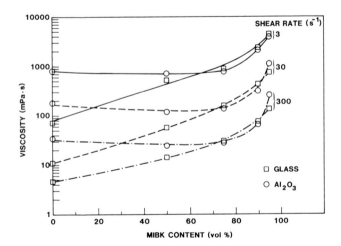

FIG. 1. Plots of suspension viscosity (at three shear rates) vs. vol% MIBK in suspension liquid phase.

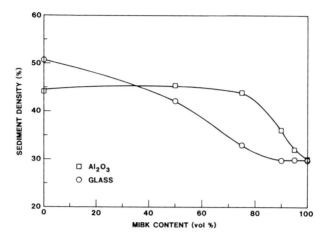

FIG. 2   Plots of sediment density vs. vol% MIBK in suspension liquid
phase.

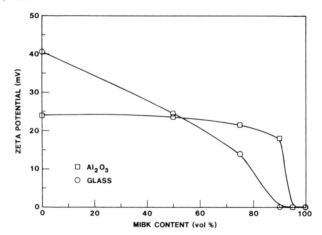

FIG. 3.   Plots of zeta potential (absolute value) vs. vol% MIBK in
suspension liquid phase.

supported by electrophoresis measurements. Plots of zeta potential
(absolute value) vs. MIBK content are shown in Fig. 3. In glass
suspensions, the zeta potential decreases continuously from a moderate value
for the MEOH suspension to zero at 9:1,MIBK/MEOH ratio. As the zeta
potential decreases, the electrostatic repulsive force between particles
decreases and increased flocculation is expected. This is consistent with
experimental observations of high suspension viscosities (Fig. 1) and low
sedimentation densities (Fig. 2) at high MIBK contents. In Al_2O_3
suspensions, the zeta potential is relatively constant over the range ~0-75
vol% MIBK, but decreases to zero at higher MIBK contents. This is also
consistent with the sedimentation and rheological behavior (Figs. 1 and 2).

    Fig. 4 shows the electrical conductivity of (1) MIBK/MEOH liquids of

316

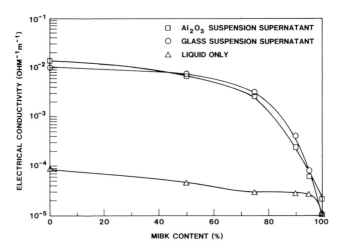

FIG. 4. Plots of electrical conductivity vs. vol% MIBK in (1) various
MIBK/MEOH liquid compositions and (2) supernatant liquids obtained
from centrifugation of Al_2O_3/liquid and glass/liquid suspensions.

various ratios and (2) supernatants of centrifuged Al_2O_3 and glass
suspensions formulated with various liquid ratios. In each case, the
conductivity decreases as the MIBK content in the liquid phase increases.
Higher electrical conductivities are observed in the suspension supernatants
(i.e. compared to the "pure" liquids) due to the presence of ions leached
into the liquid phase from the ceramic particles. The results indicate that
MIBK supports less electrolytic dissociation compared to MEOH. This is not
surprising since MIBK is an aprotic, low dielectric constant solvent (ε =
13.1 at 25°C) whereas MEOH is an amphiprotic, high dielectric constant
solvent (ε = 32.7 at 25°C) [7].

The specific mechanism(s) responsible for the development of the
particle surface charge (and zeta potential) in the various Al_2O_3/liquid and
glass/liquid suspensions were not investigated. However, the electrical
conductivity results suggest that the zeta potential decreases with increasing MIBK content because (1) dissociation of surface groups is inhibited
and/or (2) less adsorption occurs on the surface (and/or in the Stern plane
of the electrical double layer [8]) due to fewer dissociated ions in
solution. These general mechanisms are consistent with the decreased
capacity for electrolytic dissociation in MIBK.

Effect of Polymer Additions on Suspension Properties

Plots of suspension viscosity vs. shear rate are shown in Fig. 5 for
glass - 3:1,MIBK/MEOH suspensions containing various amounts (0-5 vol%) of
polyvinyl butyral resin. The suspension with no polymer addition shows
extensive low shear rate flocculation, as indicated by the high suspension
viscosities and the highly shear thinning behavior. In constrast, suspensions with small polymer additives (0.5 - 2.0 vol%) have low viscosities and
approximately Newtonian behavior. These observations indicate that suspensions with small polymer additions are well-dispersed.

The effect of polymer additions on the state of dispersion in glass -
3:1,MIBK/MEOH suspensions was also monitored by sedimentation density
measurements (Fig. 6). The suspension with no polymer addition has a very

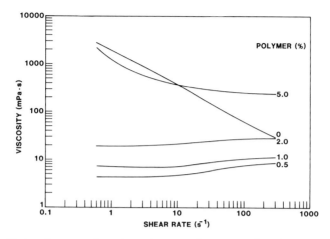

FIG. 5. Plots of suspension viscosity vs. shear rate for glass -
3:1,MIBK/MEOH suspensions containing indicated amounts of polymer.

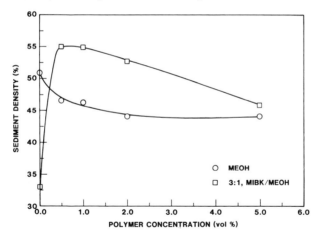

FIG. 6. Plots of sediment density vs. polymer concentration for glass -
MEOH and glass - 3:1,MIBK/MEOH suspensions.

low sedimentation density which is consistent with rheological data (Fig. 5)
indicating that the suspension is highly flocculated at low shear rates.
Small additions of polymer (0.5 - 2.0 vol%) significantly increase the
sedimentation density. This is consistent with rheological data which (Fig.
5) indicates that these suspensions are well-dispersed. Since polyvinyl
butyral is a nonionic polymer, suspension dispersion is attributed to a
steric stabilization mechanism [9] (i.e. as opposed to electrostatic
stabilization).

Suspension flocculation re-occurs at higher polymer concentrations
(e.g. 5 vol%) in glass - 3:1,MIBK/MEOH suspensions. This is indicated by
the shear thinning behavior at low shear rates (Fig. 5) and the decreased
sedimentation density (Fig. 6). The polymer concentration is apparently

318

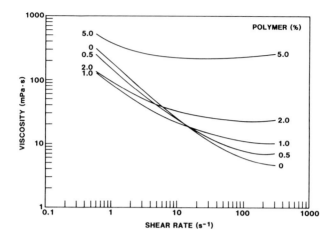

FIG. 7. Plots of suspension viscosity vs. shear rate for glass - MEOH
suspensions containing indicated amount of polymer.

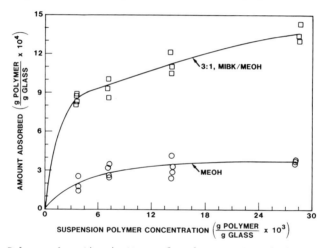

FIG. 8. Polymer adsorption isotnerms for glass - MEOH and glass -
3:1,MIBK/MEOH suspensions.

large enough for bridging flocculation to become important [9].

The effect of polymer additions on the suspension properties of glass -
MEOH suspensions was determined also. Plots of suspension viscosity vs.
shear rate (Fig. 7) show that shear thinning behavior is observed for all
suspensions. In the suspension without polymer, electrostatic repulsion
forces are too weak to prevent flocculation at low shear rates. This is
also true for suspensions containing polymer since essentially no change in
zeta potential was observed with small polymer additions (0.25 - 2.0
vol%). Furthermore, polymer additions in glass - MEOH suspensions
apparently do not promote dispersion via steric stabilization (such as
observed with glass - 3:1,MIBK/MEOH suspensions, Figs. 5 and 6). In fact,

the decrease in sedimentation density with polymer additions (Fig. 6) indicates that some bridging flocculation occurs in the glass - MEOH suspensions.

The observed differences in the rheological and sedimentation behavior between glass - MEOH and glass - 3:1,MIBK/MEOH suspensions containing polyvinyl butyral resin can be attributed largely to differences in the adsorption behavior of the polymer. Since MEOH is an excellent solvent for the polyvinyl butyral resin, polymer adsorption onto the glass particles is expected to be relatively weak. Polymer attachment is expected to occur at relatively few surface sites, as polymer chains would "prefer" to extend far into the liquid phase in order to maximize contact with the solvent molecules. This conformation would promote bridging flocculation, as opposed to steric stabilization. In contrast, stronger adsorption of the polymer onto the glass particles would be expected in 3:1,MIBK/MEOH suspensions since MIBK is a poor solvent for the polyvinyl butyral resin. The polymer would be more firmly anchored onto the particle and less extended into the liquid phase. This conformation would be more likely to promote steric stabilization.

Experimentally determined polymer adsorption isotherms support the above hypothesis. Fig. 8 shows that the amount of polymer adsorption in 3:1,MIBK/MEOH suspensions is much greater than in MEOH suspensions.

SUMMARY

The suspension behavior of Al_2O_3 and silicate glass powders was investigated using two nonaqueous liquids - methanol (MEOH) and methyl isobutyl ketone (MIBK). The effects of liquid ratio (MIBK/MEOH) and polyvinyl butyral resin additions on suspension properties were determined. Rheological and sedimentation measurements showed that all suspensions without polymer were at least partially flocculated. However, increased flocculation was observed at high MIBK contents. Electrokinetic measurements revealed that this was due to decreased electrostatic repulsion between particles. Rheological and sedimentation measurements indicated that small polymer additions promote good dispersion (via steric stabiliza-zation) in glass - 3:1,MIBK/MEOH suspension. In contrast, bridging floccu-lation apparently occurs in glass - MEOH suspensions containing polymer. These differences in suspension behavior are associated with differences in the adsorption behavior of the polymer onto the glass particles.

REFERENCES

1. R.E. Mistler, D.J. Shanefield, and R.B. Runk in: Ceramic Processing Before Firing, G.Y. Onoda, Jr. and L.L. Hench, eds. (Wiley, New York 1978) pp. 411-448.
2. J.C. Williams in : Ceramic Fabrication Processes, Vol. 9, F.F.Y. Wang, ed. (Academic Press, New York 1976) pp. 173-198.
3. M.D. Sacks and C.S. Khadilkar, J. Am. Ceram. Soc., 66 (7) 488-494 (1983).
4. M.D. Sacks and M.I. Alam, "Suspension Behavior of a Silicate Glass," submitted to Am. Ceram. Soc.
5. J.Th.G. Overbeek in: Colloid Science, H.R. Kruyt, ed. (Elsevier, Amsterdam 1952) pp. 245-277.
6. J. Lyklema, Advan. Colloid Interface Sci., 2, 65-114 (1968).
7. J.A. Riddick and W.B. Bunger, Organic Solvents (Wiley, New York 1970).
8. J. Lyklema in: The Scientific Basis of Flocculation, Part I, (Nato Advanced Study Institute, Cambridge 1977).
9. J. Gregory, ibid.

MICROSTRUCTURAL TRANSFORMATIONS IN ALUMINA GELS

F. W. DYNYS, M. LJUNGBERG AND J. W. HALLORAN
Case Western Reserve University, Cleveland, Ohio 44106

ABSTRACT

Microstructural development in alumina gels is
dominated by a series of phase transformations between
the hydrous oxides to transition alumina phases, between
transition phases, and transformation to alpha-alumina.
These microstructural transformations are illustrated in
boehmite AlOOH and bayerite Al(OH)$_3$ gels.

INTRODUCTION

The gel process naturally produces an exceedingly fine and uniform microstructure in the xerogel precursor of crystalline ceramics. The fine size of
the microstructural features promise exceptional reactivity and sinterability.
However the course of evolution from the desiccated xerogel to the final dense
ceramic is not necessarily a uniform process of densification and grain growth.
Rather there are important examples in which the evolution is punctuated by
one or more phase transformations which can profoundly alter the microstructure. Indeed, microstructure development can be dominated by the processes
of phase transformations. This situation occurs whenever there is a microstructural change caused by dehydroxylation of a hydrous oxide xerogel, or
when the oxide xerogel is an amorphous or metastable crystalline phase. Many
xerogels are in amorphous or metastable forms [1], and so these transformations are of general importance. In this paper we will discuss the case of
alumina gels, in which the transformation behavior is especially rich. We
will consider several phase transformations in psuedoboehmite and bayerite
derived gels, with emphasis on microstructural alterations.

Experimental Proceedures

Psuedoboehmite gels (AlOOH) were prepared by hydrolysis of aluminum
sec-butoxide using proceedures similar to those of Yoldas [2]. Hydrolysis
was accomplished by mixing a 0.35 molar solution of the alkoxide in sec-
butanol with a 16.5 molar solution of water in sec-butanol, such that the
molar ratio of water to alkoxide was 100. Nitric acid, present at 10^{-2} molar
in the water solution, served to peptize the psuedoboehmite to form a clear
sol. The psuedoboehmite sol was placed in a drier at 85°C. where it formed a
translucent gel. Upon drying, the gel cracked into pieces about a centimeter
in size. The X-ray diffraction pattern was consistent with poorly crystalline
psuedoboehmite.

Some gels were aged at room temperature to crystallize bayerite [3].
After hydrolysis the sol was allowed to age at room temperature for six hours,
during which a gel formed which occupied 2/3 of the original sol volume.
Excess supernate was decanted and replaced with water coating 10^{-5} molar
nitric acid. This proceedure was repeated after 24 hours. After six days the
batch was a thixotropic opaque white gel. This gel was dried at 85°C to
produce 5 mm size fragments. X-ray diffraction showed well-crystallized
Al(OH)$_3$ as bayerite. Boehmite and the trihydrate polymorphs nostradite and
gibbsite were not detected.

Mat. Res. Soc. Symp. Proc. Vol. 32 (1984) Published by Elsevier Science Publishing Co., Inc.

No crystalline hydroxide phases could be detected in either gel after heating to 400°C. The X-ray diffraction patterns after the 400°C treatment agreed with poorly crystalline gamma-Al$_2$O$_3$ for the psuedoboehmite gels and eta-alumina for the bayerite gels, consistant with the literature [4]. Thermo-gravimetric analysis showed 5 wt.% residual volatiles at 400°C, and about 1 wt.% residual volatiles at 600°C for both gels.

The microstructure of the gels was observed using TEM. Preparation of the hydrated gels involved suspending crushed gel particles in acetone to disperse thin fragments in a copper grid. The heat-treated gels were mechanically polished to a thickness of 40 microns, followed by ion-thinning to produce electron-transparency.

Gel Microstructures - Psuedoboehmite Gels

The microstructure of the dried psuedoboehmite xerogel is shown in Figure 1. The gel has a distinct fibrillar texture with fibrils 1-2 nm in cross section. This texture is consistent with previous observations of pseudoboehmite formed from Al-sec butoxide [5]. A fibrillar structure is frequently observed in boehmite gels, although boehmite fibrils are considerably larger (5-10 nm) [6,7]. Note however that non-fibrillar particulate gels have also been reported for psuedoboehmite [8] and boehmite [7].

Annealing of the gel at 600°C resulted in the crystallization of gamma-Al$_2$O$_3$, as has been previously reported [7]. The gamma-alumina xerogel retains a somewhat fibrillar texture but is considerably coarser than its psuedoboehmite precursor. Figure 2 shows that the fibrils in this case clearly consists of aligned particulate features about 5-10 nm in diameter. The psuedo-boehmite to gamma-Al$_2$O$_3$ transformation is in this case clearly accompanied by some coarsening. This is in contrast to the behavior in massive (micron-sized) boehmite, in which the gamma-Al$_2$O$_3$ develops a finer microporous texture [9].

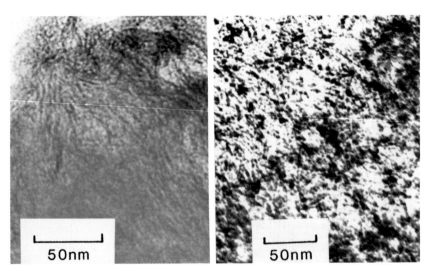

Figure 1: Psuedoboehmite gel dried at 85°C.

Figure 2: Gamma alumina xerogel after annealing at 600°C.

The gels annealed at 1000°C displayed X-ray diffraction patterns consistent with theta-Al_2O_3. The microstructure had been substantially coarsened, and the fibrillar texture had been obliterated. Figure 3 shows that the theta-Al_2O_3 xerogel consists of particulate features about 20 nm in diameter, with some particles as large as 50 nm. At higher magnifications it is apparent that the theta-Al_2O_3 particles are interconnected in a vermicular structure. Electron diffraction ring patterns demonstrated that the theta-Al_2O_3 gel was polycrystalline. In this study we have not addressed the behavior between 600 and 1000°C, so we can not determine to what extent of the coarsening was due to the transformation from gamma-Al_2O_3 and what fraction was due to subsequent sintering. It is possible that the theta-Al_2O_3 structure shown in Figure 3 is in fact the product of two transformations; gamma-Al_2O_3-to-delta-Al_2O_3, followed by delta-Al_2O_3-to-theta-Al_2O_3. However, while delta alumina is known to form from well crystallized boehmite [10], it has not been reported to form in poorly-crystallized psuedoboehmite [5], and so is not expected here.

The transformation to alpha-Al_2O_3 produces the most significant alteration in the gel microstructure. Alpha alumina formation is associated with substantial coarsening of the microstructure and the formation of large pores [11]. The coarsening is sufficient to arrest densification during sintering [7]. The alpha-alumina appeared in two distinct morphologies. Very slow transformation at relatively low temperatures (e.g., 50 days at 1000°C) produced what we term "discrete" alpha-alumina grains shown in Figure 4. These consist of dense crystals of alpha-alumina, 250-500 nm in cross-section, surrounded by the much finer theta-alumina matrix. At 1050°C or in H_2O-containing atmospheres the discrete grains grow more rapidly. As these dense grains consume the porous theta-alumina, they form voids which eventually become nearly the size of the alpha-alumina grains themselves. Voids as large as 500 nm have been observed in gels partially transformed at 1050°C, representing a more than ten-fold increase in pore size, see Figure 5.

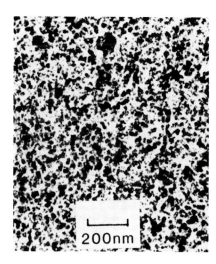

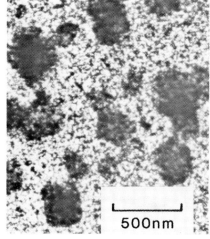

Figure 3: Theta alumina produced by annealing psuedoboehmite gel at 1000°C.

Figure 4: Alpha alumina in the "discrete" morphology growing in the theta-alumina at 1000°C.

In the temperature range 1150-1200°C the transformation occurs much more rapidly and produces a distinctly different morphology in which large porous colonies of alpha-alumina grow into the theta-alumina matrix. These colonies are as large as 1000 nm, and consist of vermicular shaped particles 75-100 nm in cross-section, with interpenetrating porosity of a similar size. An example is shown in Figure 6. Electron diffraction and dark-field imaging shows that the entire colony usually consists of one single crystal [7,11]. As the alpha-alumina grows into the porous alumina matrix, it develops an interconnected vermicular morphology which allows it to re-distribute porosity while maintaining contact with the surrounding matrix. This process is analogous to cellular precipitation [12]. The kinetics and mechanisms of the theta-to-alpha alumina transformation, including transitions between the discrete and vermicular morphologies, are discussed in detail elsewhere [13]. For the present purposes it is sufficient to note the microstructural alterations caused by this phase transformation.

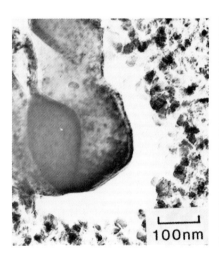
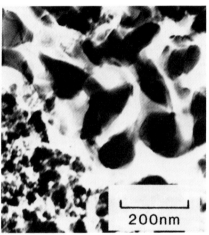

Figure 5: Voids formed at the interface between theta-alumina and alpha alumina in the "discrete" morphology 1050°C

Figure 6: Portion of a colony of alpha alumina in the vermicular morphology. Theta-alumina matrix at lower right. (1150°C).

Bayerite Gels

Aging the psuedoboehmite sol causes a recrystallization to form bayerite $Al(OH)_3$. This transformation apparently occurs via classical dissolution and reprecipitation [5]. Under the present conditions the bayerite formed as the large oblong blocky crystals. The bayerite particles appeared distinctly bimodal in size, with most particles having a small dimension of 100 nm, with about 10 vol% as large particles with small dimension around 500 nm. All particles were oblong with typical aspect ratios of 2-3.

The distribution of the larger particles were not random. Frequently clusters containing 10-20 large particles were observed, often with a preferred alignment reminiscent of tactoids in face-face flocculated clay.

Dehydroxylation at 400°C for 4 hours produced poorly crystalline eta-alumina, which was distinguishable from gamma-Al_2O_3 by its X-ray diffraction pattern [10]. Annealing at 1150°C for 2 hours in air produced a mixture of theta-alumina and alpha-Al_2O_3. These could be clearly distinguished in the microstructure.

The theta-Al_2O_3 preserved a topotaxial relationship with the original bayerite. After decomposition and transformation to theta-Al_2O_3, the relicts of the bayerite blocks were virtually unchanged in external shape, becoming very porous relicts of 10-20- nm theta-Al_2O_3 particles (Figure 7). Single crystal selected area diffraction patterns and dark field imaging demonstrated that the theta-alumina particles within relict blocks had a very close orientation. This demonstrates that the theta-Al_2O_3 preserved a topotactic relationship with the original bayerite.

Some of the appropriately oriented relict blocks showed distinct layers with a spacing of about 25 nm, perpendicular to the long side of the blocks. A good example is Figure 8. Selected area diffraction patterns, indexed as cubic, demonstrated that the layers are parallel to the (111) cubic planes. Bayerite is known to decompose via cleavage of basal planes [10]. This would create a very fine layered structure in the eta-alumina, about an order of magnitude finer than the layering observed here in theta-alumina. This coarsening due to the eta-to-theta alumina transformation is consistent with the surface area decrease observed by Aldcroft et al. [3].

Transformation to alpha-alumina occured via nucleation and growth of vermicular colonies of alpha-alumina. In spite of the dramatic differences in the microstructure of the bayerite-derived gel, the kinetics were quite similar to the psuedoboehmite gel [13]. The size (500-100 nm) and morphology of the alpha alumina was also similar to psuedoboehmite gels. Again the formation of alpha-alumina dramatically coarsened the gel microstructure.

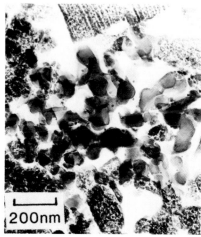

Figure 7: Theta-alumina from bayerite gel showing porous relicts of the original blocky crystal of bayerite.

Figure 8: Theta-alumina from bayerite gel being consumed by a colony of vermicular alpha-alumina. Note the layering in the bayerite relict at top center 1150°C

In partially-transformed gels it was apparent that the growing alpha grains directly consumed the blocky relicts, maintaining a sharp interface between the alpha grain and the porous theta-alumina. Figure 8 shows the advancing alpha colony in a partially transformed region of the bayerite-derived gel. The central bayerite relict in Figure 8 also illustrates the tendency for the alpha-alumina to grow more rapidly into the layers of theta-alumina within the bayerite relict. In some cases the alpha grain advanced rather differently in processes in which the new phase grew more rapidly around the surface of the bayerite relict, forming a shell of alpha-alumina.

The fully-transformed bayerite-derived gel had a microstructure which was virtually indistinguishable from the transformed psuedo-boehmite gel. The coarsening accompanying the formation of vermicular alpha-alumina obliterated the texture of the bayerite gel.

SUMMARY

In alumina gels the major microstructural activity is a result of a series of phase transformations. These convert the original microstructure from a tangle of 1 nm psuedoboehmite fibrils to micron-sized colonies of interconnected vermicular particles of alpha-alumina 100 nm in cross-section. This represents a coarsening of at least one hundred fold. Subsequent microstructure development upon further sintering of the alpha-alumina gel is a comparatively uneventful process of conventional densification and grain growth, which has been described by Badkar and Bailey [7]. Clearly in the alumina gel system, microstructural development is dominated by the factors which control these phase transformations. We suggest that effects which are similar, although perhaps less exaggerated, are to be found in other crystalline ceramic gel systems.

Acknowledgement

This work was funded by the National Science Foundation under Grant No. DMR-8103066. M. Ljungberg was an exchange student from the Technical Institute, Uppsala University, Sweden. F. W. Dynys is now with General Electric Co., Lighting Research, Cleveland, Ohio.

REFERENCES

1. S.A.Tiechner et al., Adv. Colloid Interface Sci., 5, 245 (1976).
2. B.E.Yoldas, J. American Ceramic Soc., 65, 307 (1982).
3. D.Aldcroft, G.C.Bye and C.A.Hughes, J. Appl. Chem., 19, 167 (1969).
4. D.Aldcroft et al., J. Appl. Chem., 18, 301 (1968).
5. D.Aldcroft and G.C.Bye, Proc. Brit. Ceramic Soc., No. 13, 125 (1969).
6. L.Moscou and G.S. van der Vlies, Kolloid Z. 163, 35 (1959).
7. P.A. Badkar and J.E.Bailey, J. Materials Sci., 11, 1794 (1976).
8. G.C.Bye and J.G.Robinson, Kolloid Z. 198, 53 (1964).
9. S.J.Wilson, J. Sol. State Chem., 30, 247 (1979).
10. B.C.Lippens and J.H.DeBoer, Acta Cryst., 17, 1312 (1964).
11. F.W.Dynys and J.W.Halloran in: Ultrastructure Processing of Ceramics, Glasses, and Composites, L.L.Hench and D.R.Ulrich, ed., (Wiley, N.Y. in press).
12. F.W.Dynys and J.W.Halloran, J. American Ceramic Soc., 65, 442 (1982).
13. F.W.Dynys, Ph.D. thesis, Case Western Reserve University (1984).

NOVEL MATERIALS THROUGH CHEMICAL SYNTHESIS

ORGANICALLY MODIFIED SILICATES BY THE SOL-GEL PROCESS

H.SCHMIDT
Fraunhofer-Institut für Silicatforschung, Neunerplatz 2
D-8700 Würzburg, F.R.G.

ABSTRACT

The introduction of organic groups into inorganic networks
by the sol-gel process opens the possibility for the preparation of new materials, and typical properties resulting
from inorganic as well as from organic components may be
combined. Some general aspects and different examples of
material developments are reviewed.

1. INTRODUCTION

Generally solid materials are prepared by the sol-gel process by hydrolysis and condensation of metal-organic network forming and network modifying monomers [1-10] . In order to get pure inorganic polymers it is necessary to remove every organic residue, as hydrolyzing groupings or organic solvents. Therefore generally only starting compounds with totally
hydrolyzable organic groups are used. In the case of silicon the ortho-
esters are used, since the organic residues are $\equiv$Si-O- bonded and the
$\equiv$Si-O-C$\equiv$ bond easily can be hydrolyzed. The use of unhydrolyzable $\equiv$Si-C$\equiv$
bonded ligands leads to organic modification of the $\equiv$Si-O-Si$\equiv$ network. If
-CH$_3$ or -C$_6$H$_5$ are used as modifying components, silicones are obtained
[11-14] . The introduction of organic groups may change the material properties strongly: Substitution of one $\equiv$Si-O- bridge (e.g. in fused silica) by
one $\equiv$Si-CH$_3$ group (monomethylpolysiloxane resin) increases the thermal
expansion coefficient from $0.5.10^{-6}$ to about $100.10^{-6}.K^{-1}$. The question
arises, whether it will be a general possibility to develop new materials
by introducing other organofunctional groupings into siliceous or homologous networks. It was shown by Gulledge and Andrianov that basically it
is possible to obtain solids by cocondensation of organo silicon with other
network forming components like Ti, Al, P and others [15, 16] . Organically
modified or additional network modified silicates are called heteropolysiloxanes in this paper.

The preparation of multicomponent systems is strongly affected by the
reactivity of the starting monomers and reaction conditions, since this may
influence the condensation, the way of crosslinking and the material properties [17-20, 1] . The strong influence of the type of catalyst (HCl or
NH$_3$) could be demonstrated in [21] .

In connection with questions of application, technical aspects of the
synthesis of materials and the manufacturing of products from these have to
be taken into account. Thus, sometimes a complete line, starting with the
synthesis of special monomers and ending up with the manufacturing procedure has to be built up, if new materials for practical use should be developed. In the following chapters some examples will be given to demonstrate
how heteropolysiloxanes may be used to fulfill demands for very special materials.

Mat. Res. Soc. Symp. Proc. Vol. 32 (1984) Published by Elsevier Science Publishing Co., Inc.

2. RESULTS

2.1. General aspects of the introduction of organic groups

The $\equiv$Si-C- bond opens the possibility to introduce any organic group Y into an inorganic network by the sol-gel process, if suitable compounds of the type $(YR')_n Si(OR)_{4-n}$ (n = 1-3; e.g. R = alkyl; R' = alkylene) are available or preparable. If Y is a polymerizable ligand, a second type of crosslinking becomes possible: polymerisation or polyaddition. This opens an almost infinite number of variations, only from the point of view of organic modification.

As known from the preparation of glasses by the sol-gel process, if one wants to receive monolithic materials, shrinking is a serious problem. In principle, these problems should exist with heteropolysiloxanes, too. But the introduction of an organic group as network modifier could help to overcome these problems by generating a more flexible network. How far such effects can be used for suitable material preparation must be subject to investigation. As a function of the degree of condensation brittleness should be able to be varied in a wide range. If in analogy to inorganic gels, with high degrees of condensation porous intermediates appear, they might be used as materials, too [22, 23]. On the other hand, low degrees of polymerization, in connection with organic residues should result in good solubilities in organic solvents and, if the shrinking problem is overcome, useful materials for coatings should be preparable.

2.2. Examples of material development

2.2.1. Special coatings

The covalent coupling of biochemical compounds is of interest from different reasons: Immobilization of enzymes is widely used in the field of bioengineering. The immobilization of antibodies is a powerful tool in medical diagnosis. Immobilized antibodies under certain circumstances are able to bind specificly the antigen against which they are generated. If hormones to be analyzed are used as antigens, they can be separated from other serum proteins by the antibody binding and simultaneously be determined, as described in [24-26]. The important step of covalent antibody immobilization requires proper surfaces. Reactive surfaces could be prepared by cocondensation of $R_2Si(OC_2H_5)_2$, $Si(OC_2H_5)_4$ and $YR'Si(OC_2H_5)_3$. Y is an organofunctional group. Therefore $-(CH_2)_3NH_2$, $-(CH_2)_3NHCO\text{-}\langle O \rangle\text{-}NH_2$, $-(CH_2)_3S(CH_2)_2CHO$, $-(CH_2)_3S(CH)_2CH_2OH$ and $-(CH_2)_3SH$ were used. R is $-CH_3$ or $-C_6H_5$, R' = alkylene. The reaction was carried out according to eq.(1).

$$ m\ R_2Si(OC_2H_5)_2 + n\ Si(OC_2H_5)_4 + p\ YR'Si(OC_2H_5)_3 \xrightarrow[-ROH]{+H_2O;\ H^+} \quad (1) $$

$$ \underset{(I)}{\qquad} \underset{(II)}{\qquad} \underset{(III)}{\qquad} $$

$$ \left[\begin{array}{c} R \\ | \\ Si-O \\ | \\ R \end{array} \right]_m \left[\begin{array}{c} O \\ | \\ Si \\ | \\ O \end{array} \right]_n \left[\begin{array}{c} R'Y \\ | \\ O-Si- \\ | \\ O_{1/2} \end{array} \right]_p $$

with R = $-CH_3$: few water
with R = $-C_6H_5$: excess of water
has to be used

(IV)

(IV) is soluble in organic solvents and was used as a coating material for the inner walls of small glass tubes. These tubes then were coated with antibodies and the immunoassay carried out. (I) is necessary to get homogeneous dense layers, (II) in order to get insoluble solids after coating and (III) provides the reactive surface groups. The presence of Y at the surface was proved by surface analysis. In the case of R = -CH$_3$ reactive, non fully condensed prepolymers had to be formed by using understoichiometric addition of water. Excess water led to insoluble materials. After coating hot water treatment accomplished condensation; -C$_6$H$_5$ groups led to water resistant materials soluble in most organic solvents. Best results for the bonding of T3 (triiodine thyronine) antibodies were achieved by use of -CHO and -CH$_2$OH group containing coatings.

Since the inorganic sol-gel process requires high curing temperatures for forming dense films, this procedure cannot be applied for building up layers onto organic polymers in order to improve mechanical surface properties. It is an important question, how far low temperature curing coatings with improved properties like scratch resistance may be prepared. In [27] it could be shown that the introduction of titania reduces the shrinkage problem of a SiO$_2$/CH$_2$OHCHOHCH$_2$O(CH$_2$)$_3$SiO$_{3/2}$ cocondensate substantially. Understoichiometric (with respect to hydrolysis) addition of water leads to viscous liquids which can be used as coating materials and cured at 90-120°C to clear films showing a good scratch resistance in the Erichson test (the surface is scratched by a Vickers diamond with different loads).

Curing temperature and time are very important with respect to receiving good scratch resistance. Thus, it was possible to develop coatings which did not show any scratches with loadings up to 200 g when applied as thick layers (>50 µm). The dependence of scratch resistance as a function of curing temperature is shown in figure 1. For example, organic polymers like acryl glasses or polycarbonates are scratched by loadings >15 g. Lowering the content of inorganic compounds leads to softer, less scratch resistant

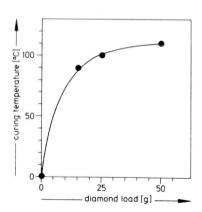

FIG.1. Dependence of scratch resistance on curing temperature (Erichsen scratch test); the curve represents the maximum load without causing scratches; curing time: 30 min; material composition: SiO$_2$: 30 mole%; TiO$_2$: 20 mole%; ⌐OCH$_2$CHCH$_2$O(CH$_2$)$_3$SiO$_{3/2}$: 50 mole%; thickness of the coating: 30 µm; coating procedure: dip coating; substrate: different organic polymers (e.g. polycarbonate, PMMA, CR 39)

materials. It is not clear how structural factors, especially the incorporation of titania, may be related to mechanical properties. Since the bulk
material is flexible (modulus of elasticity about 30.10^2 MN.m^{-2}), it seems
reasonable that this type of scratch resistance may be a compromise between
hardness and flexibility. The best results were obtained with epoxide ligands so far, but the role of the epoxide grouping is not clear, too. If
the epoxide is cleaved by addition of water, wettable materials with an
antifogging effect are achievable [28]. Measurements show that the contact
angle in a simple system decreases to about 20°. However, these systems are
not optimized yet.

As pointed out above in connection with the coatings for immobilization
of antibodies, the use of diphenylsiloxanes as network formers leads to
materials soluble in organic solvents. This is due to the steric hindrance
of the phenyl groups lowering the degree of polymerization and leading to
reduced chain lengths and, as a consequence, to soluble polymers [14].
Moreover, it could be proved that by use of phenyl groups, $\equiv$SiOH containing
thermoplastic silicates may be developed [29]. This type of thermoplastic
polymer forms moisture resistant seals to glass surfaces, if they are
sealed onto glass surfaces by a hot melt process. Therefore, aluminum foils
are coated with the polymer (thickness about 10 μm). In order to get proper
seals the OH group content of the material has to be adjusted by thermal
treatment (see fig.2). The behaviour is interpreted by the formation of
$\equiv$Si-O-Si$\equiv$ bridges with glass surface silanol groups.

Based on these results a technical procedure for sealing glass containers is under development.

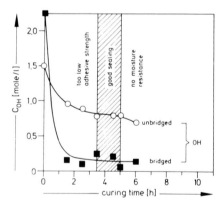

FIG.2. Content of OH groups as a function of tempering time at 150°C.
Composition: 65 mole% $(C_6H_5)_2SiO$; 35 mole% $(CH_3)(CH_2=CH)SiO$; OH groups
monitored by quantitative IR spectroscopy

For other purposes, e.g. filtration or membrane processes, porous films should be of interest. In order to get porous films a high degree of cross-linking is necessary to prevent collapsing of pores formed during condensation. Therefore the content of organic groups is limited, depending on type of group, reaction condition and type of other components. Low contents of organic groups increase the shrinking problem. Experiments using common membrane preparation techniques like casting showed that it is very difficult to get at least small pieces of thin films without cracks. Better results were obtained by using procedures like interfacial polymerization [30-32]. It is possible to receive membranes by using two immiscible liquids, one containing water and the catalyst and the other containing the silane. These materials are especially resistant against organic solvents and stable against temperatures up to 200°C depending on the organic functions. Another type of reaction was applied for the coating of porous supports, like glass fiber fleeces, with a microporous membrane using the components $(CH_3)_2SiO$ and SiO_2. Therefore the HCl containing fleece was exposed to a $(CH_3)_2Si(OC_2H_5)_2$ and $Si(OC_2H_5)_4$ containing vapor phase and the membrane reacted directly onto the fleece surface. By this it was possible to improve the filtration behaviour of the filter substantially [32].

2.2. Bulk materials

2.2.1. Dense products

As indicated above, the introduction of organic function into inorganic network may lead to more flexible products and helps to overcome the difficulties of shrinking. Furthermore, this may lead to material properties more related to organic polymers. In the sol-gel process the loss of water during condensation and porous intermediates cause shrinking during the densification process. By this high temperatures and pressures have to be applied to obtain monolithic products. Theoretically low temperature moldable products should be achievable by additional introducing of a second principle of crosslinking which allows curing without any loss of components. Thus, a multicomponent system, where the siliceous crosslinking by condensation does not lead to a brittle solid because of the presence of non-siliceous monomers, is able to be cured e.g. by polymerization according to (2), as shown in [27].

$$\equiv Si-\wedge\!\!\wedge\!-CH=CH_2$$
$$+\ O$$
$$CH_2=C-\overset{\parallel}{C}-O-CH_3 \quad \xrightarrow{\text{Cat.}}$$
$$\underset{CH_3}{|}$$
$$+$$
$$CH_2=CH-\wedge\!\!\wedge\!-Si\equiv$$

$$\equiv Si-\wedge\!\!\wedge\!-CH-$$
$$\underset{CH_2}{|}$$
$$\underset{CH_2}{|}$$
$$H_3C-C-\overset{|}{C}-O-CH_3 \qquad (2)$$
$$\underset{CH_2}{|}\ O$$
$$-HC-\wedge\!\!\wedge\!-Si\equiv$$

This principle can be realized with other polymerizable substituents, too. Thus, moldable condensates can be cured by using polymerization catalysts (heat or UV initiated). It could be experimentally proved that this method leads to solid products almost without shrinking. In addition to this increased tensile strength in comparison with the unpolymerized sample is observed (table 1).

TABLE I. Effect of polymerization crosslinking on some material properties

	Composition I [*] (non-polymerized)	Composition II [**] (polymerized)
Tensile strength $(MN.m^{-2})$;	2.15	5.15
Modulus of elasticity: $10^2 (MN.m^{-2})$	29	34
Refractive index: n_D^{20}	1.525	1.503
Mohs' hardness	3-4	3

[*] I: 20 mole% $Ti(OR)_4$ (a) and 80 mole% $(CH_3O)_3Si(CH_2)_3OCH_2\overline{CHCH_2O}$ (b);
[**] II: 5 mole% (a); 60 mole% (b); 5 mole% $(CH_3O)_3Si(CH_2)_3OOCC(CH_3)=CH_2$;
30 mole% $CH_3OOCC(CH_3)=CH_2$

2.2.2. Porous materials

Porous materials play an important role in different fields and are used for very different purposes: As insulating materials, adsorbents, carriers for catalysts and others. There are many ways to generate porosity, e.g. foaming, sintering of powders, precipitation of gels from solution. The gel formation by the sol-gel process in many cases leads to porous intermediates. Porous intermediates may be of interest, if they can be taylored for special applications. Depending on the application special demands have to be fulfilled, if adsorbents are considered: The capacity with respect to the adsorbate has to be high and the interaction strong enough, the adsorption or desorption kinetics should be quick. Organically modified gels

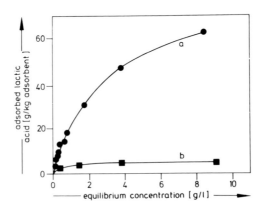

FIG.3. Adsorption isotherms of aqueous solutions of lactic acid. a: adsorbent from $Si(OC_2H_5)_4$ and $(C_2H_5O)_3Si(CH_2)_3NH_2$ with 8.10^{-5} mole $-NH_2/g$; BET surface = $350 \, m^2/g$. b: silica gel; BET surface = $500 \, m^2/g$.

allow to introduce special adsorbent-adsorbate interaction mechanisms by a simple one step reaction from the monomers to the solids. As described in [30] , the content of organic groups must not exceed certain limits. Pore size and specific surface area can be varied by reaction conditions, especially by the choice type and concentration of catalysts, e.g. HCl or NH_3 [33] . Thus, special adsorbents for adsorption of lactic acid show the effect of introducing amino groups in comparison to pure silica gel (fig.3): The results show that it is possible to obtain good adsorption of difficult adsorptives by adapting the system adsorbent/adsorbate to the special problem.

Other investigations showed that by optimizing pore size distribution and number and type of organofunctional groups, carrier for enzymes catalyzed reactions could be optimized. Thus, it could be demonstrated for different enzymes the existence of an optimal surface density of covalent coupling groups.

Microstructure does not only influence the adsorptive behaviour of porous materials, but also the mechanical properties, e.g. abrasion properties. Materials of high porosity generally are not good abrasives in the sense of a long service life. But for special applications, e.g. for grinding the human skin, soft abrasives are necessary. Therefore, especially if a well defined abrasion behaviour is required, it seems reasonable to synthesize such materials from monomers which can generate different properties in the polymer. Thus, $(CH_3)_2Si(OC_2H_5)_2$ was used as starting material for the "soft" and $Si(OC_2H_5)_4$ for the "hard" component.

By optimizing composition and reaction conditions it was possible to develop an abrasive for an acne preparation which prevents hurting of the skin reliably [21,33,34] . Figure 4 shows the abrasion behaviour as a function of composition, as it was received by a specially developed, skin adapted abrasion test.

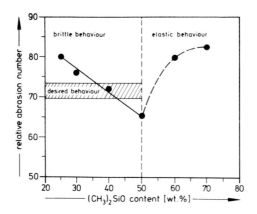

FIG.4. Abrasion test of porous materials of the composition $(CH_3)_2SiO:SiO_2$. Starting grain size: 0.3-0.4 mm.

3. CONCLUSIONS

The introduction of organic functionalities into inorganic networks by the sol-gel method seems to be a useful tool in order to prepare new materials. Examples show that materials with interesting properties for practical application can be prepared. One prerequirement for the preparation of taylormade materials is the knowledge of the possible influences of composition and reaction conditions on the material properties. A very helpful mean would be the knowledge on structural details of the new polymers. Inspite of these limitations the construction principles of these materials open almost unlimited possibilities of synthesizing new materials for new applications.

4. ACKNOWLEDGEMENT

The author appreciates gratefully the helpful discussions with Prof.Dr. H.Scholze in connection to the different investigations. Furthermore, he wants to thank his coworkers, Dr.H.Böttner, Dr.A.Kaiser, Dr.H.Patzelt Dr.G.Philipp, and Dipl.-Min.G.Tünker, who carried out the experimental work. The author thanks the Bundesminister für Forschung und Technologie der Bundesrepublik Deutschland, the Deutsche Forschungsgemeinschaft and the many industrial plants for the financial support.

5. REFERENCES

[1] H.Dislich, Angew.Chem. 83, 428-435 (1971).

[2] R.Roy, J.Amer.Cer.Soc. 52, 344 (1969).

[3] K.S.Mazdiyasni, R.T.Dolloff, and J.S.Smith,II, J.Amer.Cer.Soc. 52, 523-526 (1969).

[4] S.P.Mukherjee and J.Zarzycki, J.Mater.Sci. 11, 341-355 (1976).

[5] B.E.Yoldas, J.Mater.Sci. 12, 1203-1208 (1977).

[6] K.Kamiya and S.Sakka, Res.Rep.Fac.Eng.Mie Univ. 2, 87-104 (1977).

[7] M.Nogami and Y.Moriya, J.Non-Cryst.Solids 37, 191-201 (1980).

[8] G.Carturan, V.Gottardi, and M.Graziani, J.Non-Cryst.Solids 29, 41-48 (1978).

[9] S.Sakka and K.Kamiya, J.Non-Cryst.Solids 43, 403-421 (1980).

[10] J.D.Mackenzie, J.Non-Cryst.Solids 48, 1-10 (1981).

[11] A.Ladenburg, Ann. 173, 143-166 (1874).

[12] E.G.Rochow in: An Introduction to the Chemistry of the Silicones. J.Wiley and Sons, New York 1951.

[13] K.A.Andrianov in: Organic Silicon Compounds (State Scientific Publishing House for Chemical Literature, Moscow 1955).

[14] W.Noll in: Chemie und Technologie der Silicone. 2.Auflage (Verlag Chemie, Weinheim 1968).

[15] H.C.Gulledge, US.Pat. 2,512,058, 20.6.1950.

[16] K.A.Andrianov and A.A.Zhdanov, J.Polym.Sci. XXX, 513-524 (1958).

[17] E.Åkerman, Acta Chem.Scand. 11, 298-305 (1957).

[18] R.Aelion, A.Loebel, and F.Eirich, J.Amer.Chem.Soc. 72, 5705-5712 (1950).

[19] H.Schmidt, H.Scholze, and A.Kaiser, J.Non-Cryst.Solids 48, 65-77 (1981).

[20] L.H.Sommer, C.F.Frye, M.C.Muslof, G.A.Parker, P.G.Rodewald, K.W. Michael, Y.Okaya, and P.Pepinski, J.Amer.Chem.Soc. 83, 2210-2212 (1961).

[21] H.Schmidt, H. Scholze, and A. Kaiser, J. Non-Cryst. Solids 63, 1-11 (1984)

[22] G.Carturan, G.Facchin, V.Gottardi, M.Guglielmi, and G.Navazio, J.Non-Cryst.Solids 48, 219-226 (1982).

[23] G.Carturan, J.Non-Cryst.Solids (1984) in press.

[24] H.Schmidt and O.v.Stetten, DP 27 58 507, 28.12.1977.

[25] H.Schmidt and H.Scholze, Ger.Offen. 2,758,414, 12.7.1979.

[26] H.Schmidt, O.v.Stetten, G.Kellermann, H.Patzelt, and W.Naegele, Proc. Radioimmunoassay and Related Procedures in Medicine 1982, 111-121, Vienna 1982.

[27] H.Schmidt and G.Philipp, J.Non-Cryst.Solids 63, 283-292 (1984).

[28] H.Schmidt, G.Philipp, and C.F.Kreiner, Dt.Offen. 31 43 820, 11.5.1983.

[29] H.Schmidt, G.Tünker, and H.Scholze, DP 30 11 761, 20.3.1980.

[30] H.Schmidt and H.Scholze, DP 27 58 415, 12.7.1979.

[31] H.Scholze, H.Schmidt, and H.Böttner, Ger.Offen. 29 25 969, 29.1.1981.

[32] A.Kaiser and H.Schmidt, J.Non-Cryst.Solids 63, 261-271 (1984).

[33] H.Schmidt, A.Kaiser, H.Patzelt, and H.Scholze, J.Physique 43, 275-278 (1982).

[34] H.E.Kompa, H.Franz, K.D.Wiedey, H.Schmidt, A.Kaiser, and H.Patzelt, Ärztliche Kosmetologie 13, 193-200 (1983).

PROCESSING OF ADVANCED CERAMIC COMPOSITES

Roy W. Rice
Naval Research Laboratory, Code 6360
Washington, D.C. 20375

ABSTRACT

The field of ceramic composites is reviewed as an outstanding example of where more sophisticated chemistry can aid ceramic processing. Processes reviewed, mostly from work of the author and colleagues, include use of solution, polymer and high temperature chemistry, e.g. in sol-gel, polymer pyrolysis, solid state, and chemical vapor deposition processes. Future needs and opportunities for chemistry in ceramic processing are also discussed, e.g. coating part or all the matrix or its precursor onto the particulates or fibers of the resultant composite.

INTRODUCTION

Ceramic composites have been attracting increasing interest and study; particularly composites for improved mechanical behavior, but the opportunities for electrical and electronic ceramic composites are at least as great.[1] While composites used for their mechanical performance may pose some new design challenges, the substantial opportunities they offer for improved mechanical reliability is the driving force for their development, i.e. they have resulted in mechanical properties that were felt unlikely, or undreamed of, 5 to 15 years ago. Ceramic composites are an outstanding example of both the need for, and the results achievable from, applying more sophisticated chemistry to their processing.

This paper surveys a number of ceramic composite developments illustrating a variety of contributions of chemistry to their processing along with some future processing needs and opportunities. Examples are drawn mostly from the experience and current work of the author and his colleagues, with a few key examples drawn from other investigators. While the focus is on composites for mechanical properties, most of the processing approaches are also pertinent to electrical and electronic composites.

II. SURVEY OF CERAMIC COMPOSITE PROCESSING AND RESULTS

A. Particulate Composites

Our first major application of chemistry to composite processing was the processing of Al_2O_3-ZrO_2 composites by mixing an Al_2O_3 and a ZrO_2 sol then gelling, crushing, calcining, and subsequently hot pressing the resultant powder. As discussed elsewhere this sol-gel approach resulted in a far more homogeneous composite than could be achieved by conventional mixing and hot pressing of Al_2O_3 and ZrO_2 powders (Fig. 1), and was apprently the first demonstration of fracture strengths which increased along with fracture toughness in this system.[2,3] Such sol-gel processing is only one example of a broad spectrum of applications of lower temperature, more traditional aqueous (or non-aqueous) chemistry to ceramics.

This potential for lower temperature chemistry should not obscure the equally important opportunity for higher temperature chemistry applicable to many composite needs, as illustrated by BN particulate composites.

Originally composites using Si_3N_4, MgO and Al_2O_3,[4,5] with later development of mullite[5] and Si_3N_4[6], matrices were made by conventional mixing
and hot pressing of matrix and BN powders. However, besides the issue of
homogeneity and especially the spatial distribution and the size of the BN
particles that are endemic to powder processing, there were practical
issues of the cost of BN powder and the densification and especially shape-
ability of components in processing (sintering does not appear feasible).
A basic chemical approach that shows excellent promise of being a solution
of all of these problems for several matrices is to generate the BN insitu
by the following reactions:[7]

$$2AlN + B_2O_3 \rightarrow Al_2O_3 + 2BN \qquad (1)$$

$$Si_3N_4 + 2B_2O_3 \rightarrow 3SiO_2 + 4BN \qquad (2)$$

$$18AlN + 2Si_3N_4 + 11B_2O_3 \rightarrow 3\ [3Al_2O_3 \cdot 2SiO_2] + 18BN \qquad (3)$$

Such insitu formation of BN uses less costly raw materials, thus addressing
the practical issue of the cost of BN, as well as its spatial distribution.
Further, the transient B_2O_3 based liquid phases appear to provide excellent
promise for practical and versatile densification and shaping which can be
followed by high temperature treatment for completion of the solid state
(BN forming) reaction (Fig. 2).

Even higher temperature chemistry is clearly of importance as demonstrated by the development of precipitation toughened partially stabilized
ZrO_2 (PSZ materials). There is the development pioneered by Garvie and
colleagues of sintering, then high temperature solution heat treating,
followed by nucleation and growth of precipitates at lower temperatures.[8]
More extreme temperatures are involved in melt processing we have successfully used (with CERES Corp.) in achieving spectacular properties of PSZ
single crystals.[9,10] This, in turn, has lead to investigation of possibilities of melt processing of polycrystalline PSZ and related materials
where chemistry is used to control grain and/or precipitate sizes.

Another potentially important opportunity that we are investigating
for processing ceramic particulate composites is self-propagating synthesis
(SPS), some times also referred to as self-propagating high temperatures
synthesis. This basically consists of transfer reactions, wherein an anion
associated with one metal is transferred to another metal illustrated by
the typical thermite reaction:

$$2Al + Fe_2O_3 \rightarrow Al_2O_3 + 2Fe, \qquad (4)$$

or direct elemental reactions illustrated by the following reaction:

$$Ti + C \rightarrow TiC, \qquad (5)$$

or combinations of these. All reactions must be sufficiently exothermic to
make them self propagating, i.e. so initiating the reaction in one part of
a compact of the reactants, causes it to propagate across the compact (Fig.
3). A key characteristic of interest here is the fact that these reactions
provide very high, but very transient, temperatures that may often give
non-equilibrium phase compositions and microstructures. Such processing
encompasses either high temperature solid state reactions, liquid phases,
or both. Three reactions that can lead to possible interesting composites
are:

$$4Al + 3TiO_2 + 3C \rightarrow 2Al_2O_3 + 3TiC \qquad (6)$$

$$10Al + 3TiO_2 + 3B_2O_3 \rightarrow 5Al_2O_3 + 3TiB_2 \qquad (7)$$

$$2Ti + 2B + C \rightarrow TiB_2 + TiC \; (+ B_4 C \; ?) \qquad (8)$$

Reaction[8] is a good example of the additional opportunities of chemically controlling the microstructure of reactant compacts to control the amount and spacial distribution of the product phases (TiB_2, TiC and possibly B_4C in this case) beyond effects of controlling the reactant (Ti,B and C) particle sizes. Thus, for example, mixtures of Ti with B coated C particles or of Ti with C coated B particles should each give different mixtures of TiB_2 and TiC in the product than simply mixing separate Ti,B and C particles.

Finally, another area of chemistry applied to processing of ceramic composites is that of chemical vapor deposition (CVD). This has been of considerable interest to this author for some time but opportunities to actively pursue it have been limited. However, the work of Hirai and Hayashi[11] on the Si_3N_4-TiN system is illustrative of what might be accomplished by this approach in terms of microstructures. The challenge here will be to identify successful composition leading to not only interesting microstructure but also promising properties.

B. Fiber Composites

Ceramic fiber composites using SiO_2 based glass matrices (which might be subsequently crystallized) with graphite fibers showed spectacular improvements in fracture toughness. However, this opportunity attracted limited attention until similar results were demonstrated with SiC based fibers[12] that removed the severe oxidation limitations associated with graphite fibers. Both of these systems were processed by drawing fiber tows (bundles) through a slurry or, more preferably a stable colloidal suspension, i.e. slip, of the ceramic powder matrix with appropriate stabilizers, binders, etc. and then laying up a preform e.g. by filament winding (Fig. 4). Such slip preparation is a direct extension of much standard ceramic powder processing, but does call upon the use of chemistry as an aid in obtaining suspensions with the appropriate particle sizes, stability and wetting as well as possibly subsequent drying and handling characteristics. This technique has been extended by this author and colleagues to several more refractory, generally crystalline, matrices such as mullite, ZrO_2, and MgO. Preliminary investigations have also been made using this technique for fiber composites with reaction sintered silicon nitride (RSSN) matrices.

The above slurry or slip technique has indeed proved itself of considerable versatility, but is generally restricted to those situations for which filament winding operations are essentially applicable. This means not only shape limitations but also generaly a limitation to continuous fibers, since such slurries and slips are often not sufficiently fluid to be effectively and uniformly infiltrated into fiber preforms. Thus, one of the alternate means investigated by the author and his colleagues has been that of sol-gel processing, wherein one indeed has considerable potential versatility to infiltrate various fiber preforms allowing use of a wider range of shapes and fiber types, i.e. use of chopped fibers and whiskers. Use of SiO_2 sols and ZrO_2 sols to make ceramic fiber composites has indeed been shown to be successful (Fig. 4B), resulting in quite homogeneous fiber matrix distribution. However, like all but the RSSN matrix from the slurry-slip technique, such sol-gel processing also typically requires hot pressing (or HIPing in a can) as a means of final consolidation, hence is limited in its potential for low cost.

Polymer pyrolysis has previously been indicated as a very important opportunity to make ceramic fiber composites in a versatile and efficient fashion.[13] Results in both this author's laboratory and at Dow Corning indeed show that there is very substantial promise for this approach which avoids hot pressing. J. Jamet, in collaboration with this author and colleagues has shown that SiC or Al_2O_3 based fiber tows with a SiC producing polymer plus a ceramic filler can indeed lead to very promising composites. Here the approach has been to draw fiber tows through the polymer plus particulate filler, lay them up (e.g. as in Fig. 4A) mold them as a typical polymeric matrix composite, then pyrolyze this insitu to an all ceramic composite with good strength and toughness. Chi et al (at Dow Corning) have shown comparable strengths (~ 400 GPa) and toughnesses by similar processing, though they have also investigated multiple polymer impregnation of the initial composite in analogy with much carbon-carbon composite processing. The relatively low temperatures (e.g. 1000°C) of such pyrolysis processing is an important advantage, e.g. allowing use of Al_2O_3 fibers not amenable to the high temperature of hot pressing. While the absence of hot pressing in the above polymer processing is of considerable importance for obtaining lower costs and versatile shapes, this is presently bought at the cost of lower density which limits strength and may also provide environmental problems for some composites. Initial experiments by the author, colleagues, and Chi indicate that composites made by polymer pyrolysis cannot be significantly densified by subsequent hot pressing.

Another important chemical approach to processing ceramic fiber composites is chemical vapor deposition (CVD). Besides obtaining adequate infiltration with the high volume fractions of fibers typically desired, e.g. 30 to 60%, an important challenge is to carry out the deposition at temperatures where the aggressive gas species often associated with CVD will not seriously degrade the fibers, e.g. as shown by preliminary experiments the author (with Deposits and Composites Inc). Attempts to deposit a SiC matrix at ~ 1300°C resulted in either total loss or gross degradation of the SiC based fibers depending in part upon both process conditions as well as "vintage" and hence quality of the SiC fibers. That these problems can indeed be solved is clearly demonstrated by French investigators at SEP, who generated by CVD a SiC matrix around SiC based fibers to give a composite with strengths over 350 GPa.

It is generally agreed that the toughness of ceramic fiber composites results from limited fiber-matrix bonding allowing extensive fiber-pull out, which clearly suggests the use of fiber coatings to control bonding to the matrix. The author and colleagues have recently corroborated this. For example, SiC based fibers with ~ 0.1 μm thick of a coating material selected to limit bonding to a SiO_2 matrix gave a 4 and ~ 100 fold increases respectively in strength and toughness (as measured by area under the load-deflection curve), over uncoated fibers.

III. FUTURE DIRECTIONS FOR CERAMIC COMPOSITES PROCESSING

The results of ceramic composites research and development today have indeed been significant, if not in fact, revolutionary. The result has been to not only open up broad new areas for scientific research and technological development of composites themselves and to significantly aid the growing interest in greatly expanding high technology uses of ceramics, but to also provide both a stimulus and opportunity for much more imaginative processing. Many improvements will result from more conventional powder based ceramic processing with most of these improvements coming from the application of improved chemistry to the preparation and the processing of such powders. However, the focus will instead be on a variety of less conventional processing techniques since these represent a very significant opportunity, but are often under-represented in terms of both their support and investigation, in part because they require more chemistry than is common to much existing ceramic education and practice.

A very generic way of producing highly uniform and controlled particulate composites that is particularly illustrative of the impact and
opportunity of chemistry on ceramic processing is coating particles of the
dispersed phase (or its precursor) with the matrix (or its precursor) and
then subsequently consolidating such coated particles to form the composite
(Fig. 5). Clearly such coating is a basic way to meet the important composite requirement of avoiding dispersed particle agglomerates and significant
matrix regions having few or no dispersed particles.[4] Obviously, such
coating has to be consistent with subsequent composite processing, e.g.
large coated particles may be difficult to densify. In such (as well as
other) cases coating the particles only to the extent to determine a
minimum particle-particle spacing, then introducing added matrix material
between the coated particles may lead to somewhat more variable particle-
particle spacing (bounded on the lower end by the coating thickness), but
should significantly aid consolidation.

Another generic, method of particulate composite processing is developing the dispersed particles by solid state reactions and precipitation.
The intermediate temperature processes such as that illustrated for
mullite-BN system should have many other analogs. Correspondently the idea
of using precipitation from high temperature solid solutions is of considerable potential for ceramics, and indicates a number of possibilities of
melt processing of ceramics. These include using high temperature chemistry and handling of the solidification process to directly develop fine
microstructures from the melt (also followed by possible post processing),
or by forming molten droplets for subsequent consolidation and heat treatment. The SPS process may also offer a number of possible extensions of the
above processes.

Finally, I feel the area of chemical vapor deposition, deserves substantially more attention because of the possible opportunity to homogeneously mix different phases, that may often be difficult or impossible
to mix by other means. A particular, but by no means the only, possibility
is to form a fully dense two-, or multi-, phase body at temperatures well
below those that would normally be required for consolidation of mixtures
of the powders of the different constitutents. Thus, one can envision
using CVD to make composites that would not retain the necessary particulate phase identity at the high temperatures of more conventional powder
based processing.

Ceramic composites utilizing continuous fibers offer little or no
opportunity for processing via sintering. Thus such composites must either
be processed by powder based methods using hot pressing, or possibly HIP-
ing, or by non-traditional methods. Because of the size and especially
shape limitation of hot pressing, one indeed needs to look more to the
possibility of either being able to HIP the material or to develop matrices
by means that do not require hot pressing. One of several possibilities to
be investigated could be the use of sol-gel, polymer pyrolsis, or other
chemical techniques to be discussed below to seal the surfaces of composites so they can be HIPed without cans. An important adjunct of such processing would be coating fibers analogous to that discussed above for
particulate composites, i.e. coating fibers with all, or part, of matrix,
or its precursor. (This may be in addition to fiber coating for other purposes such as controlling fiber-matrix bonding). Fibers coated with the
matrix or its precursor may also be amenable to other chemical methods of
processing, with RSSN as a matrix being a particular example. Again it
should be noted that the coating on the fiber need not be the entire source
of the matrix but, in fact, could often provide just enough of the matrix
to give a desired minimum of interfiber spacing.

Other chemical techniques of preparing matrices for ceramic fiber composites are of significant interest besides the RSSN process. An important example is the formation of AlPO$_4$, e.g. entirely from phosphoric acid and aluminum hydroxide solutions or by phosphate bonding Al$_2$O$_3$. Such phosphate processing is of course applicable to other materials and represents only one example of a variety of possible applications of ceramic cement and castable technology to forming a variety of ceramic composites. Again such composites will typically be porous so that there may not be as high a performamce as possible; however, again the opportunity for futher densifying these e.g. but sealing and HIPing etc. or multiply impregnations warrants investigation.

The polymer pyrolsis process has already shown excellent promise for composites, with extensive opportunities for further exploitation. One important opportunity is to investigate a much broader range of polymers. Another is to pursue the important observations of Jamet that particulate fillers used to control shrinkage in the polymer can also improve the ceramic yield from the polymers and that pyrolysis under pressure may be very beneficial. Finally, there should be many more important opportunities to further apply chemical vapor deposition to the forming of ceramic fiber composites, not only in generating the matrices, but also as a means of sealing composites for HIPing.

The area of short fiber composites, i.e. using chopped fibers or whiskers, provides needs and opportunities that often involve aspects of both particulate and continuous fiber composites. Coating fibers or whiskers with part or all of the matrix (or a precursor) is felt to be of particular importance to achieve uniformity and volume fractions needed. Thus, for example, forming tapes of short fibers or whiskers that are coated with part or all of the matrix or its precursor provides extremely versatile processing opportunities. Alternatively one can infiltrate whisker (fiber) mats, e.g. formed by filtration or other felting techniques, coat them with part, or all, of the matrix precursor and then consolidate them. This is expected to be much more practical and successful than simply attempting to form a sufficiently dense fiber preform then attempting to infiltrate and consolidate such preforms.

ACKNOWLEDGEMENT

The author wishes to acknowledge partial support of the Naval Air Systems Command, the Office of Naval Research and the Defense Advanced Research Projects Agency respectively for support of work on ceramic fiber composites, particulate composites, and SPS processing which have been an important basis for doing much of the work here and stimulating much of the author's thoughts on these topics. Dr. Dave Lewis is thanked for comments on the manuscript.

REFERENCES

1. R. W. Rice, CHEMTECK, 13, pp 230-239, Am. Chem. Soc., 1983.
2. Paul F. Becher, Am. Cer. Soc., 64, (1) pp 37-39, 1981.
3. R. W. Rice, Proc. Am. Cer. Soc., 2 (7-8) pp 493-508, 1981.
4. R. W. Rice, P. F. Becher, S. W. Freiman and W. J. McDonough, Proc. Am. Cer. Soc., 1, (7-8) pp 424-443, 1980.
5. D. Lewis, R. P. Ingel, W. J. McDonough and R. W. Rice, Proc. Am. Cer. Soc., 2 (7-8), pp 719-727, 1981.
6. K. S. Mazdiyasni and Robert Ruh, J. Cer. Soc., 64, (7), pp 415-419, 1981.
7. W. S. Coblenz and D. Lewis III, Bull. Am. Cer. Soc., 62 (9), 968, 1983.
8. R. C. Garvie, R. H. J. Hannink, and C. Urbani, Ceramurgia Int., 6 (1) 19-24, 1980.

9. R. P. Ingel, R. W. Rice, and D. Lewis, J. Am. Cer. Soc., 65 (7), C108-9, 1982.
10. R. P. Ingel, D. Lewis, B. A. Bender, and R. W. Rice, J. Am. Cer. Soc., 65 (9), C150-2, 1982.
11. T. Hirai, S. Hayashi, J. Mat. Sci., 17, 1320-8, 1982.
12. K. M. Prewo and J. J. Bennan, J. Mat. Sci., 15, 463-68, 1980.
13. B. E. Walker, Jr., R. W. Rice, P. F. Becher, B. A. Bender, and W. Coblenz, Bull. Am. Cer. Soc., 62 (8), 916-23, 1983.

PROCESSING OF Al_2O_3-ZrO_2

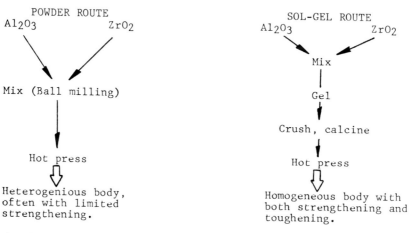

POWDER ROUTE
Al_2O_3 ZrO_2

Mix (Ball milling)

Hot press

Heterogenious body, often with limited strengthening.

SOL-GEL ROUTE
Al_2O_3 ZrO_2

Mix

Gel

Crush, calcine

Hot press

Homogeneous body with both strengthening and toughening.

Fig. 1. Schematic of the powder and the sol-gel routes used for Al_2O_3-ZrO_2 composites

PROCESSING BN PARTICULATE COMPOSITES

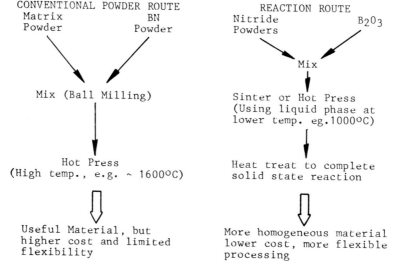

CONVENTIONAL POWDER ROUTE
Matrix   BN
Powder   Powder

Mix (Ball Milling)

Hot Press
(High temp., e.g. ~ 1600°C)

Useful Material, but higher cost and limited flexibility

REACTION ROUTE
Nitride B_2O_3
Powders

Mix

Sinter or Hot Press
(Using liquid phase at lower temp. eg.1000°C)

Heat treat to complete solid state reaction

More homogeneous material lower cost, more flexible processing

Fig. 2. Schematic of the conventional powder and the insitu reaction routes to processing BN particulate composites

SPS PROCESSING

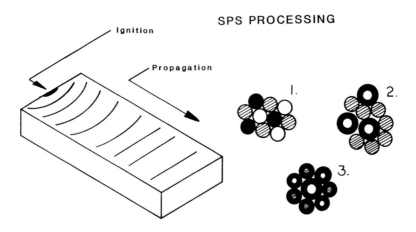

Fig. 3. Schematic of the SPS process. The left sketch illustrates how a reaction propagates (as a high temperature zone moving) through a pressed compact of the reactants when ignited a one end. The three right hand sketches schematically illustrate different arrangements of the reactants: 1) mixing powder particles (circles) of different reactants (e.g. Ti, B, and C), 2) coating one reactant on one of the others (e.g. Ti on B) mixed with the third reactant, and 3) coating two of the ractants on the third (e.g. coating Ti on particles of both B and C).

PROCESSING CERAMIC FIBER COMPOSITES

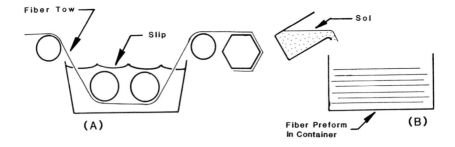

Fig. 4. Schematic of ceramic fiber composite processing techniques. A) can be used for processing with sols, but is particularly appropriate for slips, while B) can be used some with slips, but is more generally applicable for use with sols.

 345

Processing of Composites by Coating
The Dispersed Phase

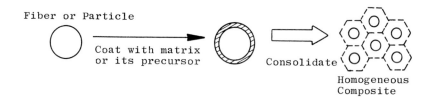

Fig. 5. Schematic of processing composites by coating the particulate or fiber phase with part, or all, of the matrix or precurssor. Upon consolidation of such coated particles or fibers, one can most closely approach the ideal homogeniety sketched at the right.

MULTI-PHASIC CERAMIC COMPOSITES MADE BY SOL-GEL TECHNIQUE

RUSTUM ROY, S. KOMARNENI AND D.M. ROY
Materials Research Laboratory, The Pennsylvania State University, University Park, PA 16802

ABSTRACT

Instead of aiming to prepare homogeneous gels and xerogels, this paper reports on work done to prepare deliberately diphasic materials. This has been achieved by three different paths: (1) mixing 2 sols; (2) mixing 1 sol with 1 solution; and (3) post formation diffusion of either one or two solutions.

By the last named process we have made SiO_2, mullite and alumina based composites, with silver halides, $BaSO_4$, CdS, etc., as the dispersed phase. The crystal size can be confined to the initial pores by rapid diffusion giving rise to extremely fine second phases in the submicron range. Subsequent reduction of appropriate metallic salts can be used to give finely dispersed metals (e.g. Cu, Ni) in essentially any xerogel matrix. The open porosity makes these metal atoms very accessible.

By the first two processes we have made both single phase and di—phasic gels of the same composition (prototype: mullite) and shown that though they cannot be distinguished by XRD, SEM, and TEM, by DTA and thermal processing, they are radically different. Such di—phasic gels store more metastable energy than any other solids.

INTRODUCTION

We developed the sol-gel technique starting in 1948 for two purposes. First, to make ultrahomogeneous glasses and avoid the tedious standard method of the time of making glasses from oxide melts by crushing, remelting and so on several times. Second, Roy and Osborn had embarked on a low-temperature hydrothermal study of the system $Al_2O_3-SiO_2-H_2O$ [1], and the glass forming region in the system $Al_2O_3-SiO_2$ was limited to ~0-25% Al_2O_3. Other starting materials such as silica glass + $\gamma-Al_2O_3$ or boehmite reacted to give corundum so early that equilibrium was unattainable. One needed a new method to make noncrystalline phases over a much wider compositional range. Ewell and Insley had already introduced the coprecipitated gels made from Na_2SiO_3 and $Al(NO_3)_3$, but even after tedious electrodialysis these always contained substantial (~1%) Na^+. The use of alkoxide precursors specifically tetraethoxysilane and aluminum isopropoxide proved to be a key to the new generalizable sol-gel process. In the next several years we used the new sol-gel method to make several hundreds of compositions as the homogeneous minimally structurally biased starting materials and making homogeneous glasses [1] for phase equilibrium studies both dry and 'wet', in dozens of binary, ternary and quaternary systems involving all the major ceramic oxides [2-6].

The solution-mixing and sol-gel route was first utilized technologically to make nuclear fuel ceramic pellets by ingeniously shaping the final product while it was a gel at Oak Ridge and Harwell [7]. Scaling up of homogeneous glass making was done at Owens—Illinois [8]. In these studies and in the revival of interest in the sol-gel process consequent

Mat. Res. Soc. Symp. Proc. Vol. 32 (1984) © Elsevier Science Publishing Co., Inc.

upon the spectacular success of Sowman and colleagues at 3M [9] in making fibers and abrasive grain work has focussed on exactly the same goal: make homogeneous gels, and derive from them, ceramics.

This paper, following the first reports by Roy and Roy [10], describes the reorientation of our original goal of homogeneity on the finest possible scale for the sol-gel method. We set as our new goal the making of new materials with controlled micro- or nano-heterogeneity. The potential value of such di- or more generally, multi-phasic materials is considerable as they allow us to make entirely new families of opto-electro-elastic composites. Further they allow us to make solids which store very large amounts of metastable energy.

EXPERIMENTAL

In the making of any composite we may either make the dispersed phase first (say Al_2O_3 fibers) and build the matrix phase around that (set a gel or melt a glass around the fibers) or one can make the matrix phase and subsequently grow the dispersed phase within it (e.g. all precipitation hardened alloys). In this work we have avoided the first or straightforward method and used either simultaneous creation of the two phases, or the growth of the dispersed phase into a pre-existing matrix.

Opto-Electro-Elastic Composites

1. Materials Studied. Our first objective was to attempt to prepare optically active composites in a dielectric matrix. For the matrix SiO_2, Al_2O_3, and $3Al_2O_3 \cdot 2SiO_2$ suggested themselves as the simplest candidates. Dispersed phases suitable for photochromic applications (such as photography) are the silver halides, and for photoreceptor application (such as xerography) include cadmium chalcogenides.

Method Developed

Two methods have been used to make the di-phasic composites. Of those, the one that is relevant to the opto-electro-elastic composites is a two-stage process.

Various silicate or aluminosilicate gels may be used although SiO_2 appears to be the easiest to work with. Tetramethoxysilane (4 parts) dissolved in alcohol (15 parts) and hydrolyzed in water (1 part) at 65°C in 2 cm diameter test tubes. The stiff gel can be removed from the test tubes after drying a little as monoliths. These monoliths are then soaked in the metal nitrate solution for 48 hours. They are removed and washed in distilled water and then soaked in dilute HCl or H_2SO_4, etc. as desired. By controlling the concentration of the above Cl or SO_4 anions in solution one can form very large numbers of very small nuclei. These diphasic materials can then be dried at various temperatures or with microwave radiation.

Two classes of di-phasic materials result from this approach, in one both phases are noncrystalline; in the other, one of the phases is crystalline.

Diphasic Ceramic-Ceramic Noncrystalline Xerogels

The second generalized method for making diphasic ceramic xerogels is shown schematically in Fig. 1. The upper portion of the figure represents the normal process of making a homogeneous single phase xerogel and also shows the fact that single phase xerogels may of course (depending on the total composition) yield one, two, or more crystalline phases depending on the phase rule as the composition approaches equilibrium. This is, of

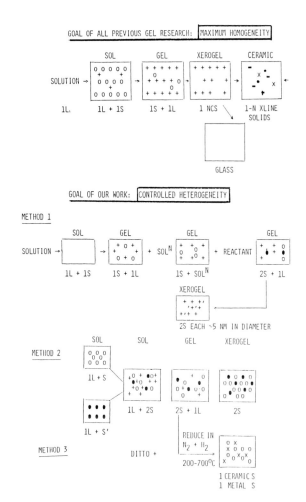

FIG. 1.   Stages in the preparation of single phase and di-phasic gels.
Three different classes of the latter are illustrated.

course, precisely the way in which such single phase gels were used in our
early studies [4] both under dry and hydrothermal condition as the easiest
route to the equilibrium crystalline assemblage whether one, two, or
multi-phasic.

    In the lower portion of Fig. 1 we show two different classes of
processes for making diphasic gels. The first is basically mixing two
separate sols and gelling the mixture, and the second is dispersing one
sol in a solution of a second composition (of one or more components) and
then gelling the latter. In the latter case it is difficult to know how
much diffusion of the ions of the latter solution may have entered the
'solid' particles of the first sol. The gels are dried by conventional or
microwave heating to yield diphasic xerogels.

Diphasic Ceramic-Metal Xerogels

To make ceramic-metal composite materials a third variant is used. We start with either of the following ways for making the xerogel.

1. Method 1. For method 1, in which all components of the xerogel are mixed simultaneously, aluminum nitrate and zirconyl chloride were formed in aqueous, acidic (pH ~1) solution in the case of alumina and zirconia systems. In the case of silica, tetraethoxysilane was dissolved in excess ethanol with an aqueous solution containing the heavy metal ion added simultaneously. The metallic precursor materials were the same for xerogels made by method 1 or 2.

2. Method 2. Pre-made sols containing the constituent oxide component were obtained either as commercial ZrO_2 or $Al(OOH)$ sols; or by hydrolyzing and polymerizing Al-nitrate, tetraethoxysilane or tetraisopropyl orthotitanate. The sols were allowed to sit for various periods of time after being prepared before the heavy metal solution was added.

An additional step of reduction of the diphasic (or monophasic) xerogel product from either of the processes listed above is needed since the xerogel components consist of or contain a relatively easily reducible ion such as Cu, Ni, Co, etc., then the product is reduced usually in a gas phase at 200-700°C usually in N_2+H_2 mixtures, or in solution in contact with the gel. We generate thereby a noncrystalline oxide xerogel matrix which contains probably a crystalline metallic phase, although the units of the latter are so small that we may be approaching noncrystalline metals phases also.

RESULTS

The different classes of diphasic composites prepared so far are described below together with their structure and properties.

Photosensitive SiO_2-AgCl, SiO_2-CdS Materials

Thin layers of the SiO_2-AgCl diphasic material were prepared by making the gel in a Petri dish, with a range of AgCl concentration from 1-10 mg AgCl per 5 cm^3 of gel.

Powder x-ray diffraction shows a crystalline line-broadened AgCl pattern superimposed on a broad amorphous SiO_2 band with samples containing greater than 4% AgCl. Scanning electron microscopy of an opaque sample containing 4% AgCl showed evenly dispersed AgCl crystals of less than one micron diameter. The transparent samples of SiO_2-AgCl, with lower concentrations of AgCl, contain smaller crystals presumably less than 0.1 micron diameter. Indeed even high resolution TEM studies failed to locate definite AgCl crystals in the low concentration gels (see Fig. 2) suggesting that these are well below 5nm in size.

Infrared absorption spectra of SiO_2-AgCl samples dried at 60°C, 500°C and 700°C show that the spectrum of the sample dried at only 60°C is identical to that of the sample heated to 500°C. No nitrates or residual organic phases were detected. Absorption bands at 1200, 1120, 800 and 460 cm^{-1} were observed, which correspond to those typical of silica glass. The band at 950 cm^{-1} has previously been reported in gel glasses and was attributed to Si-OH vibrations. The band at 3500 cm^{-1} is attributed to Si-OH stretching and adsorbed water. The 950 and 3500 cm^{-1} bands were greatly reduced in intensity by heating to 700°c.

During the drying process at 60°C the SiO_2-AgCl samples shrink considerably, reaching a density of 1.85 g/cm^3. Additional heat treatment at

5̲0̲ nm

(a)

1̲0̲ nm

(b)

FIG. 2. Nanostructure of AgCl in SiO$_2$ composites studied by TEM showing that size of AgCl crystals is 5 nm or less.

higher temperatures does not appear to further increase the density.

Samples with greater than 10 mg AgCl per 5 cm^3 silica gel are opaque and white, those with lesser amounts of AgCl are transparent. Upon exposure to sunlight all the materials darken. Samples that were initially transparent remain transparent while darkening, those initially opaque become opaque and black upon exposure to sunlight. The darkening can be almost completely reversed in the samples prepared with 4 mg or less AgNO$_3$ per 5 cm^3 silica gel depending on the preparation parameters of the present samples (not optimized in any way) by storing them in the dark for hours to days. Heating to 400°C also clears the SiO$_2$-AgCl material, which can then be darkened again by re-exposure to sunlight. Examples of such photosensitive SiO$_2$-AgCl xerogels are shown in Figs. 3-4.

The addition of the crystalline phase also significantly alters the mechanical properties of the silica gel before densification. For example, gels containing a small amount of AgCl show a much lower tendency

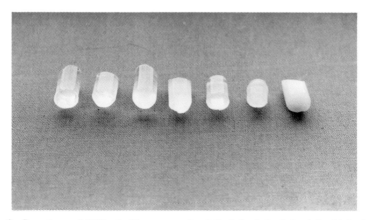

FIG. 3. Samples of SiO$_2$-AgCl prepared with 0.4, 1.0, 2.0, 4.0, 10.0, 20.0 and 250 mg of AgNO$_3$, respectively, before drying.

352

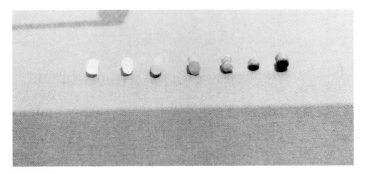

FIG. 4.   Same as in Fig. 3 after drying at 60°C and exposing to sunlight.

to crack during densification than pure silica gels.   It appears that the
introduction of low concentrations of an additional phase may be a useful
step in preventing cracking during heat treatment, which is a common
problem in the production of monolithic articles via the sol-gel process,
the latter being the major focus of much sol-gel research.   Details on
this study are given elsewhere [11].

SiO_2 Gel Composites with Crystalline CdS, $BaSO_4$, $PbCrO_4$

The same method has successfully produced many other nano-composites
of 10-50 nm size phases of SiO_2 and another composition [12].   CdS-in-SiO_2
composite thin films have been made using $1N$ $Na_2S_2O_4$ at 65°C as the
solution for precipitating the CdS. $BaSO_4$ and $PbCrO_4$ have also been
incorporated as the dispersed phase.

X-ray powder patterns show the presence of crystalline CdS, $BaSO_4$ and
$PbCrO_4$ in samples with concentrations greater than ~10 mg per 5 cm^3 of
gel.   Photoconductive properties have not been measured yet but it is
clear that they can be controlled by controlling the composition (e.g. Se
to S ratio) and doping in the crystalline phase, and its concentration.
The experimental data are summarized in Table I.

Noncrystalline Diphasic Composites

Table I also lists data on other compositions where the second phase
remains noncrystalline as in $CePO_4$, $AlPO_4$, Nd_2O_3 and Ho_2O_3.   Whereas in
the latter some diffusion of the Nd ions into the SiO_2 may be expected
giving gradients in the composition, in the two phosphate examples the two
phases should be quite separate.   Unfortunately SEM-EDAX fails to resolve
the separate compositional areas, which is consistent with our expectation
that these are on the order of 10-50 nm.

Comparison of Mullite ($3Al_2O_3 \cdot 2SiO_2$) Mono- and Di-Phasic Composites

Single phase xerogels had been prepared by dissolving tetraethoxysilane and aluminum nitrate 9-hydrate in absolute ethanol, then gelling
the solution by heating to 60°C in a water bath for several days.

Diphasic xerogels were prepared by different methods.   First was by
mixing aqueous silica sol (Ludox 'AS,' $2\underline{M}$ in SiO_2) with boehmite sol.
Opaque gels were formed by reducing the solvent volume by evaporation at
room temperature.

TABLE I. Parameters for Preparation of Diphasic Materials by the Introduction of a Second Phase into 5 cm^3 of Silica Gel

Diphasic System	Source of Cation		Source of Anion	Phases Detected XRD
SiO_2-CrPO_4	4.0 mg	$Cr(NO_3)_3 \cdot 6H_2O$ in 25 ml H_2O	0.5 $\underline{M}$ H_3PO_4	NCS
''	40 mg	''	''	
''	400 mg	''	''	
SiO_2-BaSO_4	0.1 mg	$Ba(NO_3)_2$ in 25 ml H_2O	0.5 $\underline{M}$ H_2SO_4	$BaSO_4$
''	1.0 mg	''	''	
''	10.0 mg	''	''	$BaSO_4$
''	100 mg	''	''	$BaSO_4$
$SiO_2-PbCrO_4$	1000 mg	$Pb(NO_3)_2$ in 20 ml H_2O	0.5 $\underline{M}$ chromic acid	$PbCrO_4$
SiO_2-CePO_4	500 mg	$Ce(NO_3)_3$ in 50 ml H_2O	0.5 $\underline{M}$ H_3PO_4	NCS
SiO_2-AlPO_4	500 mg	$Al(NO_3)_3 \cdot 9H_2O$ in 50 ml H_2O	''	NCS
$SiO_2-Nd_2O_3$	1000 mg	$Nd(NO_3)_3$ in 25 ml H_2O	none used*	NCS
$SiO_2-Ho_2O_3$	200 mg	$Ho(NO_3)_3$ in 25 ml H_2O	none used*	NCS

*The metal-oxide was formed by heating the samples to 400°C.

A second method was to disperse a boehmite sol in an alcohol solution of the tetraethoxysilane and causing gelling by heating. A third method was to mix aluminum nitrate solutions with Ludox, with subsequent gelation.

Silica-alumina xerogels were also prepared by immersing a piece of silica gel (formed by the hydrolysis of tetraethoxysilane in ethanol-water) in an aqueous aluminum nitrate solution.

Xerogels were formed by drying the gels in air at 60°C and subsequent heating. The xerogels were ground to powders using an agate mortar and pestle. X-ray diffraction, SEM and electron probe studies showed homogeneous noncrystalline materials with no differences between the single and diphasic materials. DTA curves, however, tell a very different story. Such curves for single phase and diphasic materials are shown in Figs. 5 and 6, respectively.

All of the single phase dried gels exhibit a broad endotherm below 400°C due to the loss of water, ethanol and nitrates. All mullite xerogels show only one sharp crystallization exotherm at 960°C being attributed to mullite crystallization and exactly reminiscent of the transition from metakaolinite to mullite + silica seen on the standard DTA pattern of kaolinite.

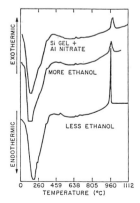

FIG. 5. DTA heating curves for single phase xerogels of aluminum to silicon atom ratios 3:1. Bottom: xerogel prepared by gelling 10.2g $Al(NO_3)_3 \cdot 9H_2O$ and 2 ml TEOS in 10 ml ethanol. Middle: xerogel prepared by gelling 10.2g $Al(NO_3)_3 \cdot 9H_2O$ and 2 ml TEOS in 30 ml ethanol. Top: xerogel prepared by immersing a TEOS-ethanol gel in 1.5$\underline{M}$ aluminum nitrate solution for 4 days.

354

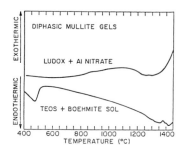

FIG. 6. DTA heating curves for diphasic xerogels of 3:1 Al:Si atom ratio from TEOS-boehmite sol and Ludox-aluminum nitrate.

Figure 5 also shows the effect of preparation conditions on the crystallization exotherms in dried and xerogels having the 'ideal' mullite composition (aluminum to silicon atom ratio 3:1). Parallel to Yoldas' observation on pure Al_2O_3 gels [13], increasing the gel volume changes the structure of the xerogel sufficiently to be reflected in a much less sharp crystallization exotherm, but does not change its position at all. The enthalpy of crystallization was estimated from the DTA peak area to be on the order of 20 cal/gram (9 kcal/mole mullite) for single phase xerogels of 3:1 aluminum to silicon atom ratio.

DTA of the silica gel immersed in Al-nitrates showed a small, broad crystallization exotherm at 960°C characteristic of a single phase silica-alumina xerogel (Fig. 5), showing that the Al^{3+} ions had penetrated and reacted with the Si-O solid units in the preformed silica gel. However, the homogeneity of the solid units is clearly very different from the best single phase material.

The diphasic xerogels exhibit radically different behavior upon heating from single phase xerogels of the same stoichiometric composition. DTA heating curves for diphasic xerogels are shown in Fig. 6. The endotherm at 400°C is associated with the decomposition of the boehmite phase. No trace of the 960°C exotherm seen in single phase xerogels was observed. Instead a very broad exotherm associated with mullite formation occurs over the entire region from about 700 to above 1250°C. X-ray powder diffraction data on diphasic gels made from Al_2O_3-sols shows the presence of a discrete boehmite phase at low temperatures. Cristobalite crystallization precedes mullite formation when the gels are heated, confirming the fact that the two discrete phases are reacting 'independently' up to ~1000°C.

The diphasic xerogels prepared by gelling tetraethoxysilane with boehmite sol, and by gelling Ludox and aluminum nitrate provide the most interesting data. The latter is not complicated by the boehmite dehydration and shows no trace of the mullite formation exotherm at 960°C. Instead there is apparently continuous reaction over a several hundred degree range, with a much larger total ΔH of reaction which is, however, very difficult to quantify.

Samples of both single phase and diphasic xerogels were subjected to ultrafast heating on a platinum strip furnace, by presetting the temperature to the desired level, turning off the furnace, placing the pinhead size sample on the strip and switching on the current. No metastable melting of the mullite composition was observable (as direct slumping) below 1750°C.

Ceramic-Metal Diphasic Xerogel Composites

Details of our work have recently been published [14] and only a summary is presented here. Some compositions studied are listed in Table II.

Table III lists typical process conditions for making a ceramic-metal xerogel. While the exact details of the process 'recipe' vary substantially over the wide range of cermet systems studied, Table III shows that all materials were initially gelled at or below 70°C. In general, when gelation was allowed to occur in an open system, it took place in times ranging from several minutes to several hours. In closed systems (no H_2O evaporation during gelation), the time ranged from 1/2 hour to several days.

Drying and heat treatment took place after the wet gel formed was found to have sufficient rigidity. The final step was to place the sample in a furnace and react with a reducing atmosphere of 'forming gas' (95% N_2, 5% H_2) temperature was varied for different samples ranging from 200°–700°C. Typical treatment times ranged from 15 to 40 minutes.

TABLE II.   Ceramic-Metal Systems Studied.

Alumina:	Cu, Pt, Ni
Zirconia:	Cu, Ni
Silica:	Cu, Sn
Titania:	Cu

Starting Materials

Oxides

Al_2O_3 – Al-nitrate, Al-isopropoxide, AlOOH sols
ZrO_2 – Zirconyl chloride, ZrO_2 sols
SiO_2 – SiO_4, tetraethoxy (or methoxy) silane
TiO_2 – Tetraisopropyl orthotitanate

Metals

Cu – Cu-nitrate, Cu-sulphate, Cu-chloride
Pt – H_2PtCl_6
Sn – $SnCl_2 \cdot 2H_2O$
Ni – Ni-nitrate, Ni-sulphate

TABLE III.   Typical Process Conditions.

Gelation
    temperature – RT-70°C
    pH – <1.5
    system – open or closed
Drying
    a)   ambient several hours --> days
    b)   oven 110°–125°C, 30 minutes
    c)   microwave
    d)   reducing furnace 200°–700°C, 10 minutes

In all the systems (except TiO_2-Cu) listed above we obtain composite xerogel materials consisting of a <u>noncrystalline oxide</u> matrix which is itself either the pure original oxide or slightly doped with the metal oxide, and a finely dispersed metal phase. The nanostructure of the gel and the metal phases is shown in SEM and TEM photographs in Figs. 7 and 9 and XRD data in Fig. 8.

In sum, what these data show is that the xerogel itself is typically made up of 50 nm globules aggregated into larger 500-1000 nm globular clusters of oxide. Dispersed among these globules are metal particles ranging in size from 5nm at 400°C in the case of Pt on Al_2O_3 increasing to about 30nm at 700°C.

Figures 7 and 9 show SEM and TEM micrographs of a 10% Pt/Al_2O_3 xerogel prepared at 500°C. The pores visible as light areas in the TEM micrograph are the same size as the agglomerated particles visible in the SEM micrograph. In addition, small Pt crystallites about 10-15 nm appear as dark spots. In Fig. 9 a TEM micrograph of the xerogel 10% Pt/Al_2O_3 is compared with a sputtered film of a similar composition. Both materials contain Pt crystallites about 10 nm in size, although the sputtered film

FIG. 7. SEM micrographs of a 10% Pt/Al₂O₃ xerogel at two different magnifications.

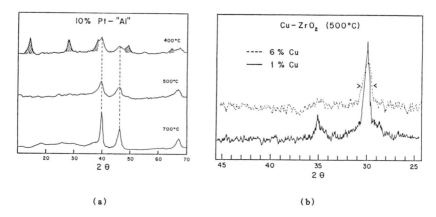

(a)   (b)

FIG. 8. XRD patterns of (a) 10% Pt/alumina gels as a function of reduction
temperature. The dotted lines indicate the position of platinum peaks.
(b) Cu/zirconia gels as a function of % Cu.

is less porous. In addition, the Pt crystallite size in the porous oxide
matrix can be varied by heat treatment in the gel, as noted previously.
In xerogels of Cu-in SiO_2, or Cu in ZrO_2 more of the copper stays in the
oxide host, whether NCS as in the case of SiO_2 or poorly crystalline as in
the case of ZrO_2. The radical difference between such diphasic xerogel
ceramic-metal composites (before they are compacted by hot pressing or
HIPing) and other materials is in their enormous accessible surface area.

This unique porosity in the xerogels was evident in a series of
experiments with $Ni-Al_2O_3$ and $Cu-Al_2O_3$ diphasic materials. Some of these
composites when reduced at 500°C-700°C were black due to the presence of a
metallic phase, although this was not always detectable by x-ray diffraction. However, after exposure to room temperature ambients for periods
from 15 minutes to one week, the metal is reoxidized and a green or blue
xerogel obtained.

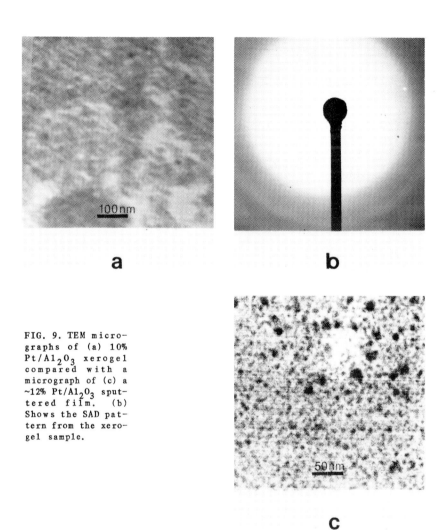

FIG. 9. TEM micro-
graphs of (a) 10%
Pt/Al$_2$O$_3$ xerogel
compared with a
micrograph of (c) a
~12% Pt/Al$_2$O$_3$ sput-
tered film. (b)
Shows the SAD pat-
tern from the xero-
gel sample.

The G-T diagram of Fig. 10 explains the theory of the actual application of each of the three mechanisms in materials processing. Metastable
assemblages of any kind whether caused by surface energy or noncrystal-
linity MUST melt at a temperature below the equilibrium melting point. By
that token diphasic xerogels, if metastable equilibrium could be attained,
will melt far below the equilibrium T$_m$. However, equilibrium (metastable
or stable) is not always attained. Indeed in the 'glass-forming' oxide
materials we are discussing, below about 1000°C equilibrium is extremely
difficult to attain at all and the reactions which occur are determined by
the kinetics. In the regime between 1000-2000°C the roles of kinetics and
equilibria are more equal in the typical experiments which run from

358

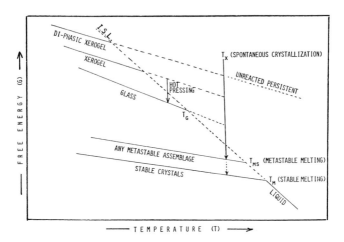

FIG. 10. Schematic G-T diagram for an isoplethal system to illustrate the variety of different reactions which can be encountered in heating metastable phases including xerogels.

minutes to days, so that one can expect to observe some of the phenomena represented schematically in Fig. 10. Other phenomena encountered include non-nucleated crystallization shown as occurring at T_x. For mullite containing compositions this seems to be ~960°C; above this temperature no NCS solid of the mullite composition can exist since it crystallizes spontaneously and very rapidly. This is why we have been unable to observe the metastable melting point of a xerogel in the $Al_2O_3-SiO_2$ system.

SUMMARY AND DISCUSSION

The preparation of di- or multiphasic xerogels has opened the way to making really new classes of materials at temperatures below 600-700°C.

1. Noncrystalline oxide xerogels containing as a second phase very small 5-50 nm crystals of a wide assortment of inorganic phases such as $AgCl$, CdS, $BaSO_4$, etc.

2. Noncrystalline or poorly crystalline oxide xerogels containing as a second phase very small 5-100 nm crystals of a wide assortment of metals Cu, Ni, Pt, etc.

3. Two noncrystalline oxide phases each of which is 5 nm or less in size. In such materials there are 3 sources of excess free energy over the stable equilibrium state

    (a) the excess surface energy of such small particles,
    (b) the difference between the stable crystalline state and the noncrystalline state,
    (c) the heat of reaction of the two phases to the stable assemblage.

Since the two phases can be very far from equilibrium with each other, they can store via (c) above in a solid far more energy than is possible by the other means (a) and (b). Typically (a) and (b) will be at most 1-3 kcals/mole of simple oxides, whereas (c) can be one order of magnitude more.

However, in the Al_2O_3-MgO system in which the 3M abrasive grains are made [9], it is likely that all or part of the gel actually melts meta-stably near 1250°C in order for the reported sintering to be accomplished at that temperature.

ACKNOWLEDGEMENTS

Present research on composite materials made via the sol-gel process is supported by Grant AFOSR 83-0212.

REFERENCES

1. R. Roy and E.F. Osborn. The System Al_2O_3-SiO_2-H_2O, Am. Min. __39__, 853 (1954).
2. R.C. DeVries, R. Roy and E.F. Osborn. Trans. Brit. Ceram. Soc. __53__, 525 (1954).
3. D.M. Roy and R. Roy. Am. Mineral. __40__, 147 (1955).
4. R. Roy. Aids in Hydrothermal Experimentation: Methods of Making Mixtures for Both Dry and 'Wet' Experimentation, J. Am. Cer. Soc. __39__, 145 (1956).
5. F.A. Mumpton and R. Roy. Geochim. et Cosmochim Acta __21__, 217 (1961).
6. R. Roy. Gel Route to Homogeneous Glass Preparation, J. Am. Cer. Soc. __52__, 344 (1969).
7. J.P. McBride. Preparation of UO_2 Microspheres by Sol-Gel Techniques, ORNL-3874 (1966).
8. I.M. Thomas. Metal-Organic-Derived (MOD) Glass Compositions. Preparation, Properties and Some Applications, Materials Research Society Abstracts, p. 370, Annual Meeting, Boston, MA (1982).
9. M.A. Leitheiser and H.G. Sowman. Non-Fused Alumina Oxide-Based Abrasive Mineral, United States Patent 4,314,827 (1982).
10. R.A. Roy and R. Roy. New Metal-Ceramic Hybrid Xerogels, p. 377, Abstracts, Materials Research Society Annual Meeting, Boston, MA (1982).
11. David Hoffman, S. Komarneni and R. Roy. Preparation of a Di-phasic Photosensitive Xerogel, J. Mat. Sci. Letters (in press).
12. D.W. Hoffman, R. Roy and S. Komarneni. Di-phasic Ceramic Composites Via a Sol-gel Method, Materials Letters (in press).
13. B.E. Yoldas, J. Mat. Sci. __14__, 1843 (1979).
14. R.A. Roy and R. Roy. Diphasic Xerogels: I. Ceramic-Metal Composites, Mat. Res. Bull. __19__, 169 (1984).

COMPOSITION AND CHEMICAL STRUCTURE OF NITRIDED SILICA GEL

R. K. Brow and C. G. Pantano
Department of Materials Science and Engineering
The Pennyslvania State University
University Park, PA 16802

ABSTRACT

Sol/gel derived silica thin films were thermally treated in NH_3 for four hours at temperatures up to 1300C. The films were analyzed by ellipsometry, X-ray photoelectron spectroscopy (XPS) and infrared spectroscopy (IR). Over 30 mol% nitrogen was incorporated in the film treated at 1300C. Using IR and XPS analyses, $-NH_x$ groups were found to be present after low temperature treatments, while nitrogen was incorporated in an oxynitride structure after the higher temperature treatments.

INTRODUCTION

It has been shown [1,2] that silica films can be deposited on silicon wafers by spinning a solution of tetraethoxysilane (TEOS), ethanol and water. The film thickness can be controlled by varying the solution composition and the spinning rate. In addition, it was found that nitrogen can be incorporated by a thermal treatment in flowing anhydrous ammonia. Chlorine pretreatments were observed to enhance this nitridation reaction; the H_2O/TEOS ratio also influenced the incorporation of nitrogen.

In this paper, the chemical nature of nitrogen incorporated in sol/gel derived silica thin films thermally treated in ammonia will be discussed. Infrared absorption spectroscopy and x-ray photoelectron spectroscopy (XPS) have been employed to describe the chemical environment of nitrogen. XPS and ellipsometry were used to determine the composition of the films as a function of treatment temperature.

EXPERIMENTAL

The solutions used to prepare the silica sol/gel samples consisted of 43% tetraethoxysilane (TEOS), 43% ethanol and 14% water (by volume). A small amount of 1M HCl was added to promote hydrolysis. The mixture was added to a reaction vessel equipped with a reflux condenser, and then heated to 60C under constant stirring and held for 30 minutes. The solution was then cooled to 40C. A portion of the solution was diluted four parts ethanol to one part solution to be used to spin-coat Si wafers. The remainder of the solution was used to make free standing silica foils for infrared analyses. The foils were created by adding 3 ml of the undiluted solution to covered 5cm diameter polystyrene petri dishes and gelling for two weeks. Foils approximately 0.1mm in thickness resulted. These were stored in a desiccator prior to heat treatments.

The substrates used for the application of the sol/gel thin films were polished n-type <100> single crystal silicon wafers (1.25 inch in diameter). All wafers were cleaned using electronic grade reagents and methods standardized by the semiconductor industry. Approximately 250 μl of the diluted solution was uniformly distributed over the surface of the

Mat. Res. Soc. Symp. Proc. Vol. 32 (1984) Published by Elsevier Science Publishing Co., Inc.

wafers using a photoresist spinner at 2000 RPM for 15 seconds. Films approximately 1000Å in thickness were produced.

Thermal treatments of the thin films and the foil samples were done in an Al_2O_3 tube furnace, equipped with a gas manifold and individual flowmeters, and connected to a rotary vacuum pump. The samples received a two-hour 400C vacuum pretreatment to remove physically adsorbed water. After the pretreatment, anhydrous ammonia or dry oxygen, was introduced into the furnace at a flow rate of 500 scc/min. The samples were heated at 1.5C/min to the temperature of interest, where they were held for four hours, after which they were cooled to room temperature in flowing, dry N_2 (500 scc/min). Treated samples were removed from the furnace and stored in a desiccator before analysis.

The infrared absorption studies of the foil samples were done using a Perkin-Elmer model 283B spectrometer, with the primary beam passing directly through the sample. The refractive index of the thin films was measured using a Gaetner ellipsometer (model L117) at an incident angle of $70°$. X-ray photoelectron spectroscopy (XPS) analysis of the thin films was done with a Physical Electronics Model 590 SAM/ESCA system, equipped with a PDP 11/04 computer for data handling. A 15 kV Mg x-ray source was used. The analyzer base pressure was maintained at ~$1x10^{-9}$ torr. The binding energies were referenced to the C1s line (@284.6eV); this line is observed on all samples due to residual hydrocarbon contamination.

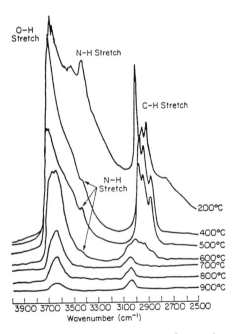

Figure 1. Infrared spectra of ammonia treated silica foils.

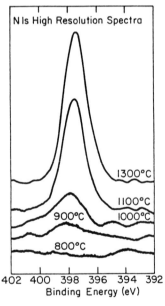

Figure 2. N1s spectra of ammonia treated silica thin films.

RESULTS AND DISCUSSION

1. Infrared Analysis

     Figure 1 shows the infrared spectra of the sol/gel derived silica
foils which were heat treated in flowing NH_3 at temperatures up to
900C. At low temperatures, significant amounts of alkoxy groups (2900-3000
cm^{-1}), physically adsorbed water (3400-3500 cm^{-1}) and hydroxyl
groups (3650 cm^{-1}) are present. In addition, a band is apparent at
3400 cm^{-1} due to $-NH_x$ species. These samples showed this
$-NH_x$ band for treatment temperatures up to 600C, but in thicker
samples the band was observed even after treatments at 900C. The
generation of $-NH_x$ species in silicates has been observed by others
after bubbling NH_3 through glassmelts at 1400C [4], as well as due to
NH_3 treatment of boroaluminosilicate gels at 750C. [3]

     An identical set of sol/gel derived silica foils were treated in
flowing O_2. These specimens did not exhibit the band due to $-NH_x$
but did show greater concentrations of adsorbed water and hydroxyl groups
than the NH_3 treated specimens for temperatures less than 700C. At
treatment temperatures in excess of 700C, the hydroxyl contents of the
specimens treated in O_2 and NH_3 were comparable.

     These observations suggest that at low to intermediate temperatures,
the ammonia interacts primarily with silanol groups to liberate water and
create $-NH_x$ species; i.e.

$$\equiv Si-OH + NH_3 \longrightarrow \equiv Si-NH_2 + H_2O \tag{1}$$

At higher temperatures, the reaction between siloxane bridges and ammonia -
or ammonia radicals - is expected to be most prevalent; i.e.

$$\equiv Si-O-Si \equiv\ + NH_3 \longrightarrow \equiv Si-OH + \equiv Si-NH_2 \tag{2}$$

Reaction 2 is consistent with the enhanced nitrogen incorporation observed
at temperatures in excess of 900C (see next section), but is inconsistent
with the observed absence of $-NH_x$ species in the specimens treated at
higher temperatures. This point is addressed further in the summary.

2. XPS Analysis

     Figure 2 shows the high resolution Nls binding energies for the thin
films treated in ammonia up to 1300C. Significant amounts of nitrogen were
not detected in samples treated below ~800C. Because of the short mean
free path of the Nls photoelectron (estimated to be 30Å in an oxynitride
matrix [5]), the spectra shown in figure 2 are representative of nitrogen
concentrations in the outer few monolayers of the film. High resolution
spectra of Si2p, Cls and Ols photoelectrons were also obtained.

     The high resolution spectra were used to estimate the chemical
composition of the thin films, following the method outlined in reference
[6]. The atomic fraction of element i was calculated from:

$$C_i = (I_i/S_i)/\Sigma(I_j/S_j) \tag{3}$$

where I is the area under a spectral peak and S is an areal elemental
sensitivity factor that depends on the x-ray flux, photoelectron
cross-section, detector efficiency, photoelectric process efficiency, the
analyzed area and the photoelectron mean free path. Values for S of 0.26,

364

0.25, 0.43 and 0.67 were used for Si2p, Cls, Nls and Ols photoelectrons respectively. The quantitative information obtained by use of these sensitivity factors agreed well with values predicted from calibration curves made by analyses of Si_3N_4 and Si_2N_2O standards.

Results of the quantitative analysis of nitrogen in the films are shown in Figure 3. Nitrogen content increased rapidly for treatments greater than 800C, and approached 35 mol% for the 1300C treatment. This can be compared with a value of 40 mol% N for stoichiometric Si_2N_2O.

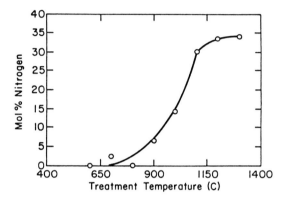

Figure 3. Nitrogen content of ammonia treated silica thin films determined by quantitative XPS

The nitrogen concentrations in the thin films are nearly an order of magnitude greater than those reported for the nitridation of bulk gels [3,7]. Two factors may account for this. First, incorporation of nitrogen in bulk gels may be limited by the diffusion of the reactant species through the pore structure, and in particular, the outer surface layers where nitridation has already occured. Likewise, water – the key reaction product due to nitridation of silica in ammonia – must be transported out of the gel structure. Wu et al. [8] reported that the formation of thermally grown Si_3N_4 was limited by the diffusion of the nitridant through the Si_3N_4 layer to the unreacted silicon substrate. Thus it may be possible that a bulk gel treated in ammonia has a high nitrogen content in its surface and a very low nitrogen content in the bulk, resulting in an overall low nitrogen content. Secondly, the thin films may be more reactive than a bulk gel because of residual stresses in the film that develop when it is dried. Morrow et al. [9] have shown that strained siloxane bonds preferentially adsorb ammonia. It may be possible that these stresses make the siloxane bonds in the film more susceptible to the chemisorption of ammonia (see reaction 2).

Nitrogen contents determined by XPS were found to correlate well with the refractive indices of the nitrided films. These results are shown in Figure 4. Also plotted is a theoretical calculation of refractive index as a function of nitrogen concentration, according to the Bruggeman approximation [10] shown in equation 4:

$$\frac{V_{SiO_2}}{V_{Si_3N_4}} \cdot \frac{n^2_{SiO_2} - n^2_{film}}{n^2_{SiO_2} + 2n^2_{film}} + \frac{n^2_{Si_3N_4} - n^2_{film}}{n^2_{Si_3N_4} + 2n^2_{film}} = 0 \qquad (4)$$

Use of equation 4 assumes that the oxynitride films are homogeneous mixtures of SiO_2 and Si_3N_4. Mol% N was determined from the calculated volume fractions of oxide and nitride in the films. This approach was used by Kuiper et al. [11] to estimate the compositions of CVD oxynitride films. Thus, ellipsometry may be a simple way to estimate the composition of these films.

In an XPS experiment, the binding energy of an inner electron is measured. Since this binding energy depends upon the element and its chemical environment, changes in the chemistry of an element can be detected by shifts in an elemental binding energy. Figure 5 plots the N1s and Si2p binding energies of the nitrided thin films as a function of nitrogen content, as determined by XPS. Both were found to decrease with increasing nitrogen contents.

The Si2p binding energy of a thermally grown SiO_2 was found to be 103.6eV, and the Si2p binding energy of CVD Si_3N_4 thin films has been reported to be 101.4eV [12]. Thus, the decrease in the Si2p binding energy shown in figure 5 can be explained as being due to an increase in the average number of nitrogen atoms coordinated to each silicon atom. Similarly, the N1s binding energy of Si_3N_4 [12] and of $-NH_x$ groups (for example, in aminopropyltriethoxysilane) [13], has been reported to be 397.4eV and 398.6 eV respectively. Thus, the shift in the N1s binding energy of the nitrided films can be explained by a reduction in the average number of H atoms bonded to nitrogen. The Si2p and N1s binding energies of the high

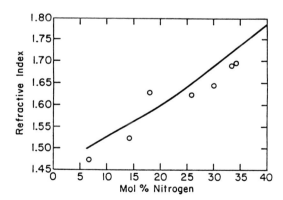

Figure 4. Refractive index of ammonia treated silica thin films. The solid line is a theoretical calculation of the effect of nitrogen concentration on refractive index

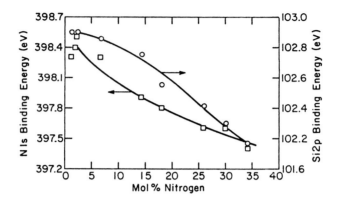

Figure 5.   N1s and Si2p binding energy shifts as functions
of nitrogen content in ammonia treated silica
thin films.

temperature (>1000C), ammonia treated films are quite close to those
reported for CVD oxynitride thin film as well as those
measured from Si_2N_2O and Si_3N_4 [12].   Therefore, the nitrogen is incorporated in these films as a nitride, i.e. is coordinated by three silicon
atoms.   Similar conclusions were drawn from an XPS study of nitrogen in
soda-lime silica oxynitride glasses [14].   The large effect that nitrogen
content has on the refractive index of the films (figure 4) is taken as
further evidence that the nitrogen is structurally incorporated as nitride
groups.

SUMMARY

     It has been found that thermal treatment of sol/gel derived silica
thin films in ammonia results in the incorporation of nitrogen in the
films.   At low temperatures, infrared analysis reveals the presence of
chemisorbed amine groups ($-NH_x$), but in films treated at temperatures
greater than 900C significant amounts of tricoordinated silicon-nitride
groups are observed.   Quantitative x-ray photoelectron spectroscopy and
ellipsometry estimate the nitrogen content to be approximately 35 mole% for
films treated at 1300C.

     Although reactions (1) and (2) can explain the incorporation of
'nitrogen' into these materials, they do not explain the formation of the
tricoordinated silicon-nitride groups.   In the case of ammonia treated
glassmelts, Mulfinger and Franz [4] have suggested that nitride species
could form due to interactions between silanol and amine groups; i.e.

$$\equiv Si-OH \ + \ \equiv Si-NH_2 \longrightarrow \equiv \underset{\underset{H}{|}}{Si}-N-Si\equiv \ + \ H_2O \tag{5}$$

and subsequently,

$$\equiv Si-OH \ + \ \equiv \underset{\underset{H}{|}}{Si}-N-Si\equiv \longrightarrow \equiv \underset{\underset{\underset{|||}{Si}}{|}}{Si}-N-Si\equiv \ + \ H_2O \tag{6}$$

One could also propose an interaction between siloxane species and amine groups; i.e.

$$\equiv Si-O-Si\equiv \ + \ \equiv Si-NH_2 \longrightarrow \equiv Si-N-Si\equiv \ + \ H_2O \tag{7}$$

These reactions become physically realistic when one recalls that these acid catalyzed silica gels are amorphous, weakly crosslinked and microporous. Nevertheless, the thermodynamic and mechanistic validity of these reactions has not yet been demonstrated. Here it can only be stated that these reactions are qualitatively consistent with the observed creation of tricoordinated nitrogen groups at T > 900C, and the corresponding elimination of $-NH_x$ species in that same temperature range.

Acknowledgements

The authors gratefully acknowledge the financial support of the National Science Foundation, Grant No. DMR-8119476, and J. Schoonover of GTE Products Corp, Towanda, Pa. for assistance with the XPS measurements.

REFERENCES

1. C. G. Pantano, P. M. Glaser and D. J. Armbrust, Ultrastructure Processing of Ceramics Glass and Composites, ed. L.L. Hench and D. Ulrich, J. Wiley and Sons, NY. 1984.
2. P. M. Glaser and C. G. Pantano, accepted for publication in J. Non Cryst. Solids.
3. C. J. Brinker, and D. M. Haaland, J. Amer. Ceram. Soc., 66 (11) 758-765 (1983).
4. H. O. Mulfinger and H. Franz, Glastechn. Ber., 38 (6), 235-242 (1965).
5. J. A. Wurzbach and F. J. Grunthaner, J. Electrochem Soc., 130 (3) 691-699 (1983).
6. C. D. Wagner, W. M. Riggs, L. E. Davies and J. F. Moulder, Handbook of X-Ray Photoelectron Spectroscopy, p. 21-22, Perkin-Elmer Corp., Eden Prairie, Mn, 1979.
7. C. J. Brinker, J. Amer. Ceram. Soc., 65 (1), C4-5 (1982).
8. C. Wu, C. King, M. Lee and C. Chen, J. Electrochem Soc., 129 (7) 1559-1563 (1982).
9. B. A. Morrow, I. A. Cody and L. S. M. Lee, J. Phys. Chem., 79 (11) 2405-2408 (1975).
10. D. A. G. Bruggeman, Ann. Phys. Leipzig, 24 636 (1935).
11. A. E. T. Kuiper, S. W. Koo, F. H. P. M. Habraken, and Y. Tamminga, J. Vac. Sci. Technol. B, 1 (1) 62-66 (1983).
12. S. I. Raider, R. Flitsch, J. A. Aboaf and W. A. Pliskin, J. Electrochem. Soc., 123 (4) 560-565 (1976).
13. J.S. Jen, Ceram. Eng. Sci Proc., Sept.-Oct. 450-457 (1982)
14. R. K. Brow and C. G. Pantano, accepted for publication in J. Amer. Ceram. Soc., 67 (4) (1984).

SOL-GEL DERIVED CERAMIC-CERAMIC COMPOSITES USING SHORT FIBERS

J. J. LANNUTTI* AND D. E. CLARK**
Ceramics Division, Materials Science and Engineering, University of Florida
Gainesville, Florida 32611 USA

ABSTRACT

Short ceramic fibers or whiskers may be ideally suited
for the fabrication of ceramic-ceramic composites using the
sol-gel process. The fibers can be uniformly dispersed in a
low viscosity sol and then frozen into the matrix through
gelation. Several ceramic composites were prepared by synthesizing an Al_2O_3 precursor from aluminum sec-butoxide and
then dispersing fibers of either zirconia, graphite or
SiC. The composites were dried at room temperature and
fired up to 1200°C. The fibers reduce the volume shrinkages
in comparison to that obtained with pure Al_2O_3 during processing. Structures with dimensions of several cm^2 can be
rapidly produced with this method without cracking.

INTRODUCTION

The value of composites for structural applications has been recognized for many years. A variety of composites including glass/polymer,
carbon/carbon and metal/ceramic are already commercially available. Although these composites offer improved toughness over the matrix material
alone, their useful temperature ranges in air are limited to below about
1200°F due to oxidation. Ceramics are more resistant to high temperature
degradation and thus provide a unique advantage over other materials. Unfortunately, the ceramic composites have not received the attention that
they deserve based on their potential advantages [1]. One of the reasons
for this lack of attention may be due to the problems encountered in their
fabrication. High temperatures and/or pressures are thought to be required
in order to produce acceptable ceramic composites.

Short ceramic fibers, or whiskers, are ideally suited for the fabrication of ceramic-ceramic composites using the sol-gel process. The fibers, or whiskers, can be uniformly dispersed in the sols and then "frozen"
into the matrix through gelation. Subsequent densification can potentially
be achieved at lower temperatures and pressures than required with conventional powder processing.

In this study composites were fabricated by casting the sol containing
the second phase into plastic trays and then gelling at 90°C. The presence
of the second phase reduces both the shrinkage and extent of macro-cracking
during drying of the gel compared to pure gel.

Although several materials were studied including zirconia and graphite short fibers, the primary focus here is on SiC whiskers. Use of glycerol as a plasticizer permitted the fabrication of tensile specimens and
other intricate shapes prior to firing. These composites were characterized with scanning electron microscopy, thermal analysis, and Fourier
transform infrared spectroscopy. Limited physical property measurements
such as shrinkages and densities are reported.

Mat. Res. Soc. Symp. Proc. Vol. 32 (1984) Published by Elsevier Science Publishing Co., Inc.

EXPERIMENTAL

Alumina sol was prepared using the Yoldas [2,3] method described in the previous paper by Lannutti and Clark [4]. The major alteration to the Yoldas method was the addition of glycerol to improve the flexibility and allow cutting of the dried gel.

The silicon carbide whiskers* are 10-80 microns in length, have a surface area of 3.0 m^2/g and a true density of 3.2 g/cm^3. These single crystals contain small concentrations of Ca, Mn, Al, and Fe as impurities. In previous experiments, these impurities had been observed to affect the pH of gel solutions which controls the gelling volume. In order to reduce those effects, the whiskers were pre-soaked for several hours in 1M HNO_3 and rinsed five times with 300 mls deionized water (1500 mls rinse total).

Another problem observed in working with these whiskers was the presence of "hard" spherical agglomerates. The agglomerates cause extensive settling prior to gelation which can result in poor homogeneity when the whiskers are mixed with the low viscosity sol. In order to reduce the extent of agglomeration, the as-washed whiskers were sonicated in an ethanol solution for approximately 20 minutes.

To prepare the alumina sol for the introduction of the whiskers, the sol was boiled down to approximately one-fourth of its as-hydrolyzed, as-peptized volume. This corresponds to a sol very near to the gelation point (see Fig. 3 in previous paper by Lannutti and Clark [4]). The ethanol and whisker slurry was then added to the sol and the mixture boiled down to its original "pre-whisker" volume. Slight variations in the pH of the whisker-/sol mixture compared to that of the unadulterated sol were observed, and these caused corresponding variations in the gelling volume.

In order to cast the whisker/sol mixtures, the sol was boiled down (with vigorous stirring) until gelation appeared imminent. The solutions were then cast into polystyrene petri dishes (coated with a non-sticking agent) and immediately placed in an 80-90°C chamber to ensure a rapid onset of gelation and thus prevent whisker settling. Approximately two hours were required to produce substantially dried gels. The use of glycerol and a mold release agent (Union Carbide R-272 organisilane ester) allowed samples to be dried more rapidly and with less cracking than with pure gels. The presence of SiC whiskers further improved the drying behavior. Three to four days are needed to dry the gels at 25°C.

The compositions prepared were 26, 35, 42 and 52 wt% SiC (corresponding to 22, 30, 37 and 46 volume %) in fired alumina.

RESULTS AND DISCUSSION

As discussed in the preceding paper by Lannutti and Clark [4] the viscosity of the sol increases significantly when the volume of sol is reduced to 0.25 of its original value (i.e. 200 cc → 50 cc). This marks the initiation of the sol → wet gel transformation. Figure 1 illustrates the viscosity changes accompanying the reduction in volume for the pure sol. It should be noted that the volume must be reduced to 0.12 of its original value before the conversion to wet gel is complete. The viscosities in this region have not been measured but most likely are several orders of magnitude greater than those indicated on the graph.

*Silar (SC-9), From Arco

The presence of glycerol, used as a plasticizer and drying agent, does not appear to affect the viscosity in the range investigated. Slight increases in viscosities, however, were observed when 52 wt% SiC whiskers (based on the weights of SiC and Al_2O_3) were added to the sol. These results could indicate that the SiC whiskers trigger the sol → wet gel transformation at a larger sol volume than is required for the pure sol. The increased viscosities, however, were not sufficiently great to prevent casting of the sol in the usual manner.

One of the major problems with sol-gel processing is the large shrinkages that occur during drying. These shrinkages are significantly reduced when SiC whiskers are present as shown in Fig. 2. The pure wet gel experiences a 25% linear shrinkage when dried under laboratory conditions of 25°C and about 60% R.H. The shrinkage of the gel with 52 wt% SiC whiskers was less than 1% under the same conditions of drying. These shrinkages were measured on samples cast into petri dishes 6.3 cm in diameter. The calculated values represent the difference between the 6.3 diameter and the diameter of the gel after drying. Since some flow of the gel may occur during drying, the shrinkage across the diameter is probably slightly less than the vertical shrinkage. The vertical shrinkage may be significantly greater than the diameter shrinkage in the pure gel. These differences are reduced when SiC whiskers are present. The mold release agent also promotes even drying by ensuring unrestricted shrinkage. The shaded region in the figure represents the range of shrinkages obtained by drying under various conditions. Less shrinkage occurs when the gel is dried under high humidity conditions, and more shrinkage occurs when the gel is dried at 90°C, than when dried in the laboratory environment. Similar trends are observed for linear fired shrinkage (Fig. 3). Nearly 40% shrinkage was obtained for the pure gel when fired to 1200°C as compared to about 1% for the gel containing 52 wt% SiC whiskers.

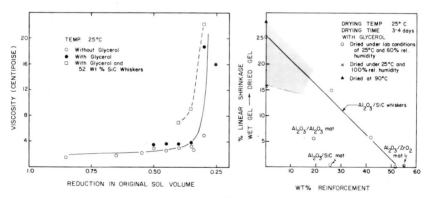

Fig. 1.   Fig. 2.

The gels containing the ceramic mats shrank slightly less during drying than the gels containing equivalent wt% of SiC whiskers. Apparently the rigidity of the mats retard the composites from shrinking during drying, but cannot prevent the shrinkage from occurring at high temperature. Thus, more shrinkage would be expected in the Al_2O_3/SiC mats composites than in the Al_2O_3/SiC whisker composites with the same % of SiC (see Figs. 2 and 3).

A summary of the thermogravimetric analysis for Al_2O_3 gel containing various percentages of SiC is presented in Table I. Weight losses occurring in three temperature ranges are shown: 25-150°C corresponds to the range where adsorbed water is removed, 150-325°C to the burnout of glycerol, and 350-550°C to the decomposition of δ-Al O(OH) into γ-Al$_2$O$_3$. The weight loss remained constant from about 550° to 1200°C. In general, the % weight losses due to glycerol burnout and to the formation of γ-Al$_2$O$_3$ decreased as the percentage of SiC whiskers increased. This result was expected since the concentration of gel decreases as the percentage of SiC increases. Also shown in this table are the theoretical values of weight loss expected as a result of the γ-Al$_2$O$_3$ transformation. These values are in fair agreement with the actual values obtained by TGA for the ones containing up to 35% SiC. Larger variations between the theoretical and actual values were obtained for composites containing higher percentages of SiC. Additionally, samples analyzed from different locations within the composites yielded slightly different results (see 42 and 52 wt% SiC data), indicating that the composites were not completely homogeneous. Although not shown, the TGA showed that the δ-Al O(OH) $\rightarrow$ γ-Al$_2$O$_3$ reaction was unaffected by the presence of SiC whiskers.

Tensile specimens similar to those shown in the preceding paper by Lannutti and Clark [4] were prepared and tested. Fracture strengths are shown in Fig. 4 for dried gel composites. The higher percentages of SiC whiskers resulted in slightly higher strengths than obtained from the pure

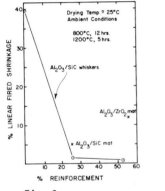

Fig. 3.

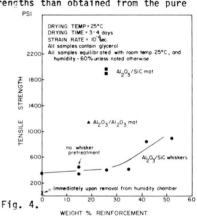

Fig. 4.

gel. The composites containing the SiC mat yielded higher fracture strengths than did the composites prepared with an equivalent weight percentage of SiC whiskers. Also shown in this figure is the strength of pure gel dried in a humidity chamber. Although much more flexible than the gel dried under room temperature and room humidity conditions, it is considerably weaker. The mechanical properties of the dried gels appear to be very dependent on their moisture content with drier samples being stronger but less flexile. The composite dried under room temperature and humidity conditions provide a satisfactory combination of flexibility and strength to permit easy handling. Fired tensile specimens were also tested, but slight warpage led to gripping and alignment problems and resulted in unreliable data.

Infrared diffuse reflection spectra* are shown in Fig. 5a for pure SiC whiskers and for Al$_2$O$_3$ gels containing SiC whiskers. These spectra were obtained from the as cast surfaces of the gels and from loosely compacted SiC whiskers. A more detailed analysis of the spectra corresponding to transformations in pure Al$_2$O$_3$ gel is presented by Clark and Lannutti [5].

*FT-IR, MX-1, From Nicolet, Madison, Wisconsin
 Diffuse Stage, Barnes Analytical, Stamford, Connecticut

The reflection peak at about 1050 cm^{-1} is unique to the pure gel, while the peak at 800 cm^{-1} is unique to the SiC whiskers. During firing, the 1050 cm^{-1} peak disappears corresponding to the decomposition of δ-Al O(OH). As shown in Fig. 5b the reflection spectrum of the Al_2O_3/SiC composite appears to have about equal contribution from both Al_2O_3 and SiC indicating good distribution (i.e. no segregation of Al_2O_3 and SiC) $\rightarrow$ of the Al_2O_3 and SiC on the cast surfaces.

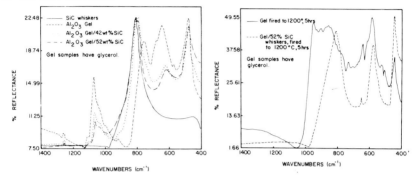

Fig. 5a.   Fig. 5b.

Scanning electron micrographs (SEMs) of fracture surfaces shown in Fig. 6 confirm that the whiskers are uniformly distributed within the composite. The agglomerates of SiC whiskers mentioned earlier are not present in these micrographs as they were in composites prepared with untreated whiskers. The final composites possess some void space after firing at 1200°C for 5 hours. Some whisker breakage observed at the tensile fracture surfaces indicated that whisker-matrix interactions do occur after five hours at 1200°C.

Al_2O_3 / 26 wt% Fired to 1200°C   Top   Bottom
Tensile Fracture Surface   Fig. 6.

ZrO_2 and carbon short fiber composites were also prepared using the same techniques discussed here.

SUMMARY

Composites of Al_2O_3/SiC can be easily fabricated by dispersing SiC whiskers in a matrix of low viscosity alumina sol and then gelling. A range of weight percentages of SiC up to 52% was investigated. The best composites based on SEM analysis contained between 26 and 42% SiC. Higher percentages of SiC resulted in higher porosity while lower percentages resulted in higher shrinkages during drying and firing. In addition to re-ducing the shrinkages, the SiC whiskers permitted the fabrication of large crack-free specimens. The use of glycerol and a mold release agent provided flexibility to the specimens and made it easier to remove them from the molds after drying. Work is now in progress to determine strength, fracture toughness and high temperature degradation properties of composites produced with sol-gel processing. Much additional research and development is required before prototype components can be fabricated and tested.

ACKNOWLEDGMENTS

This work was sponsored by the Air Force Office of Scientific Research (AFOSR) under Contract F49620-83-C-0072.

TABLE I. TGA data (% weight loss) for Al_2O_3/SiC whiskers composites synthesized by the methods described the text

Temp Range °C	0% SiC	Contain Glycerol						
		0% SiC	26% SiC	35% SiC	42% SiC		52% SiC	
25-150	5.	4.3	5.5	7.7	4.6	7.3	5.8	6.5
150-325	–	34.9	24.6	22.2	27.3	27.1	15.8	19.1
350-550	13.4	19.0	15.0	13.0	13.0	12.9	12.5	8.9
Theoretical*			14.8	13.3	12.0	12.0	10.1	10.1
Total % Wt. Loss	21.9	47.0	40.2	37.9	39.9	42.1	31.1	32.8

* Based on a 19 wt% loss during the γ-Al_2O_3 conversion and assuming the % of SiC to be exact.

REFERENCES

1. R. W. Rice, C. V. Matt, W. J. McDonough, K. R. McKinney and C. C. Wu, "Refractory-Ceramic-Fiber Composites: Progress Needs and Opportunities," Ceramic Eng. and Sci. Proceedings, published by the American Ceramics Society, 3(9-10) 714-721 (1982).

2. Bulent E. Yoldas, "Alumina Gels that form Porous Transparent Al_2O_3," J. Mater. Sci., 10(1975) 1856-1860.

3. Bulent E. Yoldas, "Alumina Sol Preparation from Alkoxides," Amer. Ceram. Soc. Bull., 3(1975) 289-290.

4. J. J. Lannutti and D. E. Clark, "Long Fiber Reinforced Sol-Gel Derived Al_2O_3Composites" in this volume.

5. D. E. Clark and J. J. Lannutti, "Phase Transformations in Sol-Gel Derived Aluminas," to be published in the proceedings of the International Conference on Ultrastructure Processing of Ceramics, Glasses and Composites, Gainesville, Florida, 1983.

LONG FIBER REINFORCED SOL-GEL DERIVED Al$_2$O$_3$ COMPOSITES

J. J. LANNUTTI* AND D. E. CLARK**
Ceramics Division, Materials Science and Engineering,
University of Florida, Gainesville, Florida 32611

ABSTRACT

A simple casting technique was used to fabricate composites comprised of long fibers in an alumina matrix. Random fiber preforms were placed in trays and subsequently a liquid alumina sol was added and gelled. Dried sheets approximately 6 cm in diameter and 0.2 cm thick were prepared without cracking. The addition of glycerol to the sol improved drying properties, made the dry gel flexible, and permitted the lamination (plying) of several sheets.

INTRODUCTION

Prewo and Brennan [1] have demonstrated the feasibility of fabricating ceramic-ceramic composites in which the matrix is glass and the reinforcement material is silicon carbide fiber yarn. In addition to having low densities, such composites exhibit excellent mechanical, chemical and thermal properties. Rice and co-workers [2,3,4] have addressed the potential problems and possible solution for developing a diverse array of ceramic-ceramic composites. They specifically noted that development of such composites has not received the attention that it deserves based on the potential opportunities that these composites offer. Some of the properties of interest in ceramic composites are strength, fracture toughness, density, stiffness and resistance to high temperature degradation. However, the major thrust of this paper is concerned with the chemistry of processing and not property evaluation.

Traditionally, the fabrication of ceramic-ceramic composites has required the consolidation of powders under high temperatures and pressures which can result in fiber damage and a reduction in performance. A second problem is the absence of good homogeneity which can lead to crazing and a reduction in strength [2]. Sol-gel processing offers a method for producing a new generation of composites. The fibers (in the form of preformed mats or weaves) may be impregnated with a low viscosity sol at 90°C without application of pressure, resulting in the formation of an intimate interface between the fibers and matrix after gelation. Additionally, very good homogeneity of matrix and fiber can be achieved. Although not proven, subsequent densification can possibly be accomplished at lower temperatures than required with conventional powder processing. Thus, the extent of matrix-fiber interaction, which is thought to influence mechanical properties, can be better controlled.

The present paper describes the processing procedures used to fabricate ceramic-ceramic composites using sol-gel derived Al$_2$O$_3$ as the matrix and preformed ceramic fibers as the reinforcement. The composites are characterized by thermal analysis and scanning electron microscopy. Physical properties such as drying and firing shrinkages, densities, and strength are discussed in a second paper in this same volume.

Mat. Res. Soc. Symp. Proc. Vol. 32 (1984) Published by Elsevier Science Publishing Co., Inc.

EXPERIMENTAL

Sol was prepared from aluminum sec-butoxide using the procedure described by Yoldas [5,6]. Excess water in the ratio of 100 moles water/mole $Al(OC_4H_9)_3$ was used to form the hydrolyzed sol which was then peptized at 90°C with 0.1 moles acetic acid per mole of $Al(OC_4H_9)_3$.††

The resulting solution is hereafter referred to as "as hydrolyzed, as peptized" and contains approximately 1.41 g of alumina (Al_2O_3) in 100 cc. Decreasing the volume of this solution (by boiling) down to approximately 25 cc produces a sol which is very close to the gelation point. Casting and drying this material results in solid bodies of δ-aluminum monohydrate, or dried gel.

The fiber preforms used in this study were SiC mat*, SiC weave*, Al_2O_3 mat**, and ZrO_2 mat.† Three of the preforms are random mats consisting of pressed long fibers and the fourth is a plain (bidirectional) weave of Nicalon silicon carbide.

The general process used for preparing the gel composites is illustrated in Fig. 1. In the case of the alumina and silicon carbide mats, a simple casting procedure was followed: placement of the silicon carbide

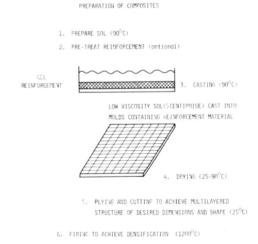

PREPARATION OF COMPOSITES

1. PREPARE SOL (90°C)

2. PRE-TREAT REINFORCEMENT (optional)

GEL
REINFORCEMENT ——— 3. CASTING (90°C)

LOW VISCOSITY SOL(5CENTIPOISE) CAST INTO
MOLDS CONTAINING REINFORCEMENT MATERIAL

4. DRYING (25-90°C)

5. PLYING AND CUTTING TO ACHIEVE MULTILAYERED
STRUCTURE OF DESIRED DIMENSIONS AND SHAPE (25°C)

6. FIRING TO ACHIEVE DENSIFICATION (1200°C)

Fig. 1.

(prefired at 700°C in air for 30 minutes to remove the polymeric binder) and the alumina (untreated) mats into polystyrene petri dishes followed by the addition of 25 cc of near-gelation sol. The mat/sol mixtures were gelled by devoluming to approximately one-tenth its original volume, then dried and fired. The low viscosity (<5 centipoise) of the sol prior to gelation facilitated the impregnation of the fiber mats.

††Aluminum sec-butoxide, From Alfa, Danvers, Massachusetts
* Nicalon, From Hexcel, Dublin, California
**Saffil, From ICI, Wilmington, Delaware
† ZYF-100, From Zircar, Florida, New York

Zirconia mats (which received no pretreatment) were judged to be too dense to be filled by this casting technique. Therefore, a predipping schedule was followed consisting of four dips in half-concentrated (50% volume reduction) as hydrolyzed, as peptized sol. From previous experience, it had been found that dried alumina gels are soluble in water. Because of this, it was thought that subsequent dips would dissolve previous gel layers. These gels had been previously observed to become insoluble when heated to above 220°C (without a change in phase). Thus, we heated the samples with 300°C heat guns in between dips. The mats achieved a noticeable rigidity after four dips, at which time the mats were cast as before.

The silicon carbide weave presented a special challenge in that previous casting attempts resulted in little or no adhesion of the matrix to the woven substrate after drying. This was attributed to the weave's inability to contract with the matrix during drying. A multiple dipping process was used to minimize this problem by using thin layers of dried gel to build up a matrix.

Multiple layers of SiC weaves were prepared by placing five single strips on top of each other in half-concentrated solution, sonicating for one minute, and removing from solution and drying. This cycle was repeated ten times. The weaves were observed to bond together after the second dip.

As produced by the Yoldas [5,6] procedure, dried alumina gel is a weak, brittle solid with very low formability. In order to more easily produce complex geometries by the sol-gel process, glycerol ($C_3H_8O_3$) was used as a plasticizer in casting to greatly improve flexibility. The added flexibility permits cutting, punching and plying of individual sheets of pure gel or gel containing fibers. It also reduces the extent of cracking during drying. The mechanisms of drying control and flexibility are thought to be due to modification of interparticle interaction in the dried gel.

The composites discussed in this paper were all prepared using 2 cc of glycerol per 100 cc of sol. Upon casting, the samples were permitted to dry under laboratory conditions (25°C, 60% R.H.) Under these drying conditions, the samples could be easily handled, cut and punched into desired shapes.

All of the samples were fired in an argon atmosphere to reduce the problem of fiber oxidation. Thermogravimetric analysis (TGA) was performed in a nitrogen atmosphere.

RESULTS AND DISCUSSION

Property changes accompanying sol-gel processing of Al_2O_3 are illustrated in Fig. 2. Starting with 200 cc of "as hydrolyzed, as peptized" sol, the volume is gradually reduced by heating at 90°C. The solution viscosity increases significantly when the volume has been reduced to 50 cc signalling the onset of gelation. Gelation is completed when the volume has been decreased to 25 cc. The viscosity variations within the sol-wet gel transformation region have not been determined, but most likely the viscosity increases by several orders of magnitude. The character of the wet gel is such that it cannot be easily handled without inducing permanent deformation. Subsequent drying of the wet gel (between 25-90°C) reduces its volume to 6 cc, at which point it becomes rigid, but is still fairly weak. The rate of weight loss (most of which is due to the evolution of water) is fairly constant during the sol-wet gel-dried gel transformations for a specified temperature. Finally, firing the dried gel to 1200°C reduces its volume to 2 cc and results in a strong and dense polycrystalline

Thermogravimetric analyses (TGA) are shown in Fig. 3 for dried Al_2O_3 gel containing various fiber mats. Three regions of behavior can be observed. Absorbed water is removed below about 150°C. The magnitude of the

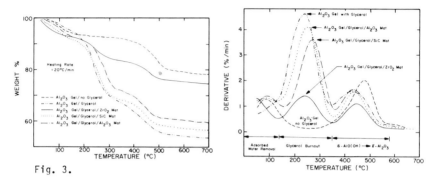

Fig. 3.

weight loss in this temperature range is very dependent on the drying procedure and exposure conditions prior to analysis. A second major weight loss occurs between 150°C and 350°C for all of the gels containing glycerol, but not for the gel not containing glycerol. Thus, it appears that glycerol decomposes in this temperature range. A third weight loss occurs between 400 and 600°C for all gel samples. This corresponds to the transformation of δ-AlO (OH) into γ-Al_2O_3 as indicated by equation (1). The percentage weight loss in this temperature range is 13% for the gel prepared without glycerol and 19% for the glycerol-containing gel. The weight losses for the gels containing the mats were less since they contained less gel. Based on weight loss measurements between 400 and 600°C the weight percentage of the various mats in the Al_2O_3 was 26% for the SiC mat, 19% for the Al_2O_3 mat and 56.3% for the ZrO_2 mat. The apparent reduction in the point of gamma transition is probably due to the continued evolution of glycerol decomposition products. The mats do not appear to affect this transition temperature. The TGA of the SiC mat alone revealed a few percent weight loss up to 1100°C, although some weight gain was observed between 400 and 800°C.

The presence of both glycerol and the mats improved the resistance to cracking during drying. Tensile test specimens could be easily fabricated from all composites in the dried conditions by simply punching with a steel die.* Representative specimens are shown in Fig. 4 after punching and also

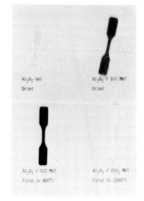

Fig. 4.

*ASTM 1822L

ceramic. Overall, there is about 99% reduction in volume when going from the starting sol to the dense ceramic.

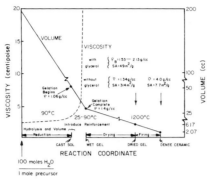

Fig. 2.

The reactions that occur when sol is transformed into a dense ceramic have been discussed in detail by Clark and Lannutti [7]. Briefly,

$$\text{Sol } [\delta\text{-A1O (OH)}] \rightarrow \text{gel } [\delta\text{-A1O (OH)}] \rightarrow \gamma\text{-A1}_2\text{O}_3 \rightarrow \alpha\text{-A1}_2\text{O}_3 \qquad (1)$$

Sol consists of isolated colloids of δ-A1O(OH) in a liquid matrix (mostly water). Evaporation of the liquid causes condensation of the colloids into a three dimensional network (wet gel). Further evaporation removes the remaining water between the colloids as well as most of the pore water (dried gel). According to Yoldas [5], the dried gel contains about 64% porosity which agrees well with the value of 66% measured in our laboratory.

One of the major problems in fabricating ceramics from sol-gels is the large volume shrinkages described above. These shrinkages usually result in cracking during drying and firing. The cracking may originate from two sources: (1) internal stresses due to differential shrinkage within the gel, and (2) interfacial stresses due to sticking of the gel to the mold material. The sticking problem was solved with the use of a mold release agent, Union Carbide R-272 organosilane ester. Glycerol was added to the sol to reduce internal stresses and provide dried gel flexibility. As drying proceeds (through liquid evaporation) the glycerol acts as a bridge between the colloids at their contact points. Deformation (through bending, stretching, compressing or shearing) of the dried gel then causes the colloids to move slightly with respect to each other without rupturing interparticle contacts. The exact mechanism leading to improved flexibility is not known, but we think that a glycerol-water layer between the colloids acts as a flexible bridge and preserves the integrity of the gel. The presence of a small amount of adsorbed water is required to achieve the maximum benefits from the glycerol. The adsorbed water present in dried gel equilibrated with the laboratory atmosphere (60% R.H.) is adequate to provide flexibility. In addition to imparting flexibility to the gel, a distinct improvement in drying properties are also obtained with the use of glycerol. The presence of glycerol increases the density of the dried gel (but not the wet gel) compared to the gel without glycerol. Surface area measurements indicated that the surface area of the dried gel containing glycerol is much less than that of the gel made without glycerol. These data suggest that the glycerol fills most of the internal pores in the dried gel.

after firing. Both the drying and firing shrinkages were significantly reduced for those gels containing fiber mats. It can be seen that all of the
tensile test specimens are nearly the same size even though two have been
fired to high temperatures. The largest shrinkages occured in the Al_2O_3
mat composites. In contrast, an Al_2O_3 gel without fibers undergoes a significant reduction in size during firing and tensile test specimens end up
about 1/3 the size of those shown. Residual carbon is observed on the surface of all glycerol-containing gels after firing in argon to α-alumina.

Scanning electron micrographs (SEMs) for some of the composites are
shown in Figs. 5-7. Several observations can be made from these micrographs. Although significant penetration of the gel into the fiber mats
did occur, the composites are porous after firing. The fired composite
with the least porosity appears to be Al_2O_3 mat. This is probably due to
the sympathetic contraction of the mat with the matrix during drying and
firing. Fiber pullout can also be seen on the fracture surface of the
fired Al_2O_3 / Al_2O_3 mat composite in Fig. 6. Firing to 800°C transforms
the gel into γ-Al_2O_3 while firing to 1200°C results in α-Al_2O_3. Thus, the
matrices of the fired samples in Figs. 5 and 6 are γ-Al_2O_3 while that in
Fig. 7 is α-Al_2O_3. The Al_2O_3 appears to be adhered to the fibers in some

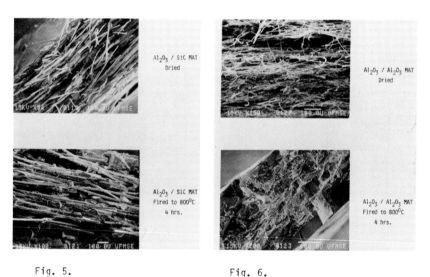

Fig. 5.   Fig. 6.

Fig. 7.

of the mats, but the extent of this adherence is not currently known. Predipping of the mats prior to their being cast in the gel improves the matrix continuity particularly for the ZrO_2 mat as shown in Fig. 7.

SUMMARY

Composites can be easily prepared by impregnating preformed fibers with low viscosity sol. The presence of the fibers reduces the extent of cracking and shrinkage during drying and firing. Additionally, glycerol can be added to the sol to provide a flexible gel after drying. The flexibility makes the gel, both with and without fibers, easier to handle and permits a wider range of forming and shaping operations than otherwise possible.

Processing and characterization are still in the early stages. Although we here demonstrated the feasibility of fabricating laboratory sized ceramic, composites based in sol-gel processing, much additional work is required to evaluate the properties and reliability of these materials.

ACKNOWLEDGEMENTS

This work was sponsored by the Air Force Office of Scientific Research (AFOSR) under Contract F49620-83-C-0072.

REFERENCES

1. Karl M. Prewo and J. J. Brennan, "Silicon Carbide Yarn Reinforced Glass Matrix Composites," J. Mater. Sci., 17(1982) 1201-1206.

2. R. W. Rice, C. V. Matt, W. J. McDonough, K. R. McKinney and C. C. Wu, "Refractory-Ceramic-Fiber Composites: Progress Needs and Opportunities," Ceramic Eng. and Sci. Proceedings, published by the American Ceramics Society, 3(9-10) 714-721 (1982).

3. R. W. Rice, "Ceramic Composites--Processing Challenges," Ceramic Eng. and Sci. Proceedings, published by the American Ceramics Society, 2(7-8) 493-508 (1981).

4. R. W. Rice, "Mechanisms of Toughening in Ceramic Matrix Composites," ibid, pp. 661-701.

5. Bulent E. Yoldas, "Alumina Gels that form Porous Transparent Al_2O_3," J. Mater. Sci., 10(1975) 1856-1860.

6. Bulent E. Yoldas, "Alumina Sol Preparation from Alkoxides," Amer. Ceram. Soc. Bull., 3(1975) 289-290.

7. D. E. Clark and J. J. Lannutti, "Phase Transformations in Sol-Gel Derived Aluminas," to be published in the proceedings of the International Conference on Ultrastructure Processing of Ceramics, Glasses and Composites, Gainesville, Florida, 1983.

THERMAL REACTION OF SILANE WITH ACETYLENE AND THE THERMAL DECOMPOSITION OF ETHYNYLSILANE

M. A. Ring, H. E. O'Neal, J. W. Erwin and D. S. Rogers, Department of Chemistry, San Diego State University, San Diego, CA 92182

Decomposition of SiH_4 in Acetylene

The volatile products from the thermal reaction (414°C) of silane in excess acetylene are hydrogen, ethylene, vinylsilane, ethynylsilane, vinylethynylsilane (possibly divinylsilane) and ethynyl-divinylsilane (1,2). We have reexamined this reaction using a 3 C_2H_2/1 SiH_4 reaction mixture and have obtained product yield curves for these products versus percent silane loss. We have also found that product curves are unaffected when propylene at pressures equal to that of acetylene is also present. Since only trace quantities of propylsilane are produced in the presence of propylene, we can rule out reactions involving silyl radicals. Thus the SiH_4-C_2H_2 reaction involves silylene and silene intermediates. The products can be explained by a mechanism similar to one proposed by Barton and Burns (3).

$$SiH_4 + (M) \longrightarrow SiH_2 + H_2 + (M)$$

$$SiH_2 + HC{\equiv}CH \rightleftharpoons \left[\begin{array}{ccc} \underset{SiH_2}{\overset{HC=CH}{\diagup}} & \rightleftharpoons & \underset{SiH}{\overset{HC-CH_2}{\diagdown}} & \rightleftharpoons & \underset{Si}{\overset{H_2C-CH_2}{\diagdown}} \\ I & & II & & III \end{array} \right]$$

$$I \longrightarrow HC{\equiv}CSiH_3 \ (ES)$$

$$II \longrightarrow H_2C=CHSiH$$

$$III \longrightarrow H_2C=CH_2 + Si$$

$$H_2C=CHSiH + HC{\equiv}CH \rightleftharpoons \left[\begin{array}{ccc} \underset{HSiC_2H_3}{\overset{HC=CH}{\diagup}} & \rightleftharpoons & \underset{SiC_2H_3}{\overset{H_2C-CH}{\diagup}} \\ IV & & V \end{array} \right]$$

$$IV \longrightarrow HC{\equiv}CSiH_2C_2H_3 \ (EVS)$$

$$V \longrightarrow H_2C=CHSiC_2H_3$$

$$H_2C=CHSiC_2H_3 + HC\equiv CH \;\rightleftharpoons\; \left[\begin{array}{c} HC=CH \\ \diagdown\diagup \\ Si(C_2H_3)_2 \\ \\ VI \end{array}\right]$$

$$VI \longrightarrow HC\equiv CSiH(C_2H_3)_2 \;(EDVS)$$

In addition to the volatile products, significant yields of solid polymer are also formed. The yield of volatile silicon products was 54% at 10% silane loss and dropped to only 18% at 50% silane loss. Polymerization of the highly unsaturated products is therfore inferred.

The Thermal Decomposition of Ethynylsilane

The product yield curve for ethynylsilane reached a maximum value early in the silane – acetylene reaction (T ~ 415°C). This suggests that ethynylsilane is relatively unstable at this temperature. We have thus examined the decomposition of ethynylsilane (1% in argon) in a stirred-flow reactor at reaction pressures ranging from 5.8 to 8.5 torr between 589.6 and 783.7°K. No prior quantitative data on the thermal stability of ethynylsilane exists in the literature.

An Arrhenius plot of the kinetic data (ln k vs 1/T) showed considerable curvature with slopes which increased by as much as a factor of three between the lower and upper temperatures. The very low activation energy corresponding to the low temperature slopes strongly suggests wall induced decomposition and we have treated the data accordingly (i.e., $K_{exp} = k_{wall} + k_{gas\ phase}$). On this basis, a computer fit of the data, which fits the experimental observations quite well, gives the following rate constants:

$$k_{wall} = 1.5 \times 10^3 \times e^{-17,6000/RT}\ sec^{-1}$$

$$k_{gas} = 3.0 \times 10^{13}\ e^{-53,000/RT}\ sec^{-1}$$

The volatile products of the ethynylsilane decomposition are predominantly acetylene and ethylene, with very small yields of methane and propylene at the higher temperatures. No silicon containing products were found, and carbon balances were poor but improved with temperature (i.e., carbon yields ranged from about 16% recovery for runs below 10% conversion up to 41% recovery for runs about 70% conversion). Acetylene was formed in similar yields at all stages of conversion. By our interpretation, this infers similar acetylene yields for both the heterogeneous wall reaction and the homogeneous gas phase reaction. Ethylene, on the other hand, appears to be a product of only the homogeneous reaction. Thus its yields are a strong function of temperature and at the lowest temperatures and conversions (i.e., 0-20%, T < 685°K) are essentially zero. Paralleling prior suggestions concerning mechanism (4), it is likely that the primary process

mechanism of the homogeneous reaction to these products involves
silacyclopropene intermediates (which may or may not have biradical
intermediate precursors). Thus, overall reactions are:

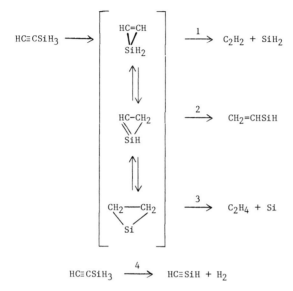

$$HC{\equiv}CSiH_3 \xrightarrow{4} HC{\equiv}SiH + H_2$$

Suggested parameters for the individual primary dissociation processes
(excluding rxn 2) are:

$$k_1 = 10^{12.9} \times e^{-47,700 \pm 3000/RT} \text{ sec}^{-1}$$

$$k_3 = 10^{13.4} \times e^{-54,600/RT} \text{ sec}^{-1}$$

$$k_4 = 10^{15.0} \times e^{-60,500/RT} \text{ sec}^{-1}$$

The activation energy for rxn 3 is based on the experimental activation
energy for ethylene formation. It agrees very well with thermochemical
kinetic estimates for a biradical reaction pathway. The back activation
energies of rxns 1 and 3 are believed to be very small and hence nearly
equal. Therefore, the activation energy difference of these reactions
must reflect the difference in their enthalpies: $E_3-E_1 \simeq \Delta H_3^\circ-\Delta H_1^\circ \simeq$
6.9 kcal. A-factors for both reactions were estimated by thermochemical
kinetic methods (5). The parameters for rxn 4 are based on estimates of
the A-factor drawn by analogy with other alkylmonosilanes, and also of
the hydrogen formation rate constant under our experimental conditions.
They are therefore less reliable and more speculative than the
parameters of the other processes.

Since rxns 1-3 dominate the kinetics of the ethynylsilane homogeneous gas
phase kinetics at static system pyrolysis temperatures, the Arrhenius

parameters of ethynylsilane are quite different, i.e., lower than those of other monoalkylsilanes studied to date (4). However, their homogeneous decomposition rates under these conditions are similar (except for those reactions subject to induced decompositions). Heterogeneous catalyis of the ethynylsilane decomposition, on the other hand, dominates the decomposition at temperature below 600°K, and at these and lower temperatures ethynylsilane is considerably more reactive than other monoalkylsilanes.

REFERENCES

(1) D. G. White and E. G. Rochow, J. Am. Chem. Soc., 76, 3897 (1954).

(2) C. H. Haas and M. A. Ring, Inorg. Chem., 14, 2253 (1975).

(3) T. J. Barton and G. T. Burns, Tetrahedron Lett., 24, 159 (1983).

(4) M. A. Ring, H. E. O'Neal, S. F. Rickborn and B. A. Sawrey, Organometallics, 2, 1891 (1983).

(5) W. W. Benson, "Thermochemical Kinetics," John Wiley and Sons, New York (1976).

Acknowledgement

    This work was supported by the Air Force Office of Scientific Research under Grant #83-0209

POLYSILANES AS POSSIBLE PRECURSORS TO SILICON CARBIDE

R. A. Sinclair, Central Research Laboratory, 3M Center, St. Paul, MN 55144
Robert West, Department of Chemistry, University of Wisconsin, Madison, WI 53706

Organometallic Polymer Routes to Ceramics

The continuing search for new types of high-strength materials, and for performance improvements in existing ceramics, has encouraged several nonconventional approaches to ceramics synthesis. Some of the advantages of polymeric routes have already been demonstrated and include new fabrication procedures leading to continuous fibers, coatings and infiltrated porous structures. At a more fundamental level, polymer pyrolysis can allow control over the microstructure of the final ceramic, with important consequences for both physical and chemical properties (1).

Although the first practical indications for polymer to ceramic conversion can be found in the production of high performance carbon fiber from organic polymers, the possibility of making ceramics from linear inorganic polymers was recognized two decades ago (2), (3). High strength continuous silicon carbide fiber, derived from polysilane starting material, has now become available commercially, largely as a result of the pioneering effort of S. Yajima (4).

Polysilane Synthesis

Polysilanes were originally synthesized by Kipping (5), but were not characterized until 1949 when Burkhard identified permethylpolysilane in the condensation reaction of dimethyldichlorosilane with sodium (6):

$$Me_2SiCl_2 \xrightarrow{\text{Na}} (Me_2Si)_n + NaCl \qquad (n = 55)$$

This highly crystalline polymer is too intractable to be processed directly but can be converted to a soluble, fusible low molecular weight polycarbosilane by thermolysis at moderate temperature (4):

$$(Me_2Si)_n \xrightarrow[\text{Ar}]{450°C} (MeHSi-CH_2)_y$$

Subsequently, the spun polycarbosilane was oxidatively crosslinked prior to a final high temperature conversion to high strength SiC fiber.

Several other types of polysilanes are being evaluated as silicon carbide precursors, including those derived from disilaryl (7) and vinylsilane (8) starting materials. Also, organosilane copolymers are a further class of compounds having desirable polymer properties useful in ceramic technology (9). The first examples of these copolymers contained only dimethylsilane and

Mat. Res. Soc. Symp. Proc. Vol. 32 (1984) Published by Elsevier Science Publishing Co., Inc.

388

phenylmethylsilane units with the $Me_2Si/PhMeSi$ ratios between 3:1 and 20:1. They were infiltrated into porous silicon nitride ceramic specimens and then pyrolyzed in situ to silicon carbide, with a resultant increase in strength for the composite (10). Polymer having a comonomer ratio near 1, known as "polysilastyrene", is soluble in common organic solvents such as toluene and tetrahydrofuran and readily characterized using GPC, NMR, MS, UV and IR techniques.

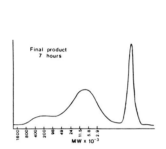

Fig. 1. Gel permeation chromatograph of polysilastyrene.

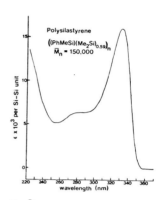

Fig. 2. Ultraviolet spectrum of a high-molecular-weight polysilastyrene fraction.

A considerable number of organosilane copolymers have now been studied (12) with a view to achieving a range in physical properties.

Conversion to Ceramics

In order to obtain high ceramic yields or pyrolysis, it is essential to control depolymerization to volatile species. The oxidative crosslinking step that has been used in the production of both carbon and silicon carbide fibers may be avoided in the case of phenyl-substituted polysilanes. These polymers are photoreactive, degrading to lower molecular weight, in solution, or crosslinking in bulk to insoluble and infusible network structures on exposure to UV radiation near 350 nm. Pyrolysis of this network structure at temperatures up to 1100°C yielded an amorphous silicon carbide composition with excellent retention of shape (9). Thus, small tubes were produced when only the surface of an extruded fiber had been photocrosslinked (Figure 3).

Polysilanes containing pendent allyl groups or siltetra-

methylene units have now been synthesized and offer scope for crosslinking by different mechanisms.

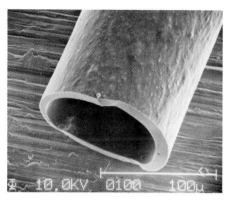

Fig. 3. Scanning electron micrograph of tube obtained by pyrolysis of photocrosslinked polysilastyrene.

Further elaboration of polymer design and conversion procedure is expected to demonstrate other advantages of these methods in materials synthesis.

REFERENCES

(1) R. N. Rice, Am. Ceram. Soc. Bull., **62** (8), 889 (1983).

(2) P. G. Chantrell and P. Popper, "Special Ceramics 1964", ed. P. Popper, Academic Press, New York (1965) p. 87.

(3) B. J. Aylett, *ibid.*, p. 105.

(4) S. Yajima, K. Okamura, J. Hayashi and M. Omori, J. Am. Chem. Soc., **59** (7-8), 324 (1976).

(5) F. S. Kipping, J. Chem. Soc., **119**, 647 (1921).

(6) C. A. Burkhard, J. Am. Chem. Soc., **71**, 963 (1949).

(7) R. Baney, U.K. Pat. Appl. 2,021,545 (1979) and 2,074,789 (1980).

(8) C. L. Schilling, Jr., J. P. Wesson and T. C. Williams, Am. Ceram. Soc. Bull., **62** (8), 912 (1983).

(9) R. West, L. D. David, P. I. Djurovich, H. Yu and R. Sinclair, *ibid.* p. 899.

(10) K. S. Mazdiyasni, R. West and L. D. David, J. Am. Ceram. Soc., **61** (11-12), 504 (1978).

(11) L. D. David, Ph.D. Thesis, University of Wisconsin-Madison (1981).

(12) X-H. Zhang and R. West, J. Polym. Sci., Polym. Chem. Ed., **22**, 159 (1984).